Wear the Change

Wear the Change examines the critical shift toward sustainability in the clothing and textile industry, a sector historically known for its significant environmental footprint. This comprehensive guide probes into the latest innovations, standards, and practical approaches that are transforming the entire value chain, from raw material sourcing to postconsumer waste management. By presenting cutting-edge research alongside actionable strategies, this book bridges the crucial gap between theoretical sustainability concepts and their real-world implementation across global textile operations.

Key features:

- Showcases successful sustainable initiatives practiced by industry leaders and consortiums worldwide
- Provides in-depth analysis of emerging circularity standards that are reshaping production models
- Offers practical guidance on implementing sustainability throughout complex supply chains
- Addresses the growing concern of microplastics pollution
- Consolidates diverse topics including cleaner production technologies, sustainable design incorporating artificial intelligence, and innovative approaches to postconsumer waste
- Presents a forward-looking perspective on opportunities and challenges facing the industry

This essential reference equips students, researchers, academics, and industry professionals with the knowledge needed to lead the textile industry's vital transformation toward environmental responsibility and sustainable business practices.

Asis Patnaik is the Head of the Clothing and Textile Technology Department at the Cape Peninsula University of Technology (CPUT), Cape Town, South Africa. He is actively involved in the teaching, research, and development of fiber-based materials suitable for manufacturing, automotive, medical, building, and footwear industries. He is the recipient of several international awards including Textile Institutes Outstanding Young Scientist Paper Award in 2008, Outstanding Contribution by an Individual award by the CSIR, South Africa in 2013, and Teaching Leader Award from CPUT Faculty of Engineering and the Built Environment in 2024.

Sweta Patnaik is an Associate Professor and Curriculum Officer in the Department of Clothing and Textile Technology at CPUT, Cape Town, South Africa. With extensive experience in teaching, learning, and curriculum development, she has led qualifications from undergraduate to Ph.D. level. Her research focuses on sustainability, waste management, entrepreneurship, e-learning, and supply chain management, with both national and international publications. Prof. Patnaik has received multiple awards, including the Nelson Mandela University Rising Star Award and DAAD UNILEAD scholarship. She serves on the editorial board of a leading journal and is a member of ISO's national standards committee. Her work has earned her various institutional and international awards.

Textile Institute Professional Publications
Series Editor: Abu Sadat Muhammad Sayem, *The Textile Institute*

Textiles and Their Use in Microbial Protection: Focus on COVID-19 and Other Viruses
Jiri Militky, Aravin Prince Periyasamy, and Mohanapriya Venkataraman

Dressings for Advanced Wound Care
Sharon Lam Po Tang

Medical Textiles
Holly Morris and Richard Murray

Odour in Textiles: Generation and Control
G. Thilagavathi and R. Rathinamoorthy

Principles of Textile Printing
Asim Kumar Roy Choudhury

Solar Textiles: The Flexible Solution for Solar Power
John Wilson and Robert Mather

Digital Fashion Innovations: Advances in Design, Simulation, and Industry
Abu Sadat Muhammad Sayem

Compression Textiles for Medical, Sports, and Allied Applications
Nimesh Kankariya and René Rossi

Microfibre Pollution from Textiles: Research Advances and Mitigation Strategies
Edited by R. Rathinamoorthy and S. Raja Balasaraswathi

Advances in Renewable Natural Materials for Textile Sustainability
Edited by Mohammad Shahid, Aranya Mallick, and Sayandeep Debnath

Biotechnology Approaches in Textile Technology: Progress and Trends
Edited by Mohd Yusuf and Mohammad Shahid

Digital Fashion: Artificial Intelligence, Virtual, and Other Emerging Applications
Edited by Abu Sadat Muhammad Sayem

Wear the Change: Sustainable Practices in Clothing and Textile Industries
Edited by Asis Patnaik and Sweta Patnaik

For more information about this series, please visit: www.routledge.com/Textile-Institute-Professional-Publications/book-series/TIPP

Wear the Change

Sustainable Practices in Clothing and Textile Industries

Edited by
Asis Patnaik and Sweta Patnaik

CRC Press
Taylor & Francis Group
Boca Raton London New York

CRC Press is an imprint of the
Taylor & Francis Group, an **informa** business

Designed cover image: © 2026 Asis Patnaik and Sweta Patnaik

First edition published 2027
by CRC Press
2385 NW Executive Center Drive, Suite 320, Boca Raton FL 33431

and by CRC Press
4 Park Square, Milton Park, Abingdon, Oxon, OX14 4RN

CRC Press is an imprint of Taylor & Francis Group, LLC

ISBN: 978-1-041-17343-4 (hbk)
ISBN: 978-1-041-17342-7 (pbk)
ISBN: 978-1-003-68922-5 (ebk)

DOI: 10.1201/9781003689225

Typeset in Times
by codeMantra

To our parents

Contents

Preface

The clothing and textile industry is among the most resource-intensive sectors globally, significantly contributing to greenhouse gas emissions, water pollution, and biodiversity loss. As global awareness around sustainability intensifies, there is a critical need to transform the industry across its entire value chain—from raw material sourcing to manufacturing, consumption, and end-of-life management.

Wear the Change: Sustainable Practices in Clothing and Textile Industries presents a comprehensive and interdisciplinary exploration of sustainability within the sector. This book begins with an overview of sustainability principles and circularity in the clothing and textile value chain, setting the foundation for understanding current challenges and opportunities. It then examines evolving circularity standards and cleaner production technologies, supported by relevant case studies and industry practices.

Subsequent chapters focus on sustainable innovations across production stages, including raw material development, yarn formation, fabric manufacturing, and garment production. These are complemented by discussions on current global sustainability initiatives shaping the industry. This book further explores innovation-driven approaches such as design for circularity, fashion design thinking, and the integration of digital technologies like AI, IoT, and blockchain to enhance transparency and efficiency.

Specialized chapters address advancements in sustainable chemical processing, technical textiles, biocomposites, and nanotextiles, highlighting emerging materials and technologies. Environmental challenges such as microplastic pollution and textile waste management are critically examined. The concluding sections explore sustainable supply chains and closed-loop systems, emphasizing collaboration, responsibility, and long-term value creation.

Each chapter presents recent developments, case studies, challenges, and future directions, offering both theoretical insights and practical applications. This book serves as a valuable resource for undergraduate and postgraduate students, researchers, academics, and industry professionals seeking to engage with sustainable transformation in the clothing and textile sector.

We extend our sincere gratitude to all contributing authors, reviewers, and collaborators whose expertise and dedication have made this publication possible. We also acknowledge the various sources that granted permission for the use of figures and tables included in this book.

We hope this book inspires critical thinking, fosters innovation, and contributes meaningfully to advancing sustainable practices in the global clothing and textile industry.

About the Editors

Prof. Asis Patnaik is the Head of the Clothing and Textile Technology Department at the Cape Peninsula University of Technology (CPUT), Cape Town, South Africa. He is actively involved in the teaching, research, and development of sustainable textiles and technical textile–based materials suitable for manufacturing, automotive, medical, building, and footwear industries. He was the recipient of several international awards, including the Textile Institute's (UK) Outstanding Young Scientist Award in 2008, Outstanding Contribution by an Individual award by the CSIR in 2013, and the Faculty Teaching Leader Award from the CPUT Faculty of Engineering and the Built Environment in 2024.

Prof. Sweta Patnaik is an Associate Professor, Learning & Teaching Coordinator, and Curriculum Officer in the Department of Clothing and Textile Technology at CPUT, Cape Town, South Africa. She has extensive experience in teaching, curriculum development, and supervising postgraduate students. Her research focuses on sustainability, waste management, technopreneurship, e-learning, and supply chain management, with publications nationally and internationally. She has secured multimillion-rand funding from the NRF, DSI, and CSIR, and holds an NRF Y2 rating. Her accolades include the Nelson Mandela University Rising Star Award (2020), DAAD UNILEAD scholarship (2021), and CPUT Teaching Excellence Award (2022).

Contributors

Christiane Siqueira de Azevedo Sá
Department of Textile Engineering
Federal University of Rio Grande do Norte
Natal, Rio Grande do Norte, Brazil

Antaleena Banerjee
Department of Fashion Technology
National Institute of Fashion Technology
Patna, Bihar, India

Clovis Medeiros de Bezerra
Department of Textile Engineering
Federal University of Rio Grande do Norte
Natal, Rio Grande do Norte, Brazil

Arul Jothi Biniya
Department of Design Space
National Institute of Fashion Technology
Bengaluru, Karnataka, India

Orebotse Joseph Botlhoko
Department of Engineering Metallurgy
Faculty of Engineering and the Built Environment
University of Johannesburg
Doornfontein, Johannesburg, South Africa

Kanti Prakash Brahma
Indian Institute of Management
Indore, Madhya Pradesh, India

Huantian Cao
Department of Fashion and Apparel Studies
University of Delaware
Newark, Delaware, United States

Kelly Cobb
Department of Fashion and Apparel Studies
University of Delaware
Newark, Delaware, United States

Kakoli Das
Department of Textile Design
National Institute of Fashion Technology
Bengaluru, Karnataka, India

Qilong Feng
Department of Computing
The Hong Kong Polytechnic University
Kowloon, Hong Kong

Huipu Gao
Textile Development and Marketing Department
Fashion Institute of Technology
State University of New York
Albany, New York, United States

Ashvani Goyal
Department of Textile Technology
The Technological Institute of Textile and Sciences
Bhiwani, Haryana, India

Patrick Hui
Department of Computing
The Hong Kong Polytechnic University
Kowloon, Hong Kong

Shamil Isaacs
Technology Station: Clothing and Textiles
Faculty of Engineering and the Built Environment
Cape Peninsula University of Technology
Cape Town, South Africa

Shafiqul Islam
Department of Textile Engineering
Dhaka University of Engineering and Technology
Gazipur, Bangladesh

Sarat Kumar Jena
Operations Management and Decision Science
Xavier Institute of Management
XIM University
Bhubaneswar, Odisha, India

Trisha Kumari
Department of Fashion Technology
National Institute of Fashion Technology
Patna, Bihar, India

Rasiah Ladchumananandasivam
Department of Textile Engineering
Federal University of Rio Grande do Norte
Natal, Rio Grande do Norte, Brazil

Lebo Maduna
Department of Clothing and Textile Technology
Faculty of Engineering and the Built Environment
Cape Peninsula University of Technology
Cape Town, South Africa

Dimakatso Makwakwa-Morajane
Centre for Nanostructures and Advanced Materials
DSI-CSIR Nanotechnology Innovation Centre
Council for Scientific and Industrial Research
Pretoria, South Africa

Vinay Kumar Midha
Department of Textile Technology
Dr B R Ambedkar National Institute of Technology
Jalandhar, Punjab, India

Arumugam Natarajan Monica
Department of Textile Design
National Institute of Fashion Technology
Bengaluru, Karnataka, India

Sudhakar Muniyasamy
Centre for Nanostructures and Advanced Materials
DSI-CSIR Nanotechnology Innovation Centre
Council for Scientific and Industrial Research
Pretoria, South Africa

Anshul Nayak
School of Mechanical Engineering
Iowa State University
Ames, Iowa, United States

Rajkishore Nayak
School of Communication and Design
RMIT University Vietnam
Saigon South, Vietnam

Miguel Higino Neto
Department of Materials Science and Engineering
Federal University of Rio Grande do Norte
Natal, Rio Grande do Norte, Brazil

Tarun Panwar
School of Fashion and Textiles
RMIT University
Melbourne, Australia

Manoj Kumar Paras
Department of Fashion Technology
National Institute of Fashion Technology
Patna, Bihar, India

Hủy Quang
The Business School
RMIT University Vietnam
Saigon South, Vietnam

Harsha Rani
Department of Fashion Management Studies
National Institute of Fashion Technology
Bengaluru, Karnataka, India

Guera Massyn Romo
Hernieuw (Pty) Ltd
Pretoria, South Africa

Rushdeen Rose
Technology Station: Clothing and Textiles
Faculty of Engineering and the Built Environment
Cape Peninsula University of Technology
Cape Town, South Africa

Mary Ruppert-Stroescu
Sam Fox School of Design and Visual Arts
Washington University
St. Louis, Missouri, United States

Bhavin J. Shah
Indian Institute of Management
Indore, Madhya Pradesh, India

Monica Sikka
Department of Textile Technology
Dr B R Ambedkar National Institute of Technology
Jalandhar, Punjab, India

Francisco Claudivan Da Silva
Department of Textile Engineering
Federal University of Rio Grande do Norte
Natal, Rio Grande do Norte, Brazil

Iris Oliveira da Silva
Department of Textile Engineering
Federal University of Rio Grande do Norte
Natal, Rio Grande do Norte, Brazil

Nadine Cynthia Sonnenberg
Department of Consumer and Food Sciences
University of Pretoria
Pretoria, South Africa

Conrad Sparks
Centre for Sustainable Oceans
Faculty of Applied Sciences
Cape Peninsula University of Technology
Cape Town, South Africa

Cyrus Tshifularo
Department of Clothing and Textile Technology
Faculty of Engineering and the Built Environment
Cape Peninsula University of Technology
Cape Town, South Africa

1 Overview of Sustainability in the Clothing and Textile Industries

Huantian Cao and Kelly Cobb

1.1 INTRODUCTION

The textile and apparel industries are one of the largest industrial sectors in the world. The global apparel market was projected to attain 1.84 trillion U.S. dollars in 2025, accounting for 1.63% of the world's GDP (Uniform Market 2025). The global textile fiber production was 132 million tons in 2024, which was more than doubled from 58 million tons achieved in 2000 (Textile Exchange 2025). The four fibers with the highest production in 2024 were polyester (77.7 million tons), cotton (24.5 million tons), nylon (7.0 million tons), and rayon (6.7 million tons) (Textile Exchange 2025). These four fibers accounted for 88% of global fiber production.

The production of textile and apparel products begins at fiber production from agricultural lands for natural fibers such as cotton and wool, forests for regenerative fibers such as rayon and lyocell, or the petrochemical industry for synthetic fibers such as polyester and nylon. The textile production includes fiber production for regenerative and synthetic fibers, yarn production, fabric production through weaving, knitting and nonwoven processes, and dyeing and finishing. The apparel production includes design, manufacturing, and finishing. After production, the apparel products will be packaged, transported, distributed, and sold by the retailers. The consumers will then use, care (wash, dry, dry clean), and dispose of the products. All of these stages intensively use water, chemicals, and energy (electricity and thermal energy), and generate polluted water and air, solid waste, and greenhouse gas (GHG).

The textile and apparel industry is also a highly globalized industry. In 2023, 98.4% of apparel products sold in the United States were imported from other countries (American Apparel and Footwear Association 2024). The United States is the latest apparel market with about 20% of global apparel market size (Uniform Market 2025) and the fourth largest cotton-producing country accounting for 12% of global cotton fiber production (U.S. Department of Agriculture n.d.). However, the textile and apparel production in the United States is low. The United States exports cotton fibers to other countries to produce yarns, fabrics, and garments and then imports the garments. The environmental impacts of textile and apparel production and use, especially agricultural production stage for natural fibers and apparel use stage, vary

DOI: 10.1201/9781003689225-1

in different countries. In Australia, at least 80% cotton agricultural land is irrigated (Roth et al. 2013), while in the United States, about 35% of cotton agricultural land is irrigated (Cotton Incorporated n.d.). Irrigated cotton production uses much more water than nonirrigated (rainfed) cotton production. To clean cotton t-shirts in China, 58.22% consumers use manual washing without electric drying and only 34.96% consumers use machine washing without electric drying; the former washing practice uses much less electricity, water, and detergent and generates much less wastewater, harmful chemicals, and GHGs compared to machine washing and electric drying (Zhang et al. 2015). The environmental impacts to clean cotton t-shirts by Chinese consumers are much lower than U.S. consumers, a vast majority of whom use machine washing and electric drying.

This chapter provides an overview of sustainability in the global textile and apparel industries, which covers the entire life cycle of textile and apparel products from textile fiber production to the end-of-life textile waste. It also discusses sustainability aspects in the textile and apparel supply chain, environmental impacts of textile and apparel production, waste management at the end of product life, use of resources in the product life cycle, sustainable models used in the industry, and opportunities and challenges in sustainable textile and apparel industries.

1.2 SUSTAINABILITY ASPECTS ACROSS THE INDUSTRY VALUE CHAIN

The textile and apparel industry is a complex system encompassing the cultivation and extraction of raw materials, fiber and fabric production, garment manufacturing, distribution and retail, purchase and consumption, and disposal (Figure 1.1). The concept of "value chain" introduced by Porter (1985) refers to the full range of activities and processes that businesses undertake to create customer value and competitive advantage through products or services developed and delivered to the end consumer. In the fashion industry, every phase of this process entails environmental and social impacts, which are increasingly influenced by supply chain disruptions, fluctuations in material prices, geopolitical tensions, global trade dynamics, intensifying competition, and evolving regulatory frameworks (Magnus et al. 2024). A sustainable fashion system, therefore, necessitates the optimization of these interconnected processes to promote long-term environmental, economic, social, and cultural well-being (Brundtland 1987; Rockström et al. 2009; Raworth 2017) and cannot be isolated to a single phase of production but must be embedded holistically across the entire value chain. The transition toward sustainability requires aligning the fashion industry's operational process with international standards. Goals such as the United Nations Sustainable Development Goals (SDGs) (n.d.) exemplify international standards as guideposts for sustainability, specifically goals that

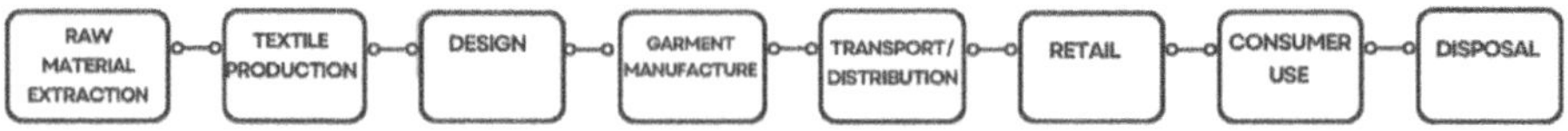

FIGURE 1.1 Stages of the linear textile and apparel value chain.

address responsible production and consumption (SDG12), clean water and sanitation (SDG6), climate action (SDG13), and decent work and economic growth (SDG8). WRAP's Textiles 2030 Circular Design Toolkit identifies the fashion industry as the third-largest global emitter of GHGs (WRAP 2023), responsible for 5% of total emissions, consuming 93 billion m^3 of water, and producing 98 million metric tons of nonrenewable material inputs annually (Ellen MacArthur Foundation 2017). The magnitude of these impacts makes the case for systemic intervention and coordinated action across all stages of the industry value chain.

1.2.1 Mapping the Value Chain

Mapping the textile and apparel value chain provides a holistic understanding of where and how environmental and social impacts occur, from raw material cultivation to post-consumer disposal and potential regeneration. The value chain encompasses a continuum of interconnected stages, each representing a system of inputs and outputs involving energy, water, chemicals, labor, and capital, with feedback loops that influence sustainability outcomes across the whole life cycle. By visualizing these interdependencies, mapping the value chain makes it possible to identify "hotspots" of environmental intensity such as fiber cultivation, dyeing and finishing, and end-of-life waste, as well as opportunities for intervention, innovation, and collaboration.

1.2.2 Raw Material and Fiber Production

The textile value chain starts with the extraction of raw materials, both natural and synthetic, to be used as the basic elements of apparel. The sustainability of this stage depends on aspects such as land use, water consumption, chemical inputs, and labor conditions linked to fiber cultivation and extraction. However, there remain significant barriers to adoption of these technologies, particularly in low- and middle-income manufacturing regions. Obstacles include limited access to regenerative crop production practices, and financial constraints among small farmers (Textile Exchange 2025). The inconsistent enforcement of certification standards, the cost of buying into standard bodies is prohibitive for smaller farms. Market price volatility for fiber products discourages investment in eco-friendly inputs. Finally, synthetic fiber production remains heavily reliant on fossil fuel infrastructure, presenting structural challenges for decarbonization (Textile Exchange 2024) despite progress in bio-based alternatives.

1.2.3 Textile Manufacturing and Processing

The manufacturing and processing of textiles involve the processes of yarn spinning, weaving or knitting, dyeing, printing, and finishing, and represent a high-impact phase of the textile life cycle due to intensive water, energy, and chemical use. Dyeing and finishing processes contribute an estimated 20% of global industrial water pollution, largely due to untreated effluent containing heavy metals, dyes, and surfactants that degrade aquatic ecosystems. Inefficient dyeing further exacerbates this impact, with approximately 15%–50% of applied dyes lost to wastewater streams (Singha et al. 2021; Shamsuzzaman et al. 2025). Textile processing is energy-intensive, contributing significantly to carbon emissions, especially when reliant on fossil-fueled thermal energy and oil heating for processes like steam production in manufacturing hubs across South and East Asia (Hasanbeigi et al. 2025). However, the adoption of cleaner production technologies in textile processing remains uneven due to high capital costs, insufficient infrastructure in many manufacturing hubs, and small and medium enterprises (SMEs), which represent a large portion of global textile production, frequently lack access to financing mechanisms required for system upgrades and upskilling, resulting in persistent environmental impacts despite available technological solutions.

Innovations such as low-liquor dyeing, digital and supercritical CO_2 dyeing, enzyme-based finishing, and zero-discharge chemical management systems have demonstrated significant reductions (de Oliveira et al. 2024; Bian et al. 2022) in environmental impact. WRAP (2021) estimated that implementing best-available technologies in dyeing and finishing could reduce water use by up to 50% and carbon emissions by 30%. To support industry accountability, several certification and monitoring frameworks—such as Bluesign®, OEKO-TEX®, and the Higg Index Facility Environmental Module—enable brands to evaluate chemical safety, emissions, and resource efficiency.

1.2.4 Apparel Design and Production

The design and production stage of apparel production is pivotal as decisions made here predetermine approximately 80% of a product's environmental footprint (WRAP 2023). Designers play an important role in determining material selection, pattern efficiency, assembly methods, and end-of-life recyclability parameters that define whether a product aligns with a linear or circular life cycle. Designers are encouraged to "commit to purity" by not mixing fiber types, to increase the demand for sustainable fiber options (such as organic and recycled materials), and to prioritize both performance and recyclability in their designs (Schumacher and Forster 2022).

Sustainable design strategies that emphasize adaptability and modularity contribute significantly to extending the lifespan of garments while minimizing waste generation. Techniques such as zero-waste patternmaking, reduced seam construction, and modular assembly improve material efficiency and facilitate repair, disassembly, and recycling processes (WRAP 2023). Designing with durability and high-quality construction in mind enables multiple use cycles. Strategic material selection, which considers factors such as compatibility, recyclability, and the minimization of mixed fiber content, exemplifies effective sustainable practices in material choice. Extending the active life of garments represents one of the most effective approaches to reducing the overall environmental impact of the fashion industry. In fact, increasing product longevity frequently yields greater environmental benefits than downstream waste management interventions, such as recycling or energy recovery (Laitala and Klepp 2021).

Integrating 3D virtual prototyping, e.g., CLO3D and Browzwear, with AI-assisted design analytics enables designers to simulate garments, optimize their fit, and minimize physical sampling. This, in turn, saves material and energy resources (WRAP 2023). Linked to material databases and life-cycle data, such digital tools support evidence-based design, balancing aesthetics with measurable sustainability outcomes.

1.2.4.1 Case Study: Circular Design Strategies in Regional Manufacturing: Lobo Mau

Lobo Mau, a Philadelphia-based apparel company founded by designer Nicole Haddad, incorporates sustainable principles in textile design and apparel production through regional manufacturing, material reutilization, and small-batch product development. The company utilizes deadstock fabrics and recycled materials, performs in-house textile printing with plastisol free water-based inks, and manufactures garments and soft goods locally within the Philadelphia region. For example, Figure 1.2 illustrates a circular furniture collection developed by Lobo Mau in collaboration with Jen Tiberi and Ryan Parker. The business approach lowers transportation emissions and allows for enhanced oversight of labor and environmental practices (Lobo Mau 2026). The operational model is consistent with slow fashion principles, circular economy strategies, and green engineering concepts (Gam et al. 2009) emphasizing reduced overproduction, efficient material utilization, and increased regional supply chain resilience (Ellen MacArthur Foundation 2017; WRAP 2023).

Lobo Mau's design process incorporates waste-led innovation through collaboration with regional partners from academia and industry to identify reuse pathways

FIGURE 1.2 Circular furniture collection (Lobo Mau, 2026).

for pre-consumer textile scraps. Research involving the company indicates that waste-led product development can convert knit production remnants into new textile forms, extending material lifespans (Yatvitskiy et al. 2024). This case illustrates a circular design model in which waste is regarded as a resource and design decisions are guided by material availability instead of linear product assumptions (Cobb et al. 2025).

As an SME, Lobo Mau demonstrates how localized production networks, small-batch manufacturing, and design-led material recovery contribute to circular apparel systems and bolster local economies (Wittchen 2022). This case reinforces the role of SMEs as Innovation Incubators in enhancing sustainable and regenerative fashion systems.

1.2.5 Distribution, Retail, and Consumer Use

Distribution, retail, and consumer use or consumption phases present a unique intersection of logistical, behavioral, and cultural sustainability challenges (Leal Filho et al. 2024). Globalized fashion supply chains depend on long-distance transport, contributing to Scope 3 emissions (McKinsey and Co. 2024). Shifting from air freight to sea and rail transport, optimizing logistics, and implementing carbon-neutral shipping programs can significantly reduce environmental impacts (WRAP 2023). Sustainable packaging or the inclusion of biodegradable, recycled, or reusable materials has become a growing area of innovation among brands such as Stella McCartney and Burberry.

Sustainable retailing encompasses both operational and behavioral change (Yang et al. 2017). Energy-efficient store design, renewable energy sourcing, and digital commerce can reduce the retail sector's carbon footprint (Yang et al. 2017). However, the consumption stage is where fashion's social and psychological dimensions intersect with sustainability. Overconsumption by consumers, fueled by fast fashion models, remains the central driver of waste and emissions. Educating consumers about product longevity, care practices, and repair can significantly extend the useful life of clothing. Consumer studies show that extending a garment's life by 50% can reduce its environmental impact by up to 40% (Jučienė et al. 2025). Initiatives such as clothing swaps, mending workshops, and awareness campaigns foster community-based sustainability cultures. Fashion simultaneously reflects and drives social changes (Burns 2019). Shifting consumer behavior is as much a cultural endeavor as it is a technical one.

1.2.6 Integrative Perspective: Toward a Sustainable Value Chain

The sustainability challenges of the textile and apparel industry cannot be solved in isolation; they require a systems-thinking approach (Meadows 2008; Fletcher 2008) that integrates environmental, social, and economic dimensions across the entire value chain. From regenerative fiber cultivation to circular design, ethical manufacturing, and responsible consumption, every stage offers leverage points for reducing impact and creating shared value.

1.3 ENVIRONMENTAL IMPACT

1.3.1 Chemical Use and Wastewater

The entire value chain of the textile and apparel industries heavily relies on chemicals and generates wastewater contaminated with these chemicals. Conventional cotton production intensively uses agrichemicals. A study in Benin found that conventional cotton farming used an average of 212kg of chemical fertilizers, 5L of chemical insecticides, and 4L of herbicides per hectare (Bonou-zin et al. 2019).

The textile industry uses about 8,000 chemicals in different manufacturing steps with a total amount of about 5 billion kg/year (Nimkar 2018). Kir et al. (2024) conducted a study on 44 facilities in different sectors of the textile industry in Turkey. They found that the average of auxiliary chemical consumption was in the range of 221–397 g/kg product and the average of dyestuff consumption was in the range of 18–30 g/kg product. The synthetic woven fabric dyeing and finishing subsector has the highest auxiliary chemical consumption of 397 ± 237 g/kg product, and cotton knitted fabric dyeing and finishing and synthetic knitted fabric dyeing and finishing sectors have the highest dyestuff consumption of 30 ± 13 and 30 ± 7 g/kg product, respectively (Kir et al. 2024). Of the 1,800 chemicals used in clothing production, 165 chemicals were classified as hazardous to human health or the environment by the European Union (EU) (Granskog et al. 2020).

A large quantity of wastewater effluent was generated from the textile industry. The textile dyeing and treatment contribute approximately 25% of all industrial water pollution (Granskog et al. 2020). China, the largest textile manufacturing country in the world, discharged a peak of about 2.45 x 10^9 tons of wastewater in 2010 from its textile industry, and then gradually decreased with an average annual decline of 6.52% from 2011 to 2015 (Zhou and Zhou 2019). Wastewater from textile wet processes have high biological oxygen demand, chemical oxygen demand, suspended solid, dissolved solid, alkalinity, salinity, ammonia nitrogen (NH_3-N), nitrogen, and colority, and contain heavy metals, phosphorus, chlorine dioxide (ClO_2), and organic halogen (Kabir et al. 2019; Zhou and Zhou 2019).

Textile production, especially the dyeing and finishing process, uses a large quantity of chemicals and water. After the process, chemical-contaminated wastewater effluent is generated, which causes a big environmental impact.

1.3.2 Energy Use and Greenhouse Gas Emissions

Cotton farming uses energy in fallow, planting, in-crop, irrigation, harvesting, and postharvest. Irrigation (about 40%–60% of the total energy use) and harvesting operation (about 20% of total energy use) are the most energy-intensive processes (Chen and Baillie 2009). A study in Australia found that energy uses from cotton farms ranged from 3.7 to 15.2 GJ/ha of primary energy, which is higher than other crops due to more intensive management practices and higher irrigation requirements. This energy use corresponds to 275–1,404kg CO_2 equivalent per hectare GHG emissions (Chen and Baillie 2009). Another study in Benin found that cotton farming emitted an average of 669.40kg CO_2 equivalent per hectare. Organic cotton farming emitted about 42% less

TABLE 1.1
Energy, Steam, and GHG Emissions Reduction from a Smart Analysis Module in Textile Dyeing to Dye 15 kg Cotton Textile

Manufacturing Stages	Electricity Consumption (kWh)	Steam Consumption (kg)	Steam Consumption (kg)
Conventional process	2.5	27.2	3.534
New process using smart analysis module	2.2	23.9	3.107

Source: Kim et al. (2024).

GHG than conventional cotton, i.e., 462.65 and 882.06 kg CO_2 equivalent per hectare for organic and conventional cotton, respectively (Bonou-zin et al. 2019).

All processes in the textile industry intensively use energy: fiber preparation consumes 21%, yarn spinning consumes 11%–18%, and weaving consumes 23% of total energy used from fiber to finished fabric (Farhana et al. 2022). The textile wet processes use energy to elevate process bath temperature, generate steams, and dry textiles. To process 1 kg of textiles, scouring without drying uses 5–18 MJ energy; bleaching and drying use 8–33 MJ energy; bleaching, dyeing, and drying use 10–35 MJ energy; dyeing, finishing, and drying use 8–18 MJ energy; and finishing uses 6–12 MJ energy (Saxena et al. 2017). A study in Turkey found that the average electricity consumption was in the range of 1.2–7 kWh/kg product and the average steam consumption was in the range of 6.8–20 kg steam/kg product (Kir et al. 2024). The wool woven fabric dyeing and finishing subsector has the highest electricity consumption of 7 ± 5.3 kWh/kg product and the highest steam consumption of 20 ± 11 kg steam/kg product (Kir et al. 2024).

The textile and apparel industries contribute about 1.7 billion tons of carbon emissions per year, which is estimated to be 6%–8% of global carbon emissions (Sharpe et al. 2022). A study on a textile finishing company in Indonesia found that 258.629 kg CO_2 equivalent was emitted in the finishing process to treat one ton of textile (Efendy and Dewi 2024). Innovative technology in the textile industry can help reduce energy use and GHG emissions. Kim et al. (2024) developed a smart analysis module to continuously monitor and analyze the dyeing parameters in real time. This technology can reduce about 12% of electricity consumption, steam consumption, and GHG emissions as shown in Table 1.1 (Kim et al. 2024).

Textile production is energy-intensive with huge consumptions of electricity and steam. Because fossil fuel is still the main source for energy generation, the energy use in textile production also results in large amounts of GHG emissions.

1.3.3 Biodiversity Loss

The decline of global biological diversity was driven by the degradation of terrestrial, freshwater, and marine ecosystems due to human activities (Navarro-Gambín et al. 2025). Textile and apparel supply chains directly caused soil degradation, conversion of natural ecosystems, and waterway pollution that negatively impacted biodiversity (Granskog et al. 2020). The five largest contributors to biodiversity loss related to textile and apparel production and consumption are cotton agriculture (intensive insecticide and water use), wood-based regenerative cellulose fibers (endangered and primary forests, water and soil pollution from chemicals used in forest plantation and pulp processing), textile dyeing and treatment (hazardous chemical use and water pollution), microplastics (laundry of synthetic textiles), and solid textile waste (habitat loss due to landfills, pollutants from incineration and landfills) (Granskog et al. 2020). In addition, the GHG emissions from the textile and apparel industry accelerate climate changes that raise sea levels leading to coastal area and wetland habitat reduction and change temperature affecting species behavior and ecosystem functions (Navarro-Gambín et al. 2025). Sandin et al. (2013) compared the production of wood-based regenerative cellulose fibers and cotton fibers and found that the transformation of natural land (e.g., transforming from natural forests to cotton farming) has a much higher impact on biodiversity loss than the occupation of land. Tian and Cheng (2024) found that provinces in China where leather and textile industries create higher export value had higher biodiversity loss.

The entire life cycle of textile and apparel products including natural and regenerative fiber production, textile dyeing and finishing, apparel laundry, and disposal causes environmental problems that result in biodiversity loss. The environmental problems related to biodiversity loss include hazardous chemical use, water and soil pollution, GHG emissions and climate change, and landfills.

1.4 WASTE MANAGEMENT AND MATERIAL CIRCULARITY

1.4.1 The Post-Consumer Waste Problem

The textile and apparel industries generate significant waste, making them one of the most environmentally and socially harmful industries. In 2018, the United States generated about 17.03 million tons of textile waste, out of which 66.4% was landfilled, 18.9% was incinerated for energy recovery, and 14.7% was recycled (Cao et al. 2022; Shamsuzzaman et al. 2025). The disparity indicates that a persistent imbalance exists between the volume of materials generated and the capacity of existing recovery infrastructure to manage them. This indicates the presence of structural inefficiencies within the fashion value chain. Post-consumer textile waste refers to garments that individuals or organizations, such as hospitals, hotels, and industrial laundries, have worn and subsequently discarded. Post-consumer waste constitutes the largest category of textile waste (U.S. Government Accountability Office 2024).

1.4.2 Problems with Fiber Blends and Contamination

The growing complexity of post-consumer waste streams poses significant barriers to the efficient recycling of textiles. Mass-market apparel incorporates blended fabrics, synthetic fibers such as elastane, functional surface treatments, and chemical additives, which collectively complicate the recycling process. Cotton-polyester blends are widely used in apparel manufacturing worldwide. However, the unique thermal, mechanical, and chemical properties of the fibers pose challenges in their separation, which can reduce the efficiency of both mechanical and chemical recycling methods (Schumacher and Forster 2022).

The National Institute of Standards and Technology (NIST) states that blends constitute one of the primary obstacles to achieving circularity. Recyclers state that elastane contamination exceeding 2%–3% constitutes a critical failure threshold in mechanical reprocessing (Schumacher and Forster 2022). Concerns regarding contamination following use exacerbate the technical difficulties further.

The U.S. Government Accountability Office (2024) reports that collected textiles are often not suitable for reuse because their fiber composition is "unknown or mislabeled." This indicates that more advanced fiber identification technologies are required.

1.4.3 Global Consumption Patterns and Fast Fashion

The patterns of global consumption intensify the textile waste problem. Fast fashion models entail significantly shortened garment lifespans with consumers wearing items for shorter periods, i.e., an item fewer than 7–10 times before disposing of it (Shamsuzzaman et al. 2025). The Accelerating Circularity (2021) indicates that micro-use phases generate substantial quantities of low-quality, low-value material that cannot be easily converted back to fiber, thereby complicating sorting and recycling efforts. Global reporting confirms this pattern. The UNIDO report on Circular Economy for Textiles (2017) emphasizes that the rapid turnover of fast fashion items generate material flows with insufficient structural integrity, rendering them incapable of enduring the mechanical forces required for fiber recovery, thus fostering reliance on virgin materials to satisfy performance requirements. Likewise, life-cycle assessment (LCA)-based research (Peña et al. 2021; Edirisinghe et al. 2024) indicates that without substantial advancements in fiber purity and traceability, circular pathways will remain constrained in scale.

1.4.4 Material Circularity Strategies

The transition from a linear "take-make-dispose" model to a circular materials economy is critical to sustainable transformation in the fashion industry. Circularity emphasizes designing systems that keep materials in use for as long as possible, maintaining their value through strategies such as reuse, repair, remanufacturing, and recycling (Ellen MacArthur Foundation 2017). Textile reuse represents the most resource-efficient form of waste reduction. Rental platforms, resale markets, and repair services are examples of emerging reuse innovations that demonstrate environmental benefits while encouraging new business logics that decouple profit from production volume (WRAP 2023). Upcycling, or transforming discarded materials into higher-value products, offers further potential, merging creativity with sustainability (Bhatt et al. 2019).

1.4.5 Recycling Technologies

Recycling represents the key to material circularity, from mechanical to chemical or even hybrid recovery systems. Mechanical recycling includes shredding and re-spinning textiles into new fibers. This process reduces fiber length and diminishes tensile strength; the results are often downcycled into products like insulation or industrial wipers. Chemical recycling enables fiber-to-fiber regeneration through depolymerization and re-polymerization of waste textiles. The technologies used include Lenzing's Refibra™ and Teijin's Eco Circle™, which use dissolution to convert waste cotton and polyester into near-virgin-quality regenerated cellulosic or synthetic fibers, respectively (Cao et al. 2022).

Hybrid systems that integrate mechanical preprocessing with chemical depolymerization are emerging as key solutions for blended fabrics, considered the most common and problematic waste stream in worldwide apparel production. The rise of fiber-to-fiber technologies and industrial symbiosis models redefine material value chains. Fiber regeneration platforms, such as Renewcell's Circulose®, enable cotton and polyester waste to become high-value input materials within the supply chain (Ellen MacArthur Foundation 2021a). Textile waste is increasingly used as a feedstock for non-textile applications, including composite panels, automotive insulation, and construction materials, indicating a trend toward closed-loop systems beyond apparel production. As of Shamsuzzaman et al. (2025), various innovations are examples of how circular design can eventually inspire new sustainable industrial ecosystems characterized by reduced waste and resilience.

1.4.6 Policy, Infrastructure, and Stakeholder Roles

Achieving circularity in fashion necessitates strong policy frameworks, coordinated collection and sorting systems, and active collaboration among stakeholders at global, national, and local levels. New policy mechanisms are reallocating the responsibility for textile waste to the producers who launch products into the market. Extended Producer Responsibility (EPR) legislation requires brands to finance or oversee collection, sorting, and recycling systems, thereby aligning economic incentives with environmental results. The EU's Green Deal, Ecodesign for Sustainable Products Regulation (ESPR), and Green Claims Directive collectively enhance standards for transparency, reparability, and recyclability in apparel (Siliņa et al. 2024). California's Responsible Textile Recovery Act of 2024 instituted the inaugural state-level EPR policy in the United States, mandating producers to participate in stewardship programs for the effective collection and recycling of post-consumer textiles (Siliņa et al. 2024).

Notwithstanding advancements in policy, infrastructure continues to pose a substantial obstacle. The UK Textiles Pact (2026–2030 Roadmap) identified insufficient sorting systems and the absence of large-scale fiber-to-fiber recycling facilities as significant deficiencies in attaining circularity. In 2025, merely 16% of fibers utilized by Pact signatories were recycled, marking a substantial rise from 2% in 2021 (WRAP 2025). National pilot initiatives on advanced sorting—coordinated by WRAP and comprising 18 industrial collaborators—have evidenced the economic viability of large-scale material recovery, supported by collaborative data, technological integration, and intersectoral investment (WRAP 2023).

Individual entities cannot attain circularity; it necessitates a holistic approach. The Textiles Action Network, organized by WRAP, links regional initiatives throughout

Europe, Australia, and Africa to standardize practices and disseminate knowledge (WRAP 2023). Academic institutions and research laboratories are essential in creating novel material assessment frameworks and life-cycle data that guide circular design (Schumacher and Forster 2022). Brands and manufacturers implement these findings via product innovation and transparent sourcing, while consumers increasingly engage as active participants in the circular economy through conscientious purchasing, repair, and resale practices. The Textiles 2030 Circular Design Toolkit underscores that attaining circularity necessitates a collaborative effort involving internal teams, robust supplier relationships, and innovative partnerships throughout the industry (WRAP 2023). This interdisciplinary collaboration highlights the evolution of fashion from a linear, extractive model to a regenerative, design-oriented ecosystem.

In summary, waste management and material circularity in the fashion and textile sector signify both technological and cultural shifts. The scale of post-consumer waste requires both sophisticated recycling technologies and a reconfiguration of design principles, manufacturing processes, and international policy structures. Circular models—rooted in fiber regeneration, EPR, and stakeholder collaboration—illustrate how waste can transform into a renewable resource instead of a finality. In this evolving framework, design serves as a mechanism of environmental governance, while circularity signifies both innovation and accountability within the 21st-century fashion industry.

1.5 USE OF RESOURCES

1.5.1 Land Use

Natural fibers such as cotton, wool, and flax are agricultural products that use agricultural land in their production. Global agricultural land area, including rainfed cropland (1,216 million hectares), irrigated cropland (343 million hectares), and pastureland (3,201 million hectares) for grazing livestock, was 4.76 billion hectares by 2020, which accounted for 32% of the world's total land area (Fuglie et al. 2024). The global production of natural fibers in 2024 is summarized in Table 1.2.

TABLE 1.2
Global Production of Natural Fibers in 2024

	Fibers	Production (Million Tons)	% Among All Fibers	% Among Natural Fibers
Plant fibers (cellulose)	Cotton	24.5	19	75
	Flax	0.3	0.3	0.9
	Hemp	0.3	0.2	0.9
	Other plant fibers	6.2	4.7	19
Animal fibers (protein)	Wool	1.1	0.9	3.4
	Silk	0.09	0.1	0.3
	Other animal fibers	0.05	0.04	0.2

Source: Textile Exchange (2025).

TABLE 1.3
Irrigated and Dryland Cotton Lint Productivity in Different Regions in Texas

Region	Irrigated Cotton Lint Productivity (lbs/acre)	Dryland Cotton Lint Productivity (lbs/acre)
South Plains	500–1,000	300
Permian Basin	500–1,000	300
Rolling Plains	750–1,300	350
Winter Garden	750–1,250	400

The production of 24.5 million tons of cotton fibers in 2024 accounted for 19% of all fiber production and 75% of natural fiber production (Textile Exchange 2025). The worldwide cotton harvest area was about 32.4 million hectares in 2024/2025, which was near the average of the last 5 years (Meyer and Dew 2024). The global land used for cotton fiber production was about 0.7% of all agricultural land and about 2% of all croplands. Cotton can grow and be harvested in both rainfed cropland (dryland) and irrigated cropland. Table 1.3 summarizes cotton lint production in irrigated and rainfed croplands in different regions in Texas. The productivity of irrigated cotton is about three times that of dryland cotton. Cotton plants can cause soil erosion in the cropland. Studies in Australia found that soil loss from irrigated cotton was 93–243 kg/ha, which was much lower than up to 4–5 tons/ha soil loss from dryland cotton (Nachimuthu and Webb 2016).

Sheep use pastureland and require more land than cotton plants. In Australia, the leading wool-producing country, 44.04 ha of land are needed to produce one bale of wool. Compared with 0.12 ha of land to produce one bale of cotton, wool uses 367 times more land per bale than cotton (Feldstein et al. 2021). Pasturelands cleared and grazed for non-native animals such as sheep result in severe land degradation such as soil erosion and desertification (Feldstein et al. 2021). It should be noted that only a portion of the used land and soil erosion by sheep can be allocated to wool fiber production since other valuable products such as meat and skin are also produced by sheep.

1.5.2 Water Use

Cotton farming requires intensive water use. The global water use for cotton crop production is 198.4 billion m^3/year, among which 99.0 billion m^3 is "blue water" (ground or surface water for irrigation) and 99.4 billion m^3 is "green water" (infiltrated rainwater) (Chapagain et al. 2006). The global water consumption related to cotton products is 256 billion m^3/year, which accounts for 2.6% of the global water footprint (Chapagain et al. 2006).

The textile wet processes, including preprocessing such as desizing, scouring, and bleaching; processing such as dyeing and printing; and post-processing such as finishing, consume a significant amount of water: 80–150 L of water for processing 1 kg of fabric (Kabir et al. 2019). A study in Turkey found that the average water consumption was in the range of 55–345 L/kg product (Kir et al. 2024). The wool woven fabric dyeing and finishing subsector has the highest water consumption of 345 ± 262 L/kg product (Kir et al. 2024).

1.5.3 Resources for Energy

The textile industry has two categories of energy consumption: electricity used for machinery, lighting, office, air conditioning, and motor drives; and thermal energy steam used for scouring, bleaching, dyeing, finishing, and washing (Farhana et al. 2022). Steam used in the textile industry is generated from oil, liquefied petroleum gas, coal, and natural gas, and the rapidly depleting resources of fossil fuels account for about 80% of global industrial energy usage (Farhana et al. 2022). The share of fossil fuels among all energy sources in major textile and apparel production countries was higher than the global average: 99.4% in Bangladesh, 93.9% in Indonesia, 91.0% in India, 85.1% in China, and 84.8% in Vietnam (Sharpe et al. 2022). The textile industry in China, the largest textile manufacturing country in the world, used about 1500 Petajoules (PJ) of energy in 2019 (Hasanbeigi and Zubeiri 2022).

1.5.4 Resources for Synthetic Fiber Feedstock

The global synthetic fiber production was 91.0 million tons in 2024, which accounted for 69% of all fiber production (Textile Exchange 2025). The significant increase in global fiber production in the past few decades was mainly driven by the increase of polyester fiber production: the production of polyester fibers was more than tripled, while the production of all fibers was more than doubled from 2000 to 2024 (Textile Exchange 2025). A vast majority of synthetic fibers use petrochemicals derived from crude oil as the feedstocks for their production. The production of synthetic fibers used around 350 million barrels of oil in 2017, which was 1.35% of global oil production (Trunk et al. 2021).

To address the depletion of crude oil resources used in synthetic fiber production, the textile scientists and industry developed synthetic fibers from feedstocks derived from recycled plastics and textiles, renewable bio-based resources, and captured CO_2 emissions (Textile Exchange 2024). The recycled polyester production was 9.3 million tons, accounting for about 12% global polyester production, in 2024 (Textile Exchange 2025). About 98% of recycled polyester feedstocks were made from polyethylene terephthalate plastic bottles, and the rest of them were derived from discarded polyester textiles and other post-consumer plastics such as ocean waste and packaging waste (Textile Exchange 2025). Polyester, nylon 6, nylon 6,6, polylactic acid, and polybutylene succinate fibers can be produced from bio-based feedstocks (Textile Exchange 2024). However, the production of biosynthetic fibers is low at this time. About 0.01% of total polyester fibers and 0.4% of all nylon fibers were produced from bio-based feedstocks (Textile Exchange 2024). The captured CO_2 can be converted to ethanol through gas fermentation or ethylene through electrochemical conversion, and both of them can be used to produce feedstocks for polyester synthesis (Textile Exchange 2024). Biological processes and chemical synthesis can also convert CO_2 to feedstocks for synthetic fiber polymers (Textile Exchange 2024).

1.6 SUSTAINABLE BUSINESS MODELS

1.6.1 From Linear to Circular: Rethinking the Fashion Economy

A linear production model, while economically advantageous in the short term, is ecologically and socially unsustainable, resulting in significant waste, overproduction, and GHG emissions (WRAP 2023). The circular fashion economy broadens value creation by developing products and systems that sustain materials in use at their maximum utility and value for longer periods of time (WRAP 2023). This involves the reconfiguration of production and consumption systems to promote reuse, resale, repair, remanufacturing, and recycling. Circular models redefine customer relationships by encouraging access, sustainability, and shared value across diverse product life cycles, rather than exclusively focusing on point-of-sale transactions.

The transition to a circular model necessitates a cohesive transformation across all organizational levels—strategic, operational, and cultural. It requires collaboration among designers, manufacturers, consumers, and policymakers to collectively develop a sustainable textile economy. WRAP's Textiles 2030 initiative outlines a systems-oriented approach that includes designing circularity, implementing circular business models, developing material recovery infrastructure, and decarbonizing the supply chain (WRAP 2023).

1.6.2 Innovative Circular Business Models in the Fashion Sector

Circularity is exemplified by various unique business model archetypes that transform the creation, distribution, and preservation of value in the fashion industry. Product-Service Systems (PSS) rental, subscription, and leasing systems facilitate access to apparel without the conveyance of ownership (ThredUp 2024). Rental and subscription services like Rent the Runway and My Wardrobe HQ enable consumers to access diverse options while minimizing unused wardrobe space. These models attract consumers desiring innovation with reduced environmental impact and provide brands with a consistent revenue stream through asset reutilization. Studies demonstrate that renting or leasing clothing for a minimum of 20 uses can reduce their carbon footprint per wear by over 30%, provided that logistics and cleaning processes are efficiently administered. Resale and recommerce platforms are among the most rapidly expanding sectors of the circular economy. Platforms such as ThredUp, Vestiaire Collective, and The RealReal are making pre-owned fashion a prevalent and attainable option for many people. ThredUp's (2024) resale report indicates that the global secondhand market is anticipated to attain 350 billion U.S. dollars by 2030. Moreover, brand-supported resale initiatives like Patagonia's Worn Wear and Levi's Secondhand prolong product longevity while fostering deeper customer relationships by linking sustainability to authenticity and the brand narrative.

1.6.3 Repair Solutions

Repair-focused business models restore product functionality and preserve material value. Companies such as Nudie Jeans and Arc'teryx provide lifetime repair warranties, while independent repair centers and repair cafes are emerging as community-based responses to overconsumption. The UK Textiles Pact (2026–2030) acknowledges the growth of repair and alteration services as a crucial strategy for reducing waste and creating employment opportunities (WRAP 2023).

1.6.4 Closed-Loop and Regenerative Manufacturing

Closed-loop manufacturing aims to reclaim fibers and fabrics at the conclusion of their life cycle and reintroduce them into the production cycle. Examples encompass Adidas's Made to Be Remade program, which enables customers to return worn shoes for disassembly and remanufacture, and Stella McCartney's collaborations with Renewcell and Evrnu to advance circular fiber innovation. By shifting production toward material regeneration, these models exemplify the principle of "designed to be remade" (Ellen MacArthur Foundation 2021b).

1.6.5 Enabling Technologies and Digital Transformation

Technological innovation is essential for the scalability of sustainable business models. Digitalization enhances traceability, efficiency, and customization, which are fundamental for achieving circularity. Blockchain technology enables full transparency by incorporating verifiable product histories into digital ledgers. The forthcoming EU Data Product Passport (DPP) framework will mandate that fashion brands disclose environmental data, material composition, and recyclability metrics for every product introduced to the market (Siliņa et al. 2024). Innovative brands like Chloé and EON are developing blockchain-enabled labeling to enable consumer tracking of garments from fiber origin to end-of-life recovery.

AI and machine learning are now more common in predicting demand, managing inventory, and optimizing designs. This has cut down on overproduction, which is the main cause of waste. AI-driven design systems make patterns more efficient, materials better, and style forecasting more accurately. This makes sure that supply and consumer preferences are in sync. One drawback about AI is that it takes a lot of energy to run AI storage centers.

Three-dimensional (3D) virtual prototyping platforms such as CLO3D, Browzwear, and Optitex enable designers to digitally render garments, thereby reducing reliance on physical samples. This change saves materials and energy, speeds up time-to-market, and is in line with the goals of sustainable design and resource efficiency (WRAP 2023). Some people who do not like virtual prototyping comment that it makes fashion move faster.

Smart and responsive textiles, including fibers that change temperature or biodegradable conductive materials, exemplify how technology and sustainability intersect. Although nascent, these materials innovations are expected to enhance product longevity and merge sustainability with functionality, in accordance with the long-term goals of the circular economy (Shamsuzzaman et al. 2025).

1.6.6 Life-Cycle Assessment and Environmental Footprint Analysis

LCA offers a quantitative methodology for evaluating environmental impacts throughout a product's lifespan, encompassing raw material acquisition through end-of-life management (Burns 2019). Numerous brands utilize LCA tools, including the Higg Index and the Product Environmental Footprint, to assess resource consumption, emissions generation, and water-related impacts. These evaluations inform ecodesign strategies and direct investments in circular innovation.

1.6.7 Environmental, Social, and Governance (ESG) Frameworks

ESG reporting frameworks and voluntary certifications (e.g., B Corporation, Global Recycled Standard, and OEKO-TEX) are becoming more integrated into corporate accountability systems. ESG reporting facilitates stakeholder comparison of performance based on social and environmental criteria, thereby promoting adherence to global sustainability objectives. The Textiles 2030 Roadmap incorporates science-based targets—a 50% reduction in carbon footprint and a 30% reduction in water footprint by 2030—as collective performance benchmarks for participating organizations (WRAP 2023).

In addition to conventional accounting practices, the triple bottom line framework (people, planet, profit) and the emerging regenerative accounting frameworks advocate for the inclusion of ecological restoration and social welfare as fundamental organizational objectives (Elkington 1997; Parkhurst et al. 2024). This shift underscores a philosophical transition from companies solely focused on profit maximization to entities dedicated to minimizing harm and generating positive environmental and social value.

1.6.8 The Future Trajectory of Sustainable Business Practices within the Fashion Industry

The confluence of design, policy, and digital transformation portends a significant revaluation of value within the fashion industry. The forthcoming decade will likely witness hybrid business models that integrate resale, rental, and repair services within cohesive brand ecosystems. EPR and similar policy instruments will enhance accountability, whereas developments in chemical recycling, digital traceability, and regenerative agriculture will strengthen the material basis of circularity (WRAP 2023). Ultimately, sustainable business models should be assessed not only on the basis of profitability but also according to their ability to mitigate environmental consequences, foster social equity, and maintain material life cycles. According to the Ellen MacArthur Foundation (2021b), the circular economy represents "not merely an economic system, but a cultural reorientation toward restoration, resilience, and responsibility." Within the fashion and textile industries, this reorientation constitutes both an ethical obligation and a creative prospect: to harmonize innovation with environmental limits and design with ecological principles.

1.6.9 Summary of Sustainable Business Models

Table 1.4 shows summary of sustainable business models (shown in the next page).

1.7 OPPORTUNITIES AND CHALLENGES

The textile and apparel industries intensively use resources such as agricultural land, water, and energy, and are highly globalized. The rapidly depleting resources, high water and air pollution, and huge amounts of solid textile waste caused by big quantities of textile and apparel production and consumption cause severe environmental challenges. Among them, resource depletion, including the depletion of water, energy, and raw materials, is a high-priority challenge that also brings opportunities in sustainable development.

Technical and management innovations provide abundant opportunities in sustainable textile and apparel industries. Drip irrigation in cotton cropland could save 20%–30% water compared with surface irrigation systems, and innovative technologies for soil-moisture monitoring and accurate crop coefficients for irrigation scheduling could also save irrigating water (Roth et al. 2013). Waterless dyeing technologies such as waterless air-dyeing process, supercritical CO_2 dyeing process, and plasma-assisted waterless dyeing process would save water and energy consumption and achieve better colorfastness (Jahan and Islam 2025). For example, the air-dyeing process uses 90%–95% less water and saves 86% energy than water dyeing (Jahan and Islam 2025).

Steam and thermal energy are intensively used in the textile industry, especially the dyeing and finishing process. The efficiencies of combustion boiler, thermal oil boiler, and electric boiler are 80%, 85%, and 98%–99%, respectively (Hasanbeigi and Zubeiri 2022). It is estimated that the annual potential energy saving of the textile industry in China will be 84 and 35 PJ if the steam boiler and fossil-fuel-fired thermal oil boilers, respectively, are electrified by 2050 (Hasanbeigi and Zubeiri 2022). In addition, electricity can be generated from renewable sources such as hydropower, solar, and wind, and the global renewable sourced electricity generation was nearly 10,000 TWh in 2024 (Ritchie et al. n.d.).

Developing regional circular textile and apparel supply chains to turn textile and apparel waste into new textile products provides opportunities to solve the textile waste and resource depletion problems (Thomas et al. 2025). Interviews with textile and apparel industry stakeholders, including representatives from brands and retailers, waste collectors, recyclers, nonprofit organizations, academic institutions, and government agencies, identified challenges and opportunities for textile circularity: technologies for materials identification, sorting, and preprocessing of post-consumer textile waste to enable recycling; innovations in recycling technologies to retain the quality of recycled materials; new design and manufacturing approaches considering feedstock variability and consumer needs; and prevention of greenwashing (Thomas et al. 2024).

TABLE 1.4
A Comparative Summary of Different Sustainable Business Models

Business Model Type	Description	Benefits	Limitations/Challenges	Examples
Product-service systems (rental, leasing, subscription)	Consumers access garments temporarily rather than owning them.	Extends product utilization; recurring revenue; reduced environmental impact per wear.	Logistics complexity; cleaning impacts; profitability challenges; adoption barriers.	Rent the Runway; Nuuly; My Wardrobe HQ
Resale and recommerce	Secondhand garments resold through platforms or brand programs.	Extends garment life; scalable; growing consumer acceptance.	Sorting/authentication costs; logistics challenges; rebound consumption risk.	ThredUp; Vestiaire Collective; Patagonia's Worn Wear
Repair and maintenance services	Repair, alteration, and refurbishment services to maintain product use.	Extends lifespan; builds loyalty; supports local jobs.	Labor-intensive; limited scalability; consumer willingness varies.	Nudie Jeans; Arc'teryx's ReBird; Repair Cafes
Closed-loop/take-back and remanufacturing	Products collected after use and materials recovered for new production.	Reduces virgin resource use; supports circularity; brand value.	Infrastructure needs; technology limitations; high costs.	Adidas' Made to Be Remade; Stella McCartney + Renewcell
On-demand and small-batch production	Products manufactured after purchase or in limited quantities.	Reduces overproduction; minimizes waste; supply–demand alignment.	Higher unit costs; longer delivery; scalability challenges.	Made-to-order SMEs; localized production brands such as eShakti; Imogene + Willie; Nooworks
Digital and AI-enabled optimization	AI forecasting, blockchain traceability, and 3D prototyping tools.	Reduces overproduction; improves efficiency; enhances transparency.	High investment; energy use; skills gap.	CLO3D; Browzwear; EON Digital IDs
Regenerative and bio-based material systems	Renewable materials and regenerative agriculture integration.	Reduced environmental impact; ecosystem benefits.	Higher costs; supply limitations; certification complexity.	Regenerative cotton programs; Fibershed; bio-based fibers; New York Textile Lab.
Hybrid circular ecosystems	Combination of resale, repair, rental, and recycling within one system.	Maximizes life-cycle value; diversified revenue; systemic impact.	Operational complexity; coordination challenges.	Patagonia's Worn Wear; Eileen Fisher Renew, Renewal Workshop

REFERENCES

Accelerating Circularity. 2021. Accelerating circularity chains of custody, claims, and traceability briefs 2021. https://www.acceleratingcircularity.org/accelerating-chain-of-custody [Accessed 26 November 2025].

American Apparel and Footwear Association. 2024. ApparelStats 2024. https://www.aafaglobal.org/AAFA/Solutions/Tools_Resources_Nav/AAFA/Solutions_Pages/Tools_Research.aspx?hkey=15779e99-86fe-435f-b154-a6723544e8e9 [Accessed 26 November 2025].

Bhatt, D., Silverman, J. and Dickson, M. A. 2019. Consumer interest in upcycling techniques and purchasing upcycled clothing as an approach to reducing textile waste. *International Journal of Fashion Design, Technology and Education* 12 (1): 118–128.

Bian, X., Fan, S., Xia, G., Chae, Y., Yu, H. and Xin, J. H. 2022. Sustainable ultrasonic dyeing of unscoured and unbleached cotton sliver using low liquor ratio. *Journal of Cleaner Production* 379: 134853.

Bonou-zin, R. D. C., Allali, K. and Fadlaoui, A. 2019. Environmental efficiency of organic and conventional cotton in Benin. *Sustainability* 11: 3044.

Brundtland, G. H. 1987. *Our Common Future: The World Commission on Environment and Development*. Oxford University Press, Washington, DC.

Burns, L. D. 2019. *Sustainability and Social Change in Fashion*. Fairchild Books, New York.

Cao, H., Cobb, K., Yatvitskiy, M., Wolfe, M. and Shen, H. 2022. Textile and product development from end-of-use cotton apparel: A study to reclaim value from waste. *Sustainability* 14: 8553.

Chapagain, A. K., Hoekstra, A. Y., Savenije, H. H. G. and Gautam, R. 2006. The water footprint of cotton consumption: An assessment of the impact of worldwide consumption of cotton products on the water resources in the cotton producing countries. *Ecological Economics* 60: 186–203.

Chen, G. and Baillie, C. 2009. Development of a framework and tool to assess on-farm energy uses of cotton production. *Energy Conversion and Management* 50 (5): 1256–1263.

Cobb, K., Cao, H., Yatvitskiy, M., Ludwig, K. and Gupman, S. 2025. The ReSpool fiber research (RFR) model: A protocol for the evaluation of mechanically recycled textile materials towards "second life" product applications. *Sustainability* 17: 10753.

Cotton Incorporated. n.d.. Why irrigate cotton? https://www.cottoninc.com/cotton-production/ag-resources/irrigation-management/why-irrigate-cotton/ [Accessed 24 November 2025].

de Oliveira, C. R. S., de Oliveira, P. V., Pellenz, L., de Aguiar, C. R. L. and da Silva Júnior, A. H. 2024. Supercritical fluid technology as a sustainable alternative method for textile dyeing: An approach on waste, energy, and CO_2 emission reduction. *Journal of Environmental Sciences* 140: 123–145.

Edirisinghe, L. G. L. M., de Alwis, A. A. P. and Wijayasundara, M. 2024. Sustainable circular practices in the textile product life cycle: A comprehensive approach to environmental impact mitigation. *Environmental Challenges* 12: 100985.

Efendy, R. P. and Dewi, K. 2024. Greenhouse gas inventory on textile finishing industry PT X. *Journal of Community Based Environmental Engineering and Management* 8 (1): 1–8.

Elkington, J. 1997. *Cannibals with Forks: The Triple Bottom Line of 21st Century Business*. New Society Publishers, Gabriola Island, BC.

Ellen MacArthur Foundation. 2017. A new textiles economy: Redesigning fashion's future. https://www.ellenmacarthurfoundation.org/a-new-textiles-economy [Accessed 24 November 2025].

Ellen MacArthur Foundation. 2021a. Recycling textiles to prevent waste: Renewcell. https://www.ellenmacarthurfoundation.org/circular-examples/recycling-textiles-to-prevent-waste-renewcell [Accessed 24 November 2025].

Ellen MacArthur Foundation. 2021b. Circular business models: Redefining growth for a thriving fashion industry. https://content.ellenmacarthurfoundation.org/m/4bd1902f3acbac0a/original/Executive-Summary-Circular-Business-Models.pdf [Accessed 24 November 2025].

Farhana, K., Kadirgama, K., Mahamude, A. S. F. and Mica, M. T. 2022. Energy consumption, environmental impact, and implementation of renewable energy resources in global textile industries: An overview towards circularity and sustainability. *Materials Circular Economy* 4: 15.

Feldstein, S., Hakansson, E., Katcher, J. and Vance, U. 2021. Shear destruction: Wool, fashion and the biodiversity crisis. https://www.collectivefashionjustice.org/shear-destruction [Accessed 3 November 2025].

Fletcher, K. 2008. *Sustainable Fashion and Textiles: Design Journeys*. Earthscan, London.

Fuglie, K., Morgan, S. and Jelliffe, J. 2024. Global changes in agriculture production, productivity, and resource use over six decades. U.S. Department of Agriculture, Economic Research Service, Amber Waves. https://www.ers.usda.gov/amber-waves/2024/september/global-changes-in-agricultural-production-productivity-and-resource-use-over-six-decades [Accessed 30 October 2025].

Gam, H. J., Cao, H., Farr, C. and Heine, L. 2009. C2CAD: A sustainable apparel design and production model. *International Journal of Clothing Science and Technology* 21: 166–179.

Granskog, A., Laizet, F., Lobis, M. and Sawers, C. 2020. Biodiversity: The next frontier in sustainable fashion. https://www.mckinsey.com/industries/retail/our-insights/biodiversity-the-next-frontier-in-sustainable-fashion [Accessed 30 October 2025].

Hasanbeigi, A., Springer, C. and Wei, D. 2025. Low carbon thermal energy roadmap for textile industry. global efficiency intelligence. https://www.globalefficiencyintel.com/low-carbon-thermal-energy-roadmap-for-textile-industry [Accessed 16 November 2025].

Hasanbeigi, A. and Zubeiri, M. J. S. 2022. Electrification of steam and thermal oil boilers in the textile industry: Techno-economic analysis for China, Japan, and Taiwan. *Energies* 15 (23): 9179.

Jahan, R. and Islam, M. R. 2025. An overview of waterless dyeing in textile industry: Advancements, challenges, and environmental impact analysis. In *Advancements in Textile Coloration: Techniques, Technologies, and Trends*. (Eds. Shahid, M., Maiti, S., Khan, S. A. and Adivarekar, R. V.), Springer, Singapore, 75–102.

Jučienė, A., Gurauskienė, I. and Kruopienė, J. 2025. Calculating the environmental impact reduction due to extended lifespan of clothing through clothing swaps. *Sustainability* 17 (10): 4411.

Kabir, S. M. F., Chakraborty, S., Hoque, A. M. A. and Mathur, K. 2019. Sustainability assessment of cotton-based textile wet processing. *Clean Technologies* 1: 232–246.

Kim, M., Shin, J. Y., Lim, S., Lee, H., Kwon, S. C., Hong, S. and Ryu, S. 2024. Reduction of greenhouse gas emissions by optimizing the textile dyeing process using digital twin technology. *Fashion and Textiles* 11: 17.

Kir, A., Ozturk, E., Yetis, U. and Kitis, M. 2024. Resource utilization in the sub-sectors of the textile industry. Opportunities for sustainability. *Environmental Science and Pollution Research* 31: 25312–25328.

Laitala, K. and Klepp, I. G. 2021. Clothing longevity: The relationship between the number of users, how long and how many times garments are used. *Proceedings of the 4th PLATE 2021 Virtual Conference*, Limerick, Ireland, 26–28 May 2021.

Leal Filho, W., Dinis, M. A. P., Liakh, O., Paço, A., Dennis, K., Shollo, F. and Sidsaph, H. 2024. Reducing the carbon footprint of the textile sector: An overview of impacts and solutions. *Textile Research Journal* 94 (15–16): 1798–1814.

Lobo Mau. 2026. *Lobo Mau*. https://lobomau.com/ [Accessed 16 February 2026].

Magnus, K.-H., Ibáñez, P., Scalco, A. and Hügl, J. 2024. Reimagining the apparel value chain amid volatility. McKinsey and Company. https://www.mckinsey.com/industries/retail/our-insights/reimagining-the-apparel-value-chain-amid-volatility/ [Accessed 18 November 2025].

Meadows, D. H. 2008. *Thinking in Systems: A Primer*. Chelsea Green Publishing, Chelsea, VT.

Meyer, L. and Dew, T. 2024. Cotton and wool outlook: July 2024. Report No. CWS-24g. U.S. Department of Agriculture, Economic Research Service. https://ers.usda.gov/sites/default/files/_laserfiche/outlooks/109537/CWS-24g.pdf [Accessed 30 October 2025].

McKinsey and Company. 2024. What are scope 1, 2, and 3 emissions? https://www.mckinsey.com/featured-insights/mckinsey-explainers/what-are-scope-1-2-and-3-emissions [Accessed 25 November 2025].

Nachimuthu, G. and Webb, A. A. 2016. On-farm soil conservation measures in cotton farming systems of Australia: A sustainability analysis. *Journal of Soil and Water Conservation* 71 (3): 75A–80A.

Navarro-Gambín, P., Bonetti, M., Villa, M., Brunori, G. and Vergamini, D. 2025. Paving the ground for biodiversity-positive transformative change in fashion: An exploration of drivers of biodiversity loss and barriers to transformative change in the textile, apparel, and fashion sector. *International Journal of the Commons* 19 (1): 322–341.

Nimkar, U. 2018. Sustainable chemistry: A solution to the textile industry in a developing world. *Current Opinion in Green and Sustainable Chemistry* 9: 13–17.

Parkhurst, T., Standish, R. J., Prober, S. M., Kobryn, H. and Vardon, M. 2024. Balancing the books of nature by accounting for ecosystem condition following ecological restoration. *Scientific Reports* 14 (1): 11369.

Peña, C., Civit, B., Gallego-Schmid, A., Druckman, A., Pires, A. C., Weidema, B., Mieras, E., Wang, F., Fava, J., Canals, L. M. I., Cordella, M., Arbuckle, P., Valdivia, S., Fallaha, S. and Motta, W. 2021. Using life cycle assessment to achieve a circular economy. *The International Journal of Life Cycle Assessment* 26 (2): 215–220.

Porter, M. E. 1985. *Competitive Advantage: Creating and Sustaining Superior Performance*. Free Press, New York.

Raworth, K. 2017. *Doughnut Economics: Seven Ways to Think Like a 21st Century Economist*. Chelsea Green Publishing, Chelsea, VT.

Ritchie, H., Roser, M. and Rosado, P. n.d. Renewable energy. Our world in data. https://ourworldindata.org/renewable-energy [Accessed 26 November 2025].

Rockström, J., Steffen, W., Noone, K., Persson, Å., Chapin, F. S., Lambin, E. F., Lenton, T. M., Scheffer, M., Folke, C., Schellnhuber, H. J., Nykvist, B., de Wit, C. A., Hughes, T., van der Leeuw, S., Rodhe, H., Sörlin, S., Snyder, P. K., Costanza, R., Svedin, U. … and Foley, J. A. 2009. A safe operating space for humanity. *Nature* 461 (7263): 472–475.

Roth, G., Harris, G., Gillies, M., Montgomery, J. and Wigginton, D. 2013. Water-use efficiency and productivity trends in Australian irrigated cotton: A review. *Crop and Pasture Science* 64 (12): 1033–1048.

Sandin, G., Peters, G. M. and Svanström, M. 2013. Moving down the cause-effect chain of water and land use impacts: An LCA case study of textile fibers. *Resources, Conservation and Recycling* 73: 104–113.

Saxena, S., Raja, A. and Arputharaj, A. 2017. Challenges in sustainable wet processing of textiles. In *Textiles and Clothing Sustainability* (Eds. Muthu, S. S.), Springer, Berlin/Heidelberg, 43–79.

Schumacher, K. and Forster, A. 2022. Facilitating a circular economy for textiles workshop report, special publication (NIST SP), National Institute of Standards and Technology, Gaithersburg, MD. https://tsapps.nist.gov/publication/get_pdf.cfm?pub_id=934344 [Accessed 25 November 2025].

Shamsuzzaman, M., Islam, M., Mamun, M. A. A., Rayyaan, R., Sowrov, K., Islam, S. and Sayem, A. S. M. 2025. Fashion and textile waste management in the circular economy: A systematic review. *Cleaner Waste Systems* 11: 100268.

Sharpe, S., Dominish, E. and Martinez-Fernandez, M. C. 2022. Taking climate action: Measuring carbon emissions in the garment sector in Asia, International Labour Organization (ILO) Working Paper 53, Geneva, ILO. https://www.ilo.org/publications/taking-climate-action-measuring-carbon-emissions-garment-sector-asia [Accessed 3 November 2025].

Siliņa, L., Dāboliņa, I. and Lapkovska, E. 2024. Sustainable textile industry: Wishful thinking or the new norm: A review. *Journal of Engineered Fibers and Fabrics* 19: 1–27.

Singha, K., Pandit, P., Maity, S., Sharma, S. R. and Singha, K. 2021. Harmful environmental effects for textile chemical dyeing practice. In *Green Chemistry for Sustainable Textiles: Modern Design and Approaches*, (Eds. Ibrahim, N. and Hussain, C. M.), Woodhead Publishing, Cambridge, 153–164.

Textile Exchange. 2024. The future of synthetics. https://textileexchange.org/knowledge-center/reports/the-future-of-synthetics/ [Accessed 24 November 2025].

Textile Exchange. 2025. Materials market report. https://textileexchange.org/knowledge-center/reports/materials-market-report-2025/ [Accessed 3 November 2025].

Thomas, K., Clarke-Sather, A. R., Cobb, K. and Cao, H. 2025. ReSpool: Scaling a circular supply chain for recycled textiles. *Journal of Advanced Manufacturing and Processing* 7 (2): e70000.

Thomas, K., Durrani, H., Brady, J., Ludwig, K., Yatvitskiy, M., Clarke-Sather, A. R., Cao, H. and Cobb, K. 2024. Fundamental challenges and opportunities for textile circularity. *Sustainability* 16: 11117.

ThredUp. 2024. 2024 resale report. https://cf-assets-tup.thredup.com/resale_report/2024/ThredUp_2024_Resale%20Report.pdf [Accessed 26 November 2025].

Tian, H. and Cheng, C. 2024. International trade and biodiversity: Is export a species killer? *Ecological Economics* 225: 108313.

Trunk, U., Harding-Rolls, G., Banegas, X. and Urbancic, N. 2021. Fossil fashion: The hidden reliance of fast fashion on fossil fuels. Changing Markets Foundation. https://changingmarkets.org/report/fossil-fashion-the-hidden-reliance-of-fast-fashion-on-fossil-fuels/ [Accessed 24 November 2025].

Uniform Market. 2025. Global apparel industry statistics (2025). https://www.uniformmarket.com/statistics/global-apparel-industry-statistics [Accessed 24 November 2025].

United Nations Industrial Development Organization. 2017. Circular economy. https://www.unido.org/sites/default/files/2017-07/Circular_Economy_UNIDO_0.pdf [Accessed 30 November 2025].

United Nations Sustainable Development. n.d. The 17 goals. https://sdgs.un.org/goals [Accessed 4 November 2025].

U.S. Department of Agriculture. n.d. Production: Cotton. https://www.fas.usda.gov/data/production/commodity/2631000 [Accessed 26 November 2025].

U.S. Government Accountability Office. 2024. Textile waste: Federal entities should collaborate on reduction and recycling efforts (GAO-25-107165). https://www.gao.gov/assets/gao-25-107165.pdf [Accessed 24 November 2025].

Wittchen, S. 2022. With the addition of Fabscrap, Philly's textile industry moves closer to circularity. https://gridphilly.com/blog-home/2022/01/03/circular-fashion [Accessed 6 February 2026].

WRAP. 2021. Textiles 2030 annual progress update 2022-23. https://www.wrap.ngo/resources/report/textiles-2030-annual-progress-update-2022-23 [Accessed 17 June 2026].

WRAP. 2023. Circular design toolkit for fashion and textiles. https://www.wrap.ngo/resources/guide/circular-design-toolkit [Accessed 24 November 2025].

WRAP. 2025. UK textiles pact roadmap 2026–2030. https://www.wrap.ngo/resources/report/uk-textiles-pact-roadmap-2026-2030 [Accessed 24 November 2025].

Yang, S., Song, Y. and Tong, S. 2017. Sustainable retailing in the fashion industry: A systematic literature review. *Sustainability* 9 (7): 1266.

Yatvitskiy, M., Abayode, D., Cobb, K., Thomas, K. and Haddad, N. 2024. Reviving fashion: Weaving life back into discarded textile waste. *International Textile and Apparel Association Annual Conference Proceedings*, Long Beach, CA, November 20–23, 2024.

Zhang, Y., Liu, X., Xiao, R. and Yuan, Z. 2015. Life cycle assessment of cotton T-shirts in China. *International Journal of Life Cycle Assessment* 20: 994–1004.

Zhou, L. and Zhou, H. 2019. Evaluation of textile industry wastewater treatment and reuse in China. *Desalination and Water Treatment* 144: 330–338.

2 Fashion Circularity Standards

Recent Trends and Case Studies

Rajkishore Nayak, Anshul Nayak, Huy Quang, and Tarun Panwar

2.1 INTRODUCTION

Clothing ranks as the second most essential requirement for human survival after food. Recent developments in technology and increased earnings of consumers have led to increased amounts of clothing and textiles for each household (Joung 2014). Further, the rapid growth of fast fashion has increased consumers' intention of buying trendy clothes, more often and more than they need (Niinimäki et al. 2020). This, in turn, has led to a wasteful culture, where a large quantity of clothing and textiles are discarded into the landfill prematurely despite being serviceable (Nayak and Patnaik 2021). To mitigate these problems, circular economy (CE) model, also known as fashion circularity, is one of the approaches in fashion and textiles.

Fashion circularity or circular fashion comprises a set of frameworks within the broader and widely used term "Circular Economy (CE)" (Amir et al. 2023; Papamichael et al. 2023). The frameworks for fashion circularity focus on eliminating waste, recycling, upcycling or reducing it through several approaches in the fashion value chain (Mishra et al. 2021). For example, CE can be achieved by sustainable design and using the clothing in a manner that enhances the durability of the product so that it can be used for longer time rather than landfilled (Dissanayake and Weerasinghe 2022). Modular designing is also another approach for achieving the objectives of CE (Machado and Morioka 2021). Similarly, introducing product tack-back policies, which assign responsibility to a set of organizations to manage the end-of-life (EOL) fashion and textiles, can also support the CE framework (Tran et al. 2023).

At the EOL of the product, it is given a new life by the approaches such as repair, reuse, resale, redesign and rental (Nayak et al. 2021a; Islam 2021). When the product is deemed to be useful, the material can be recovered and reused with virgin materials to produce products with lower levels of usage. Finally, the product can be responsibly reintroduced into the biosphere by suitable approaches such as composting (Nayak et al. 2025). This creates a closed-loop system, contrasting with the conventional linear model of "take-make-use-dispose" commonly followed by many

DOI: 10.1201/9781003689225-2

industries. Although it looks to be simple, successful implementation of CE model faces several challenges to achieve its objectives. The linear model and circular fashion model are shown in Figures 2.1 and 2.2, respectively.

There have been several recent publications on various approaches for fashion circularity (Baloyi et al. 2024; Saha et al. 2022). However, there has been a dearth of literature on the recent status of fashion circularity standards. This chapter focuses mainly on the recent trends in circular fashion models and various global standards

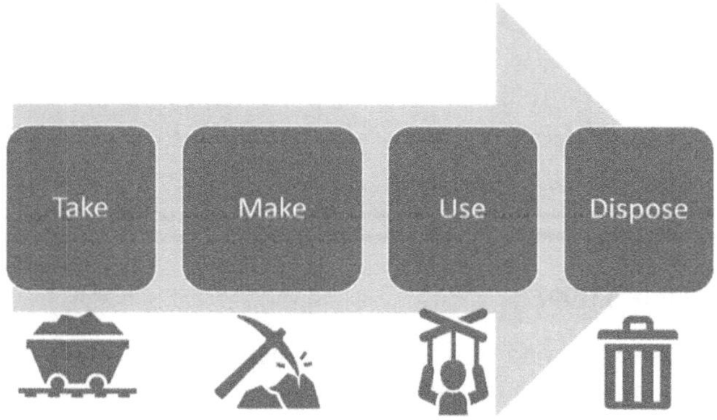

FIGURE 2.1 Linear model followed in fashion and textiles (take-make-use-dispose).

FIGURE 2.2 Circular economy fashion model: trend in recent times.

focusing on CE models. The authors have explored some of the key regulations that can enhance sustainability, recycling and circularity in fashion and textile manufacturing. The chapter was prepared using the data collected from secondary resources and the knowledge of the authors in fashion circularity. Some primary data is also included through the interview with the managers of fashion and textile industries in Vietnam. The chapter is organized as follows:

- The first section gives a brief introduction to fashion circularity, its need and some circularity data. This section covers the definitions of fashion circularity and its importance in reducing various environmental impacts of fashion. Various principles of fashion circularity are also discussed in detail.
- The next section highlights various standards relating to circular fashion models and their recent trends. Various regulations on circular fashion around the world are also included.
- Special attention has also been given to the European standards in fashion circularity. Some of the European standards such as extended producer responsibility (EPR), UK Textiles Pact and Eco-design for sustainable fashion are also covered.
- This chapter also highlights case studies from developing countries such as Vietnam on fashion circularity. Based on the secondary data for the fashion and textile industries in Vietnam, we have included a separate section. The major findings from the data analysis are presented in this chapter.
- Various opportunities and challenges are discussed in the next section. The challenges were identified from the secondary data analysis in addition to the findings from the interview. Various opportunities in fashion circularity were also identified.
- Finally, the chapter covers a brief conclusion of the major findings, followed by the future directions.
- A list of leading fashion brands working on fashion circularity and glossary relating to CE are also provided in the Appendix.

2.2 NECESSITY OF FASHION CIRCULARITY

Manufacturing of fashion and textiles in recent times has increased at an unprecedented rate. The total number of clothing items manufactured every year range from 100 to 150 billion pieces on a global scale. The total weight of these garments can be approximately 100 million tons, which leads to the usage of resources at an alarming rate. Out of these total clothing weight, about 80% end in the landfill at the EOL of the product (Nayak et al. 2021a). The remaining 20% are repurposed by various methods such as reuse, upcycling or recycling. Hence, it can be concluded that a vast majority of the resources that have the potential to be reused are wasted through incineration and/or landfill (Stella et al. 2024; Grover et al. 2025). Fashion circularity approaches can help to recover the resources for repurpose to different applications. There are several points for which the circular fashion model is necessary in recent times as discussed here.

- **Environmental prospects**: Fashion and textile sectors produce a large amount of solid waste during the manufacturing process as well as during the post-consumer stage (Azad et al. 2024; McCauley and Jestratijevic 2023). On several occasions, these wastes are either discarded into the landfills and/or incinerated leading to environmental pollution. In the landfills, these wastes pose several environmental threats such as producing greenhouse gases (GHGs) such as methane (CH_4) or chemical leaching into the soil or water (Gupta et al. 2022). Further, these wastes rapidly fill up the landfill space leading to the problem of space shortage. Incineration of fashion and textiles on the other hand produces huge amounts of carbon dioxide (CO_2) and other GHGs leading to environmental pollution. The implementation of fashion circularity can help to solve the problem of environmental pollution.
- **Economic prospects**: The use of CE models in fashion can save the industries buying new virgin materials. The recycling sector can generate large amounts of money while dealing with EOL clothing. Further, CE models in fashion can financially support government and local councils spending a substantial lower amount of money to collect, transport and dispose of the fashion and textile waste from landfills (Abdelmeguid et al. 2022). Circularity in fashion saves this expenditure by reducing the costs related to environmental remediation from landfills and incineration or repurposing of fashion and textile waste. Transitioning to a circular economic model creates economic advantages by establishing new industrial jobs in recycling and upcycling and generating value from materials that have been discarded.
- **Resource recovery**: The current practices of linear economy model led to the resources being wasted in landfills that could have been reused. Every year, millions of dollars are wasted due to the linear model in fashion and textiles (Baloyi et al. 2024). Adopting fashion circularity can help the resources being reused, which preserves the resources for future generations by reducing new resource consumption. The use of virgin materials is reduced, which helps in financial savings and lowers the consumption of nonrenewable resources.

Effective regulations and policies designed for fashion circularity must consider both supply and demand factors (Faludi 2025). On the supply side, it is essential to invest in initiatives that protect the environment, commonly known as "green finance". Transitioning from conventional business models to circular business models is crucial for a successful transition to a CE model (Huynh 2022). Policymakers should focus on the challenges of attracting investment and ensuring the financial sustainability of these initiatives to encourage further investment in the CE.

On the demand side, policymakers should work to make circular materials more appealing to consumers, including older adults, individuals with lower educational attainment and high-income groups. Fiscal policy incentives can drive the adoption of circular practices in fashion and textile sectors. Future policymakers may assess the impact of these consumer incentives on the uptake of sustainable behaviors that are consistent with CE principles.

2.3 KEY PRINCIPLES OF FASHION CIRCULARITY

The fundamental principles of fashion circularity, which are derived from the CE model, can be generally classified into four main groups: designing for circularity, using sustainable materials, product usage and EOL management or adopting recycling models for material recovery (Moorhouse and Moorhouse 2017). These principles are highlighted in the subsections below.

2.3.1 Designing for Circularity

Designing fashion and textiles with a focus on durability and recyclability, frequently incorporating modular or adjustable features to prolong their lifespan can lead to fashion circularity (D'Itria and Aus 2023). Durability helps in the use of resources for a longer period, whereas recyclability helps to recover the resources by recycling. Modular design supports reducing the number of products, ultimately saving resources (Machado and Morioka 2021). To be successful in fashion circularity, the designers should focus on CE models for producing fashion and textiles.

2.3.2 Using Sustainable Materials

While designing fashion and textile products, the focus should be utilizing recycled or renewable materials, as well as materials that can be safely reintegrated into the economy (Nayak et al. 2021b). There are several new materials that are renewable and biodegradable, widely available for manufacturing sustainable fashion (Nayak et al. 2023). For example, new fibers derived from bananas and lotus are becoming popular in fashion and textile manufacturing (Maduna and Patnaik 2023). Similarly, bio-based materials are getting wider acceptance as new sustainable materials in fashion and textile manufacturing (Nayak et al. 2024). Recycled materials are also the choice for sustainable fashion and textile manufacturing. The designers should aim to use these materials to reduce the use of nonrenewable and nonbiodegradable materials.

2.3.3 Product Usage at End of Life

Establishing business models that facilitate extended life of garments at their EOL such as rental, resale, reuse or subscription, can contribute toward fashion circularity (Amasawa et al. 2023). Creating systems for the collection, sorting, repair and responsible recycling or upcycling of textiles can lead to CE models. Implementing models such as the European Union (EU)'s EPR at the EOL of products, can help to assign the responsibility of dealing with EOL fashion and textiles rather than being landfilled (Tran et al. 2023).

2.3.4 Adopting Recycling Models for Material Recovery

The CE models should focus on complete recycling or biodegradation of materials, once they reach EOL of their life cycle, allowing them to be transformed into new products (Claussen et al. 2023). The current practice of linear model is generating millions of tons of textiles ending in landfills or being incinerated. Adopting recycling models can help in resource recovery to make new products or even support composting, creating economical value for the resources otherwise dumped to the landfills (Nayak et al. 2025).

2.4 FASHION CIRCULARITY STANDARDS

Legal authorities (global and local) and government advocate for ecolabels and sustainability standards that promote circularity in fashion and textiles. Based on these regulations, fashion and textile manufacturers and product designers are required to give special consideration to follow CE models. For any fashion and textile product to be classified as "green", the product must demonstrate ecological responsibility throughout the entire life stages including manufacturing, usage and maintenance, as well as after its useful life comes to an end or at the EOL. The triple bottom line (TBL) of sustainability, which covers environmental, social and economic factors, must be integrated into product design and manufacturing (Nayak 2020). These factors need to be certified by independent organizations to ensure products are safe for both people and the planet.

A range of organizations have developed frameworks, metrics and certifications to guide the fashion industry toward circularity. In contrast to a singular and universal standard, fashion circularity is evaluated through a blend of indicators that examine material sourcing, design, manufacturing and EOL strategies. As such, there is not a single and unified standard for fashion circularity; rather, it's a combination of widely adopted strategies and framework like the 10R approach with digital passports for easy traceability and tracking.

2.4.1 Need for Standards

Standards are essential for organizations to set their goals, develop strategies and implement operational changes to meet the objectives of CE transition (Dissanayake and Weerasinghe 2022). Various standards established within the CE framework provide guidelines and requirements for the execution of activities by all participating organizations to enhance their transition. Standards facilitate the sustainable utilization of resources and energy while safeguarding consumers, workers and the environment. Circularity standards are essential tools in fostering sustainable and circular practices during fashion and textile production. They offer guidance and encourage organizations to adopt socially and environmentally responsible practices.

2.4.2 List of Circularity Standards

Fashion circularity standards not only represent a singular, cohesive set of regulations but also include principles and developing regulatory frameworks, such as the EU's Eco-Design Regulation and EPR initiatives (Batool and Nawab 2023). The primary standards emphasize the importance of designing for longevity, repairability and recyclability; favoring safe, recycled or renewable materials; adopting extended use models such as rental and resale; and creating effective collection and recycling systems to effectively close the loop on textile waste. Some global circularity standards from international organizations and other agencies are discussed below and summarized in Table 2.1.

- **ISO 26000**: Assists organizations in incorporating social responsibility, which encompasses environmental sustainability and circularity, by advocating for ethical practices and resource efficiency within the textile industry.
- **ISO 59000 family**: Offers extensive guidelines for the adoption of a CE, addressing principles, measurement and transition pathways for enterprises, including those in fashion and textiles.
- **ISO 14040 and ISO 14044**: Global standards for life cycle assessment (LCA) that facilitate circularity by supplying frameworks to assess the environmental impacts of textile products throughout their entire life cycle.

TABLE 2.1
List of Circularity Standards for Fashion and Textiles

CE Standard Number	Title	Description
ISO 59004:2024	Circular economy – Vocabulary, principles and guidance for implementation	ISO 59004 belongs to the ISO 59000 family of standards, which is intended to promote a transition to a circular economy. These standard highlights the definition of essential terms and concepts, presents a vision for a CE model, clarifies fundamental principles and provides practical advice for implementing actionable steps toward sustainability.
ISO 59010:2024	Circular economy – Guidance on the transition of business models and value networks	ISO 59010 provides guidelines to organizations aiming to change their business models from a linear to a CE model. This standard highlight business-oriented strategies for implementing CE practices. It is a complementary standard to ISO 59004, providing comprehensive guidance on evaluating existing value creation models, mapping value chains and networks, and formulating strategies for circularity. It helps organizations promoting sustainable business practices to foster a resilient global economy.
ISO 59020:2024	Circular economy – Measuring and assessing circularity performance	ISO 59020 offers guidance for organizations to evaluate and measure their circular performance within specified economic systems. It helps to standardize the methodology by which organizations gather and compute data utilizing both mandatory and optional circularity indicators, thereby ensuring results that are consistent and verifiable. It presents a systematic framework for defining system boundaries, choosing suitable indicators and analyzing data to assess circularity performance across various levels.
ISO/TC 323	Circular economy	This standard establishes frameworks, guidelines and requirements for the execution of activities by all participating organizations, thereby enhancing their contribution to Sustainable Development.

- **EPR and eco-design regulations**: Encompass obligatory eco-design requirements and eco-modulation of fees, aimed at enhancing durability, reparability, recyclability and recycled content of textiles to promote circularity in fashion products across Europe.
- **Cradle-to-cradle certification**: Linked to circular material guidelines, this certification evaluates products based on their sustainability and circular material content specifically for the fashion sector.
- **Circular materials guidelines 1.0 by Fashion Positive**: Establishes criteria for circular materials that emphasize safer, renewable and recycled fibers engineered for durability and recyclability in textile manufacturing.
- **Make fashion circular framework from Ellen MacArthur Foundation**: While not formal standards, these frameworks significantly influence the development of CE practices and innovation within the fashion industry.
- **Circular design standard**: A guideline established by circular fashion that assists brands in designing for circularity, providing advice on material choices and the possibilities for a garment to be repaired, reused and recycled.

2.4.3 Regulations for CE

Circularity regulations play a vital role in promoting sustainability and CE in fashion and textiles. Like the standards, the regulations provide guidelines for organizations in following the ethical practices on the TBL (planet, people and profit). Regulatory bodies in many countries need to incorporate policies that help fashion and textile industries from the linear model to circular model. Many countries in EU work under stricter regulations of circularity (Bour et al. 2023). However, in majority of the Asian countries, these regulations are not strictly followed. Among all the Asian countries, China and Japan are the leading countries in implementing regulations on fashion circularity.

To promote CE in fashion and textiles, the fashion brands and manufacturers should create consumer awareness while managing the discarded clothes. Further, they should emphasize practices such as recycling, upcycling, downcycling and product take-back practices (Jing et al. 2026). The target of reducing waste through different approaches (such as zero-waste patternmaking and modular design) should also be prioritized by the fashion and textile manufacturers. As such, many of the retailers and manufacturers are still following the linear economic model, which involves take-make-use-dispose. Implementing the CE model needs wide-scale preparation, development of infrastructure for collection, sorting and recycling in addition to the new product development.

Regulations should cover the product design aspects focusing on durability and recyclability. The durability of products will ensure that they can be used for a longer period, saving the resources. Similarly, recyclability will ensure that the product can be recycled rather than landfilled or incinerated. The material labels in the product or a QR code in the product can provide necessary information to the consumers or recyclers to take the necessary action while recycling. Several agreements focused on sustainability aspects to reduce the global carbon footprint, which include the UN's climate agreements or international agreements on climate change, the Kyoto protocol, the Paris agreement, the European Green Deal and the European Climate Law. These guidelines are also applicable to various fashion and textile products.

2.5 RECENT EUROPEAN STRATEGY FOR CIRCULAR FASHION

The 28 countries listed in the EU are considered as developed economies. The consumers in these countries purchase large amounts of fashion and textiles, which lead to the rejection of around 11 kg of products per consumer annually. The huge amount of waste has several negative impacts on the planet in addition to the wastefulness of resources. To address the issue, the EU strategy for Sustainable and Circular Textiles focuses on several approaches, which are mentioned in the subsections below.

2.5.1 Brand and Consumer Responsibility

Global fashion brands are one of the greatest influences when the CE transition is considered. They can take several steps starting from product design till the product is responsibly managed at the end of its life (Ki et al. 2021). They can enhance consumer awareness through various initiatives to focus on quality, durability and longer use of clothes. Further, creating infrastructure for dealing with EOL products can be beneficial as many countries lack these facilities. At the same time, the consumers should also focus on the EOL choices for fashion and textile products. Rather than incineration or landfill, they should select the other routes to discard their EOL fashion and textiles.

2.5.2 Eco-Design for Sustainable Products Regulation

The eco-design for sustainable products regulation (ESPR) was implemented by the EU in 2024 (Commission 2024). The ESPR establishes regulatory guidelines for products marketed within the EU under sustainability drive. As per the guidelines, significant importance is given to making the products durable, prohibiting the destruction of unsold merchandise, making repairable and recyclable products, and mandating Digital Product Passports (DPPs) to furnish sustainability-related information. The objective of this regulation is to switch from buying too many products to durable products following CE model.

2.5.3 Digital Product Passports

A DPP for any fashion product serves as an electronic record outlining its complete life cycle (Colasante et al. 2025). The electronic record holds necessary information on material origin, manufacturing processes, environmental impact and recyclability. The use of digital technologies such as a QR code or a near-field communication tag or a radio frequency identification (RFID) tag. This is necessary as per the EU's ESPR for any product designed to be sold in the EU market. DPPs are designed to enhance transparency and encourage circularity within the industry by supplying comprehensive product details to brands, consumers and regulators. DPP can help to increase consumer trust, enhance brand value, foster circularity and ensure regulations are being followed.

2.5.4 Global Circularity Protocol

Global Circularity Protocol (GCP) is a novel science-based framework aimed at assisting businesses globally in measuring, managing and communicating their circular performance in a consistent manner. This protocol was developed and introduced by the World Business Council for Sustainable Development (WBCSD) in collaboration with the UN Environment Program (UNEP) in November 2025. The primary objective of the GCP is to establish a harmonized framework to facilitate global shift from a linear economy to a CE. The key objectives of GCP are as follows:

- **Standardization**: Creating consistent guidelines and metrics for circularity, parallel to the role of the GHG Protocol in carbon emissions.
- **Accountability and transparency**: Providing reliable and comparable data, reporting to foster investor confidence and channel capital toward circular solutions.
- **Impact measurement**: Assisting businesses in monitoring material flows throughout supply chains to assess their effects on climate, nature and equity.
- **Risk management and innovation**: Supporting companies in pinpointing circularity "hotspots" to prioritize interventions, stimulate innovation in business models and mitigate risks associated with material sourcing and price fluctuations.

2.5.5 UK Textiles Pact

The UK Textiles Pact, earlier known as Textiles 2030, represents the foremost voluntary initiative in the United Kingdom aimed at assisting businesses and organizations in the fashion and textile sector to shift toward more sustainable and circular practices by the end of this decade (2030). This pioneering initiative unites organizations from various sectors of the clothing and textiles industry to collaborate on achieving swift, science-driven advancements in climate action and transitioning toward a more circular system. The Pathway outlines the necessary activities and milestones to meet the UK Textiles Pact targets and will be revised based on new evidence and advancements.

Several organizations have pledged to reduce their carbon emissions by 50% and their water usage by 30% (Leal Filho et al. 2024). They are also collaborating to implement more circular fashion practices within their businesses. This roadmap serves as a guide for various organizations on how to meet these ambitious objectives and outlines significant milestones from the year 2025 to 2030. Meeting these targets will be a daunting task and necessitate substantial changes, not only from fashion brands and retailers but also from all participants in the fashion textiles value chain, including textile collectors and recyclers, reuse organizations, technology innovators, sector organizations, academic and government.

The roadmap seeks to motivate textiles businesses to collaborate effectively in order to achieve the goals of the UK Textiles Pact. Together, these businesses will experiment, learn and refine their strategies. It is a dynamic strategy that will change in future versions as they discover effective methods and actions that can create the most significant impact. As part of the roadmap, the Circularity Pathway offers more detailed guidance, particularly for signatories aiming to assume a leadership role, on how they will collaborate to develop and implement circular practices. To effectively minimize carbon impacts, the businesses must innovate and modify essential management practices, including the following:

- Designing products that maintain their aesthetic appeal for an extended period and are recyclable at their EOL;
- Experimenting with reuse business models such as resale, rental and subscriptions; and
- Establishing partnerships to source and utilize recycled fibers in new products.

2.5.6 Product Take-Back Policies

Product take-back policies or EPR is a legislation that holds the manufacturers responsible for their goods covering the entire life cycle (Tran et al. 2023). This includes the stages of raw material extraction, product manufacturing and consumer purchase of the product or even the EOL waste management of the product. The implementation of EPR supports achieving environmental objectives in reducing the total impact of a product. It can include producing durable products that can reduce the extraction. In addition to creating comprehensive data on production, products, waste creation and treatment, EPR also raises money from manufacturers to assist cover the costs of waste product collection, sorting and recycling.

EPR is a policy framework that holds producers accountable for their products throughout their entire life cycle, including the post-consumer phase (Cai and Choi 2019). This approach aids in achieving environmental objectives such as recycling goals and conserving resources that would otherwise end up in landfills. Additionally, EPR provides financial support from producers to cover the costs associated with the collection, sorting and recycling of waste materials, while also generating comprehensive data regarding production, products, waste generation and treatment. Consequently, EPR policies typically transfer the financial burden of waste management or the physical collection of waste, either partially or entirely, from local governments to the producers.

2.6 CIRCULARITY EVALUATION TOOLS AND ORGANIZATIONS

2.6.1 Circular Transition Indicators Framework

The Circular Transition Indicators (CTI) framework within the fashion industry is a quantitative and standardized system to measure and promote circularity in the fashion and textile sectors (John 2025). It was created by the WBCSD along with its members for fashion and textile sectors. The CTI framework offers universal guidelines for evaluating circular performance and accountability based on some key indicators such as those for closing loops, enhancing resource utilization, appreciating products and improving overall material productivity. Further, the CTI framework supports fashion enterprises to change as per the growing regulations, to work together on circular pathways and to accelerate the shift toward circular business models. This presents a significant economic and environmental opportunity for the fashion sector to minimize waste, reduce land utilization and lower carbon emissions while improving sustainable value generation.

2.6.2 Material Circularity Indicator

The Material Circularity Indicator (MCI) is a metric created by the Ellen MacArthur Foundation to evaluate the circularity of a product (Tetteh et al. 2025). The MCI metric is based on the principles of a CE and the ISO 59020 standard relating to circularity metrics. The MCI measures the extent to which linear material flows (such as the use of virgin materials and waste) are reduced, and restorative flows (including recycled and reused materials) are increased. Additionally, it considers the duration and intensity of a product's usage.

The MCI metric ranges from 0 to 1, where 0 indicates the product is completely linear and 1 indicates the product is completely circular. This means the score of 1 is given to products that are made from renewable materials and completely recyclable at their EOL. In contrast, 0 is given to products that are made entirely from virgin materials and end up in landfills at their EOL. The MCI is calculated from three factors such as:

- **Material input**: Percentage of virgin, recycled or renewable materials used in the product.
- **Utility factor**: Durability, lifespan and intensity of use of the product, which indicates the valuable function the product provides over time.
- **EOL recovery**: Percentage of materials that are recovered or recycled when the product reaches its EOL.

2.6.3 CircularTool by Matrec

CircularTool, developed by Matrec, serves as an assessment instrument that allows businesses to evaluate the circularity of their products and materials. It provides quantitative results based on several key performance indicators related to material origin, reparability and recyclability, which is designed to support adherence to the EU eco-design regulation. CircularTool grants access to an extensive database containing over 1,800 circular materials. This feature enables companies to compare and select materials for redesigning and improving their circularity performance.

The tool utilizes an input–output analysis methodology and promotes alignment with CE strategies and certifications. It is employed across a range of sectors, including fashion, textiles, furniture and construction. Moreover, it facilitates initiatives aimed at enhancing circularity and aids companies in demonstrating compliance with EU CE regulations. Additionally, the tool provides access to a circular materials database and supports a certification process for product circularity.

2.7 CASE STUDIES

We have investigated benefits and challenges of fashion circularity from a developing country's perspective. Secondary data sources were used for the study to examine the benefits of implementing CE models (especially focusing on EPR) in Vietnamese fashion and textile enterprises. The findings are discussed in this chapter.

2.7.1 Benefits of CE Models in Developing Countries

There are numerous benefits associated with the implementation of EPR in developing nations like Vietnam. Vietnam is emerging as a significant destination for fashion and textile sourcing for various international brands (Nayak et al. 2020). The adoption of EPR within the Vietnamese textile and garment industry can encourage fashion companies to prioritize sustainable product design, conserve resources and enhance waste management. Research indicates that the implementation of EPR policies in developing countries, including Vietnam, positively influences the environment. With the enforcement of EPR, global brands will concentrate on sustainable design strategies such as durability, renewability and recyclability. Durability is a crucial factor that is frequently overlooked due to the focus on creating inexpensive and trendy fashion items (6Wresearch 2023).

The EPR will compel brands to implement suitable strategies, including product take-back, recycling or remanufacturing at the end of a product's life cycle. Under the EPR policy, fashion manufacturers, retailers and distributors will assume the responsibility of collecting EOL products for recycling purposes. This initiative will facilitate the utilization of recycled materials that would otherwise have been disposed of in landfills, thereby conserving virgin materials. Consequently, the recovery and reuse of materials will reinforce the principles of a CE as outlined in this study.

2.7.2 Challenges in Implementing CE Standards

Many of the fashion businesses in developing countries are struggling to implement the CE standards and follow the CE guidelines. This is because small and medium enterprises (SMEs) often lack financial robustness due to their constrained budgets. For example, the CE programs such as the EPR, require both administrative assistance and technological resources to effectively manage waste throughout different phases. Should the EPR mandate the recycling of EOL clothing by SMEs, it could result in financial setbacks or potentially the shutdown of these enterprises. Implementing EPR programs presents significant challenges due to various complexities related to both the products and their management. To begin with, there exists a diverse array of fashion and textile products produced by companies that are frequently situated in different nations. Additionally, multiple levels of suppliers are involved in the supply chain process, and their identities are often not disclosed.

2.8 CONCLUSIONS AND FUTURE DIRECTIONS

2.8.1 Conclusion

There has been wider acceptance of fashion circularity standards to establish the circularity model by industries. The recent trend shows a strong industry shift toward embedding circular design principles, using sustainable materials and applying innovative recycling technologies to scale circularity across fashion and textile products. This chapter has discussed the recent trends in fashion circularity and various regulations relating to it. Global strategies for circular fashion are discussed with a special focus on the EU. Further, circularity evaluation tools such as CTI framework, MCI and circular tool by Matrec are also discussed.

New standards relating to circular fashion are being developed, which provide comprehensive and product-specific guidelines to brands in circular inputs, design for longevity or durability and recycling. Further, the certification through third-party verification to ensure compliance and credibility is also gaining significant importance. The recent trend in circular fashion that occupies the center stage includes practices such as reuse, recycling, upcycling and downcycling. Many of these practices are supported by digital tools such as product life cycle passports that boost transparency enhancing consumer association with a specific brand.

The evolution of fashion circularity standards is also aligned with increasing regulatory pressure and consumer demand for sustainability. This, in turn, is pushing brands to adopt measurable circularity indicators and LCAs. Collaboration between various brands, manufacturers, secondhand retailers, recyclers and regulators is vital in advancing these circular standards. Standards both for material innovations (such as bio-based and recyclable fibers) and digital verification systems are needed for future fashion and textiles. Together, these trends foster a holistic, credible and scalable CE within the fashion industry, aiming to reduce environmental impact while supporting ethical production and durable product design.

2.8.2 Future Directions

The future trends in fashion circularity standards will focus on the use of sustainable materials, recycling techniques and employing digital technologies for CE. Various bio-based materials that are renewable and biodegradable are paving their way in the CE model. Materials such as bacterial cellulose, oyster wool, biodegradable polyester, lotus and banana fabrics are some of the sustainable materials for future fashion and textile applications. Further, recycled materials of natural fibers, synthetics and their blends are also becoming potential candidates for future manufacturing. Several industries are researching to develop value-added products using recycled materials.

New standards need to be developed for sustainable materials and manufacturing processes in addition to addressing the challenges of EOL clothing management. These standards will become increasingly essential within regulatory frameworks and corporate responsibility reporting. With the development of global standards for CE, fashion brands should shift their business models that can meet the policy requirements and consumer expectations. The global circularity standards should offer uniform metrics for assessing environmental impacts in addition to traceability and LCA.

The use of digital technologies can help to provide real-time data on supply chain transparency, traceability and other sustainability-related information about the product. For example, the use of technologies such as RFID, block chain and AI can revolutionize the fashion and textile sectors. New policies should be developed for the use of digital technologies. The governing agencies should ensure that the policies are strictly implemented, and the industries follow the guidelines. Like the EU policies, strict measures will ensure that the policies are followed by the fashion and textile industries.

APPENDIX

List of Brands Working on Fashion Circular Economy

There are several fashion brands that are becoming popular for CE models. These brands employ various circularity aspects such as design for longevity, repair, resale, recycle, take-back programs and other circular business models that reduce waste and foster reuse across the global fashion supply chain. Here is a list of notable global fashion brands actively working on fashion circularity as of 2025.

List of Fashion Brands Working on Fashion Circularity

Fashion Brand Name	Focuses Area
Arc'teryx	Focuses on mono-materials, disassembly-ready garments and circular design standards.
Coach	Uses rescued leather and bulk materials with circular design at scale.
Decathlon	Engages in circular design pilots with durability and repairability standards.
Gant	Implements secondhand, take-back programs, and product longevity initiatives.
Gucci	Offers lifetime repair, resale through Gucci Vault and circular material use.
Levi's	Known for repair, resale and the WellThread collection made for recycling.
Lululemon	Runs a resale program "Like New" and partners for recycled nylon innovation.
OVS	Provides in-store repairs and strong circular design programs.
Patagonia	Longtime leader with repair, resale (Worn Wear) and durable products.
The North Face	Offers lifetime warranties, repair services and renewed resale line
Additional Brands with Strong CE Efforts	
Another Tomorrow	Uses recycled and renewable materials with a resale program.
Astor and Orion	Circular jewelry designed for recycling and take-back for credit.
MUD Jeans	Offers a jean leasing program and uses a high percentage of recycled denim.
Outerknown	Focuses on resale, repurposing and aims for full circularity by 2030.
RE/DONE	Upcycles vintage denim and other materials into new garments.

GLOSSARY ON FASHION CIRCULARITY

Some keywords that are widely used in fashion circularity are explained here.

Chemical recycling: An advanced form of recycling that breaks down synthetic fibers into their original monomers to be reused for new fibers and textiles.

Circular fashion (fashion circularity): An approach to fashion design and production that focuses on designing garments to be durable, repairable, reusable, recyclable or biodegradable, keeping them in use for as long as possible. This model contrasts with the traditional linear "take-make-waste" fashion system and aims to eliminate waste and pollution by ensuring garments stay in a continuous life cycle.

Closed loop: A product/material life cycle model where used garments or textiles are continuously recycled or reused to make new products without generating waste. This creates a "loop" that feeds the EOL of one product back into the production of another, often through fiber-to-fiber recycling.

Cradle to cradle: A design framework where products are created with safe, renewable materials that can either biodegrade naturally or be fully recycled, supporting a closed-loop CE.

Circular economy: An economic system designed to keep resources in use for as long as possible through repair, reuse, remanufacture, and recycling, thereby minimizing waste and environmental impact. In fashion, it means creating products that fit within cycles of use and regeneration.

Downcycling: A form of recycling where materials are converted into lower-quality products that have reduced functionality or value compared to the original, extending the material life cycle but not maintaining original quality.

Durability: A design principle prioritizing the creation of long-lasting garments that withstand multiple uses and washes, reducing the need for frequent replacement.

Ethical fashion: Emphasizes fair wages, safe working conditions and the rights of workers, often intersecting with sustainability principles.

Fast fashion: The swift and inexpensive production of fashionable clothing, associated with overproduction, waste and substandard labor practices.

Repairability: Designing clothes in a way that makes them easy to repair, extending their useful life and preventing premature disposal.

Recyclability: The ability of garments or textiles to be effectively disassembled and processed into new fibers or materials for making new products.

Rental programs: Business models where consumers rent clothing for limited periods instead of purchasing, promoting reuse and reducing the demand for new items.

Resale platforms: Online or physical marketplaces that facilitate the selling and buying of secondhand or pre-owned clothing, supporting circular fashion by extending the life of garments.

Slow fashion: A form of mindful consumerism that values quality, timeless designs and fewer purchases in contrast to fleeting trends.

Sustainable fashion: The production and consumption of clothing that conserves resources, reduces impact on planet and people and encourages longevity through the use of eco-friendly materials.

Take-back initiatives: Programs by brands or retailers where consumers can return used garments for recycling, resale or reuse, thereby keeping textiles out of landfills.

Traceability: The process of monitoring a product's journey from raw materials to the point of sale, ensuring transparency and accountability.

Upcycling: The creative process of transforming used or discarded clothing and textiles into new items of higher value or quality, reducing waste and environmental impacts compared to recycling.

Zero waste: A philosophy and design approach aiming to eliminate textile waste by redesigning production, consumption, and disposal processes to re-integrate resources into the system without loss.

REFERENCES

Abdelmeguid, A., Afy-Shararah, M. and Salonitis, K. 2022. Investigating the challenges of applying the principles of the circular economy in the fashion industry: A systematic review. *Sustainable Production and Consumption* 32: 505–518.

Amasawa, E., Brydges, T., Henninger, C. E. and Kimita, K. 2023. Can rental platforms contribute to more sustainable fashion consumption? Evidence from a mixed-method study. *Cleaner and Responsible Consumption* 8: 1–10.

Amir, S., Salehi, N., Roci, M., Sweet, S. and Rashid, A. 2023. Towards circular economy: A guiding framework for circular supply chain implementation. *Business Strategy and the Environment* 32: 2684–2701.

Azad, A. K., Haq, U. N., Khairul Akter, M. M. and Uddin, M. A. 2024. Recycling practices of pre-consumer waste generated from textile industry. In *Sustainable Manufacturing Practices in the Textiles and Fashion Sector* (Eds. Muthu, S. S.), Springer, Cham, Switzerland, 301–324.

Baloyi, R. B., Gbadeyan, O. J., Sithole, B. and Chunilall, V. 2024. Recent advances in recycling technologies for waste textile fabrics: A review. *Textile Research Journal* 94: 508–529.

Batool, G. and Nawab, Y. 2023. Regulation in recycling and circularity: Future prospective. In *Circularity in Textiles* (Eds. Baloyi, R. B., Ahmad, S., Nawab, Y. and Hussain, M.), Springer, Cham, Switzerland, 267–282.

Bour, A., Christensen, T. B., Hunka, A. D., Palmqvist, A., Skjold, E. and Syberg, K. 2023. Implications of circular textile policies for the future regulation of hazardous substances in textiles in the European Union. *Science of the Total Environment* 896: 1–16.

Cai, Y.-J. and Choi, T.-M. 2019. Extended producer responsibility: A systematic review and innovative proposals for improving sustainability. *IEEE Transactions on Engineering Management* 68: 272–288.

Claussen, L., Lloyd, A., Ruiz, D. and Havenith, G. 2023. Recycled versus virgin polyester sportswear: Can a difference be perceived in actual use? *International Journal of Fashion Design Technology and Education* 7: 1–9.

Colasante, A., D'adamo, I., Desideri, S., Iannilli, M. and Mangani, V. 2025. Environmental concerns in the fashion industry: A twin transition with the digital product passport. *Business Strategy and the Environment* 34: 9242–9256.

Commission, E. 2024. *Ecodesign for Sustainable Products Regulation*. European Commission Brussels, Belgium.

D'itria, E. and Aus, R. 2023. Circular fashion: Evolving practices in a changing industry. *Sustainability: Science, Practice and Policy* 19: 1–22.

Dissanayake, D. and Weerasinghe, D. 2022. Towards circular economy in fashion: Review of strategies, barriers and enablers. *Circular Economy and Sustainability* 2: 25–45.

Faludi, J. 2025. Sustainable fashion, circularity and consumer behavior: Systematic review and a social marketing research and policy agenda. *Social Marketing Quarterly* 31: 35–59.

Grover, T., Wang, L., Nayak, R., Padhye, R. and Khandual, A. 2025. Sustainable strategies for the valorization of cotton and cotton blended waste: Pathways to circular economy. *Cellulose* 32: 4077–4100.

Gupta, R., Kushwaha, A., Dave, D. and Mahanta, N. R. 2022. Waste management in fashion and textile industry: Recent advances and trends, life-cycle assessment, and circular economy. In *Emerging Trends to Approaching Zero Waste, Environmental and Social Perspectives* (Eds. Hussain, C. M., Singh, S. and Goswami, L.), Elsevier, Kidlington, UK, 215–242.

Huynh, P. H. 2022. Enabling circular business models in the fashion industry: The role of digital innovation. *International Journal of Productivity and Performance Management* 71: 870–895.

Islam, S. 2021. Waste management strategies in fashion and textiles industry: Challenges are in governance, materials culture and design-centric. In *Waste Management in the Fashion and Textile Industries* (Eds. Nayak, R. and Patnaik, A.), Elsevier, Kidlington, UK, 275–293.

Jing, T., Shi, G., Liu, M. T. and Zhu, M. 2026. Fast fashion vs. Luxury fashion: Who is better fitting with sustainable extension? The comparison of recycling and upcycling. *Journal of Product and Brand Management* 35(1): 148–163.

John, N. K. 2025. Micromanufacturing in Textile Industry: A transition towards circular economy and digitalization as drivers for sustainability. In *Applied Ethics and Rationality: Contemporary Indian Perspective* (Eds. Pandikattu, S. J. F. K.), Springer, Cham, Switzerland, 253–262.

Joung, H.-M. 2014. Fast-fashion consumers' post-purchase behaviours. *International Journal of Retail and Distribution Management* 42: 688–697.

Ki, C. W., Park, S. and Ha-Brookshire, J. E. 2021. Toward a circular economy: Understanding consumers' moral stance on corporations' and individuals' responsibilities in creating a circular fashion economy. *Business Strategy and the Environment* 30: 1121–1135.

Leal Filho, W., Dinis, M. A. P., Liakh, O., Paço, A., Dennis, K., Shollo, F. and Sidsaph, H. 2024. Reducing the carbon footprint of the textile sector: An overview of impacts and solutions. *Textile Research Journal* 94: 1798–1814.

Machado, N. and Morioka, S. N. 2021. Contributions of modularity to the circular economy: A systematic review of literature. *Journal of Building Engineering* 44: 1–13.

Maduna, L. and Patnaik, A. 2023. Introduction to sustainable fibres for fashion and textiles. In *Sustainable Fibres for Fashion and Textile Manufacturing* (Eds. Nayak, R.), Elsevier, London, UK, 27–48.

Mccauley, E. and Jestratijevic, I. 2023. Exploring the business case for textile-to-textile recycling using post-consumer waste in the US: Challenges and opportunities. *Sustainability* 15: 1–14.

Mishra, S., Jain, S. and Malhotra, G. 2021. The anatomy of circular economy transition in the fashion industry. *Social Responsibility Journal* 17: 524–542.

Moorhouse, D. and Moorhouse, D. 2017. Sustainable design: Circular economy in fashion and textiles. *The Design Journal* 20: S1948–S1959.

Nayak, R. 2020. *Supply Chain Management and Logistics in the Global Fashion Sector: The Sustainability Challenge*, Routledge, London, UK.

Nayak, R., Cleveland, D., Tran, G. and Joseph, F. 2024. Potential of bacterial cellulose for sustainable fashion and textile applications: A review. *Journal of Materials Science* 59: 6685–6710.

Nayak, R., Jajpura, L. and Khandual, A. 2023. Traditional fibres for fashion and textiles: Associated problems and future sustainable fibres. In *Sustainable Fibres for Fashion and Textile Manufacturing* (Eds. Nayak, R.), Elsevier, Kidlington, UK, 3–25.

Nayak, R., Nguyen, L., Patnaik, A. and Khandual, A. 2021a. Fashion waste management problem and sustainability: A developing country perspective. In *Waste Management in the Fashion and Textile Industries* (Eds. Nayak, R. and Patnaik, A.), Elsevier, Kidlington, UK, 3–29.

Nayak, R., Panwar, T. and Nguyen, L. V. T. 2020. Sustainability in fashion and textiles: A survey from developing country. In *Sustainable Technologies for Fashion and Textiles* (Eds. Nayak, R.), Elsevier, Kidlington, UK, 3–30.

Nayak, R. and Patnaik, A. 2021. *Waste Management in the Fashion and Textile Industries*, Elsevier, Kidlington, UK.

Nayak, R., Singh, A., Thai, V., Grover, T. and Tran, D. T. M. 2025. Recent sustainability trends in composting textiles: Process, factors, benefits and challenges. *Results in Engineering* 28: 107694.

Nayak, R., Van thang, L. N., Nguyen, T., Gaimster, J., Morris, R. and George, M. 2021b. Sustainable developments and corporate social responsibility in Vietnamese fashion enterprises. *Journal of Fashion Marketing and Management: An International Journal* 26: 307–327.

Niinimäki, K., Peters, G., Dahlbo, H., Perry, P., Rissanen, T. and Gwilt, A. 2020. The environmental price of fast fashion. *Nature Reviews Earth and Environment* 1: 189–200.

Papamichael, I., Chatziparaskeva, G., Voukkali, I., Navarro Pedreno, J., Jeguirim, M. and Zorpas, A. A. 2023. The perception of circular economy in the framework of fashion industry. *Waste Management and Research* 41: 251–263.

Saha, K., Dey, P. K. and Papagiannaki, E. 2022. Implementing circular economy in the textile and clothing industry. *Supply Chain Sustainability in Small and Medium Sized Enterprises,* Routledge, Oxon, UK.

Stella, F., Garofalo, S. F., Cavallini, N., Fino, D. and Deorsola, F. A. 2024. Closing the loop: Analysis of biotechnological processes for sustainable valorisation of textile waste from the fast fashion industry. *Sustainable Chemistry and Pharmacy* 38: 1–12.

Tetteh, H., Balcells, M., Bala, A., Fullana-I-Palmer, P., Margallo, M., Aldaco, R. and Puig, R. 2025. Assessing agri-food products circularity by using the material circularity and circular flow indices. *Sustainable Production and Consumption* 12: 1–18.

Tran, T. Y. A., Kieu, K. T., Herat, S. and Kaparaju, P. 2023. Implementing EPR as a tool for addressing environmental issues in Vietnam. *Environmental Science and Sustainable Development* 8: 73–92.

6Wresearch. 2023. Vietnam Textile Waste Market (2020–2026). https://www.6wresearch.com/industry-report/vietnam-textile-waste-market-2020-2026. [Accessed 7 December 2025].

3 Cleaner Production Technologies

Lebo Maduna

3.1 INTRODUCTION

The textile and clothing industry is one of the major polluters of the environment due to increased demand for new textiles, which in turn increases consumption of resources (fibers, water, chemicals and energy), resulting in more waste being generated (Jones et al. 2025; Yukseler et al. 2017). Fiber consumption has more than doubled compared to population growth. Between 2000 and 2020, the global population increased from 6.1 to 7.8 billion and textile fiber consumption increased from 60 to 100 million tons, representing an increase of 27.9% and 66.7%, respectively (Stella et al. 2024).

Conventional manufacturing technologies are still widely used in the industry, despite being inefficient and generating a significant amount of waste. Cleaner production technologies are being recommended as alternatives, because they are efficient and generate less waste. To further enhance the benefits of cleaner production, eco-friendly materials must be used (Jones et al. 2025; Xu et al. 2024; Millward-Hopkins et al. 2023). Some of the cleaner production technologies used in the textile industry include biotechnology, plasma technology, ultrasound technology, laser technology, 3D printing, ozone and waterless dyeing. Unlike conventional production methods, which often cannot target specific areas without affecting the entire material, cleaner production technologies can achieve localized treatment without causing damage.

3.2 CLEANER TECHNOLOGIES

3.2.1 Waterless Dyeing

Historically, textiles are dyed with synthetic and natural dyes to impart the desired colors. A dye like disperse dye is soluble in nonpolar solvents, whereas dyes like direct, acid and reactive dyes are soluble in polar solvents (Hassan 2025). Conventional textile treatment processes consume large quantities of water, chemicals and energy, which in turn lead to polluted water that must be extensively treated before being disposed of (Allafi et al. 2021). It is estimated that 28–285 L/kg of water is consumed during the manufacturing processes, depending on the type of fibrous materials and manufacturing processes, with about 90% of the consumption attributed to the dyeing processes, depending on the intensity of the color required (Imani et al. 2024). It was reported that using the conventional dyeing method to dye 1 kg of textile can consume about 100–150 L of water and 8–18 MJ of energy

DOI: 10.1201/9781003689225-3

(Gong et al. 2021). The dyeing process requires that the dye solution be heated to its optimal temperature (Imani et al. 2024), for example, high temperatures of up to 130°C for polyester using the disperse dye, reactive dye temperature of up to 80°C for cotton (Hassan 2025; Zhao et al. 2023). Dyeing is time-consuming; it can take up to 1–2 hours. Furthermore, the dyed textile must be washed to remove unfixed dyes and other chemicals used to improve dyeing fixation such as salts, alkali and surfactants (Hassan 2025; Imani et al. 2024).

Waterless dyeing, such as supercritical carbon dioxide (scCO_2) dyeing, is being suggested as one of the alternative dyeing methods to reduce high consumption of water, chemicals and energy. ScCO_2 dyeing uses CO_2 as a medium to dissolve dyes instead of the conventional solvents, like water used in conventional dyeing processes. A scCO_2 state occurs when CO_2 is heated and pressurized to its critical temperature and pressure; it has characteristics like those of a solvent (Zhao et al. 2023, 2025). Gong et al. (2021) indicated that CO_2 at its supercritical temperature and pressure of 31°C and 73.7 bar, respectively, is characterized by intermediate characteristics between those of a gas and a liquid. However, for practical applications, supercritical dyeing temperatures and pressures can vary. For example, the scCO_2 temperature for polyester dyeing can have a range of 100–140°C and pressure 24–26 kPa (Gong et al. 2021; Han et al. 2018). Under standard atmospheric conditions, CO_2 exists in the gaseous phase. However, at its critical temperature of 31.1°C and pressure of 73.8 bar, it enters the supercritical state, which is characterized by intermediate properties between those of gas and liquid (Huang et al. 2026; Allafi et al. 2021). This behavior is exploited as a solvent to substitute conventional dyeing methods due to its non-toxicity, cost and elimination of long processes (Huang et al. 2026). As a nonpolar medium, scCO_2 can serve as a substitute for conventional nonpolar solvents in the application of disperse dyes (Figure 3.1). Its main challenge or disadvantage for dyeing fibers is that conventional dyes used in the dyeing of natural fibers like cotton, wool and silk are not soluble in a CO_2 medium (Hassan 2025; Zhang et al. 2016). However, Long et al. (2015) investigated the use of disperse-reactive dye to dye cotton and wool, where scCO_2 was used as the main medium.

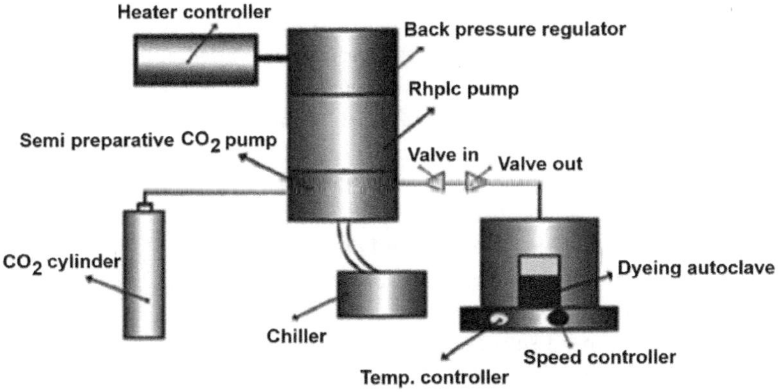

FIGURE 3.1 Schematic diagram of lab-scale scCO_2 dyeing apparatus (Mousa et al. 2024).

TABLE 3.1
Merits and Demerits of $ScCO_2$ (Huang et al. 2026; Wang et al. 2023; Li et al. 2020; Allafi et al. 2021)

Advantages	Disadvantages
No toxic solvent residue	Requires frequent cleanup to prevent fibrous materials from clogging the vents
Nonflammable	Recycling of CO_2 requires dye particles to be separated to prevent contamination of CO_2
Faster	

$ScCO_2$ can also be explored as a cleaning agent. CO_2, being a nonpolar compound with a zero-dipole moment, can be used to dissolve polar substances such as oil and grease from textile materials. It is reported that at a temperature of 150°C and pressure of 65 MPa, 98% of wax from wool can be dissolved. It is also indicated that $scCO_2$ can be used for the disinfection of bio-textile materials to prevent the growth of bacteria at temperatures between 40°C and 60°C, pressure of 10–30 MPa and time of 30–180 minutes (Allafi et al. 2021).

One of the major advantages of supercritical dyeing is that CO_2 and small remaining unfixed dye can be recovered and recycled, thereby reducing the quantities of new raw materials needed for dyeing (Table 3.1). However, it is important to ensure CO_2 and chemical dye are separated, as contaminated CO_2 gas causes poor dyeability (Gong et al. 2021; Li et al. 2020). One of the techniques used to separate dye particles from CO_2 is the cyclone separator, which separates the solid particles from gases when they are rotated at high speeds. The chemical residues are separated from the CO_2 gas due to their difference in densities, and the dye particles are collected on the wall of the separator (Wang et al. 2023; Li et al. 2020).

3.2.2 Plasma Technology

Plasma energy can be used to treat textiles. This energy can break polymer bonds, resulting in a material with a modified chemical structure (Jabbar et al. 2024). This treatment can be used to improve material properties such as hydrophilicity, flame retardancy and dye uptake (Li et al. 2025; Ayesh et al. 2024; Jabbar et al. 2024). Plasma energy depth penetration is limited to a small part of the bulk, only to the surface, and the rest of the textile is not affected (Vesel et al. 2010).

Plasma treatment is a dry process that has been explored as an alternative to modify textile surfaces without the need to use chemicals and water. It removes the drying process needed when chemical methods are used, making it eco-friendly and faster. It can be used to target a specific surface area (Li et al. 2025; Vajpayee et al. 2025; Serrano-Martínez et al. 2024). Polar functional groups such as hydroxyl, carbonyl and carboxyl can be introduced through grafting or crosslinking to improve the hydrophilicity of materials with poor hydrophilicity (Li et al. 2025; Jabbar et al. 2024; Ayesh et al. 2024; Vesel et al. 2010). Polyester fibers treated with oxygen plasma have improved surface roughness, which resulted in enhanced coefficient of friction and wettability. Furthermore, it can be used to improve the dyeability of the modified polyester when its charged groups react with the oppositely charged groups of the dye molecules (Jabbar et al. 2024; Beyene and Nalankilli 2017).

Plasma can also be used to disinfect fabrics contaminated with microorganisms without affecting textile properties, such as tensile strength and morphological properties. Surface disinfection as high as 97% has been reported; however, there are concerns due to its limited depth penetration in bulky materials: microorganisms in areas not reached by plasma will remain active (Zhao et al. 2024). Plasma-ultraviolet treatment used in bleaching reduced the consumption of H_2O_2 and NaOH by 50%, water by 33%, while energy decreased from 35 to 71 MJ and the time for batch duration decreased from 24 to 1 hour (Li et al. 2025).

3.2.3 Ultrasound Technology

Ultrasound technology using vibrating sound waves greater than 20 kHz is used in the textile industry for a range of applications such as seam joining, scouring, bleaching, retting, degumming and desizing, as well as for the preparation of dye solutions, printing and sizing paste (Jajpura and Nayak 2020). Using conventional methods for these processes is time-consuming and consumes a lot of water and energy. In conventional wet processing, chemical agents are predominantly dissolved in water to facilitate interaction with textile substrates. To enhance their transfer to textile materials, high concentrations are commonly used; however, this practice contributes to increased environmental pollution and resource depletion. Recent studies have identified ultrasound as a promising auxiliary technique that can augment these processes' efficiencies, thereby enabling reductions in chemical consumption and minimizing ecological impact (Jajpura and Nayak 2020; Ma et al. 2014).

The sizing agent that is applied to protect the yarns requires that the size paste be completely dissolved to prevent lump formation. Conventional preparation methods for size paste typically require high temperatures and prolonged processing times. The incorporation of ultrasound technology significantly reduces both temperature and processing duration, while producing a homogenized paste of superior quality. Similarly, printing paste solutions prepared using ultrasonic techniques exhibit good uniformity, eliminating clump formation. Furthermore, ultrasonic treatment can be effectively employed for desizing, that is, the removal of sizing ingredients from yarns (Jajpura and Nayak 2020).

Ultrasound-assisted scouring is used to remove impurities from fibers such as cotton, wool and silk. Conventional scouring processes consume a lot of chemicals, water and energy, and they are also time-consuming, while ultrasound-assisted scouring, resource consumption and temperature are reduced. Furthermore, there is less damage to the fibers, unlike when conventional methods are used (Jajpura and Nayak 2020). When ultrasound was introduced into the wool scouring bath, it enhanced the removal of the adsorbed dirt and grease from wool. Although the process is less energy-intensive, residues from the scouring stage pose a challenge (Allafi et al. 2021), even though it is smaller than the conventional ones.

Conventional bleaching processes are highly time-consuming, with cold bleaching requiring approximately 16 hours and boiling bleaching around 2–3 hours.

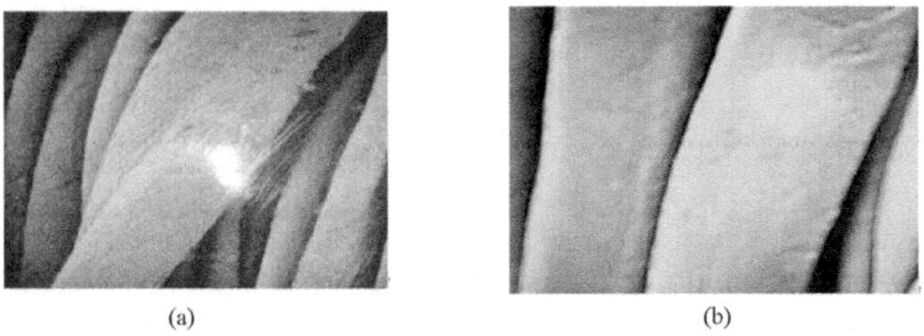

FIGURE 3.2 Micrographs of (a) control fabric bleached at 95°C without ultrasound and (b) fabric subjected to ultrasound-assisted bleaching at room temperature for 90 minutes (x1540) (Thakore and Abate 2017).

In contrast, ultrasound-assisted bleaching significantly enhances H_2O_2 utilization and improves the bleaching efficiency of cotton materials, achieving comparable cold bleaching results within just 1 hour and that of boiling in 2.5 hours (Figure 3.2). Furthermore, the integration of ultrasonic treatment with the enzyme laccase in the presence of H_2O_2 has been shown to further enhance bleaching efficiency, offering a sustainable and energy-efficient alternative to conventional methods (Jajpura and Nayak 2020).

Ultrasound-assisted laundering enhances detergency action, facilitating the effective removal of contaminants from textiles using frequencies in the range of 25–100 kHz to generate tiny bubbles in a liquid bath, and when these bubbles collapse, energy is released that shocks contaminants, causing them to be released from contaminated surfaces (Figure 3.3). Although the use of ultrasound in an aqueous detergent solution enhances the removal of contaminants, it was shown that it was accompanied by the redeposition onto textile surfaces, as observed when oil and carbon black were tested as contaminants. It is also indicated that as electric power consumption increased, both the contaminant removal and redeposition increased (Gotoh et al. 2015). Liquor-to-textile ratio is important for laundering, as a low liquor-to-textile ratio can cause damage to textiles due to increased abrasion when textile surfaces rub against each other and the walls of the bath (Gotoh and Harayama 2013). Ultrasonic treatment can be used to reduce the water and time (Ochca et al. 2019). Silk fabrics, being highly delicate, require special care during conventional home laundering to prevent damage caused by friction, which often necessitates being sent for professional dry cleaning. Similarly, wool fabrics are prone to felting and shrinkage during conventional home laundering. However, laundering with ultrasonic agitation significantly reduces felting and shrinkage. The removal of water-soluble stains poses a challenge in dry cleaning processes; ultrasound-assisted laundering addresses this through acoustic cavitation, which induces rapid liquid movement and vibration at the liquid–fabric interface, thereby dislodging stains. Compared to the intense mechanical agitation of conventional laundering, ultrasonic treatment minimizes fabric damage while maintaining cleaning efficiency (Ma et al. 2014; Choi et al. 2016; Park and Lee 2025).

In welding, ultrasonic frequencies in the range of 20–40 kHz can be used to generate heat that melts thermoplastic textile seams, creating a strong bond. For carbon-polyphenylene sulfide composite materials, frequencies as high as 619 kHz are

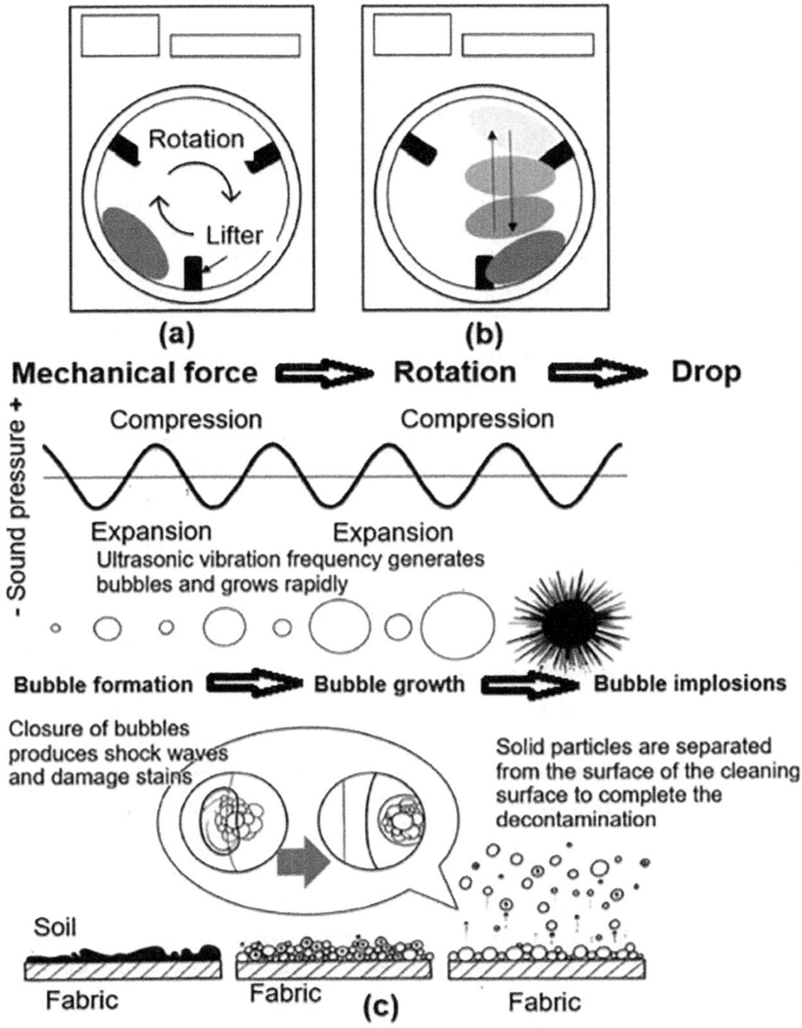

FIGURE 3.3 Principle of drum washing machine: (a) when the drum rotates due to the action of mechanical forces, the laundry inside it rotates as well. (b) The rotation causes the laundry to fall up and down. (c) Principle of washing in an ultrasonic washing machine (Park and Lee 2025)

used. Seams bonded by ultrasonic have no holes, unlike conventional seams joined by needles that create holes. This is important for textiles that must not allow the passage of liquids (Vijayakumar et al. 2025). Cutting textiles using ultrasonic welding causes less damage to textiles compared to conventional blade cutting, which generates more microfibers (Rathinamoorthy and Balasaraswathi 2024). One of the primary challenges associated with ultrasonic welding is that the resulting seam tends to exhibit increased stiffness and reduced tensile strength compared to conventional needle joining methods. Furthermore, non-thermoplastic fabric materials like cotton and wool cannot be melted to join seams.

Conventional dyeing processes typically require elevated temperatures, whereas ultrasound-assisted dyeing can be performed at comparatively lower temperatures (Thakore and Abate 2017). The application of ultrasonics enhances the homogenization of the dye bath solution, resulting in improved dye penetration and higher dye uptake. For instance, cotton dyeing can be achieved at approximately 30°C using ultrasonic assistance, while wool dyeing can occur at around 60°C. Similarly, polyester dyeing can be performed at 90°C instead of a higher conventional temperature at 130°C, demonstrating significant energy savings and process efficiency.

Ultrasound-assisted effluent treatment enhances the degradation of various organic and inorganic compounds, complementing conventional chemical and biological processes to achieve higher degradation efficiency (Bekele and Alamnie 2025; Srivastava et al. 2014). Acoustic cavitation promotes the generation of reactive free radicals, such as hydroxyl, hydrogen and hydroperoxyl, which subsequently interact with toxic contaminants, facilitating their breakdown. Using the blended sonophotocatalysis, 100% removal of Acid Orange 7 and Methylene Blue within 30 and 12 minutes, respectively, was reported, compared to conventional degradation methods like photocatalysis degradation (Mishra et al. 2025). Additionally, ultrasonic treatment can also be used for the decolorization of effluent. The free radicals can recombine with other radicals or oxygen to form an agent that can be used for decolorizing dyes, for example, the free highly reactive H and OH free radicals from the breakdown of water can react with other radicals or stable oxygen (O_2 or O_3) to form hydrogen peroxide, peroxy and superoxide that can be used for decolorizing dyes (Srivastava et al. 2014). The benefit of incorporating ultrasonic treatment in textile production processes is that it is simple to operate (Srivastava et al. 2014).

3.2.4 Laser Technology

Laser (Light Amplification by Stimulated Emission of Radiation) technology uses light waves in a laser beam that are aligned with each other, unlike normal light. Some of the sources of lasers are CO_2, Nd-YAG, Argon-ion and Helium-neon. The directed energy from the laser can be used for cutting and modifying the surfaces of textile materials (Hitz et al. 2001). Laser treatment of textiles is faster. Furthermore, it avoids the use of conventional chemical and mechanical methods, which generate harmful waste. The CO_2 laser is widely used due to its simplicity and lower operating temperatures (Atav et al. 2025; Atwee et al. 2025; Venkatraman and Liauw 2019). The energy consumed to generate laser energy can be significantly higher than the laser energy. However, its advantage is the narrow beam it generates, allowing the light energy to be focused only on the targeted area (Hitz et al. 2001). Factors such as wavelength, pulse duration, power and repetition influence the application and performance of the laser (Mittal and Lei 2018). Some of the applications of laser treatments in textiles are for surface treatment, cutting and welding (Figure 3.4).

A laser can join seams and cut simple and complex textile structures. It creates clean cuts and joins as the beam is targeted to a specific surface for cutting and welding. The conventional methods cause damage to the textiles (Rathinamoorthy and

FIGURE 3.4 Schematic representation of some laser treatments (Angelova 2020).

Balasaraswathi 2024; Wu et al. 2022; Bamforth et al. 2006). The cutting of polypropylene using a laser is reported to generate pyrolysis by-products like 1-butene, 2-methyl-2-butene, benzene, m-xylene and toluene, which pose a risk to the operator and the environment (Wu et al. 2022).

Laser-treated textiles are also being explored for the adsorption of volatile organic compounds. When a polypropylene mask with a coating of graphene oxide is treated with a laser, the smooth graphene oxide coating absorbs the energy and explodes to form a porous coating with a large surface area, which can adsorb volatile organic compounds. The heat released by the graphene oxide coating melts polypropylene and thereby enhances the fixation of graphene oxide coating to the polypropylene fabric (Cheng et al. 2024).

A laser treatment can be used to modify textile surfaces for color fading and for decorations by etching and engraving for decorative fabrics. It targets only the surface that is exposed to the laser (Atav et al. 2025; Cheng et al. 2024). The conventional methods of modifying a fabric surface into a stylish fabric use mechanical and chemical methods. Mechanical methods like sand blasting, stonewashing and sand blasting use a solid material that rubs against the surface of fabric and dust particles are released (Venkatraman and Liauw 2019; Hung and Kan 2017). Chemical methods such as bleaching and pigment printing generate chemical residues. Furthermore, they are time-consuming methods for altering fabric appearance, requiring multiple stages including drying, fixation and rinsing (Venkatraman and Liauw 2019; Hung and Kan 2017). Laser treatment exhibits a more pronounced effect in garment dyeing using reactive and direct dyes than in processes involving acid or metal complex dyes (Atav et al. 2025). Applying laser treatment on cotton-polyester fabric either pre- or post-dyeing for color fading effect can have different outcomes, with pre-laser dyeing achieving better color effect and levelness as the char formed during laser irradiation is removed during dyeing, unlike the post-laser dyeing (Hung and Kan 2017). Optimal conditions (laser resolution-dpi and residence time- μs) for color fading are reported to be in the range between 30 dpi/120 μs and 30 dpi/270 μs. Excessive heat degrades fabric; therefore, the laser energy applied to thin materials is lower compared to thick fabrics. In thicker materials, such as denim, more material remains intact after the surface has been burned. Fabric strength is also affected (Venkatraman and Liauw 2019).

Laser can also be used for disinfection and to impart antimicrobial activity to textiles. It is used to generate copper nanoparticles and titanium films on the surface of a fabric to impart antimicrobial activity (Li et al. 2025; Zhao et al. 2024; Shahidi et al. 2018). Other applications of lasers are textile fault detection, bar code scanning and accurate body measurements (Venkatraman and Liauw 2019). Some of the benefits of using laser technology are selective area treatment, less energy consumption, less chemical consumption and improved production times. However, the excessive heat generated will degrade the textile (Venkatraman and Liauw 2019; Hung and Kan 2017).

3.2.5 3D Technology

In manufacturing, 3D printing, also called additive manufacturing, is a process that manufactures a product by continuously adding material layers. Design programs like Computer-aided design (CAD) software are used to develop 3D models, which are used by a 3D printer. Using conventional manufacturing methods to create complex 3D structures is not as easy as individual parts must be joined together, and is time-consuming, whereas 3D printing is simpler and faster. 3D printing is versatile, as different products with different structures can be produced using the same 3D printing machine. Some of the benefits are reduced material waste, the ability to print complex structures, the ability to print structures in one step, and the ability to print structures directly from the technical drawing. 3D printing can also be used to apply colorants to textile surfaces, enabling 3D creation (Mukherjee and Debroy 2024; Sitotaw et al. 2020).

The clothing online shopping segment has experienced significant growth recently, with projections that it will continue to grow; however, there are challenges with this kind of shopping, as customers cannot evaluate the garment for comfort and fitting (Liu et al. 2017). In the United States, it is estimated that approximately 24% of garments purchased online in 2023 were returned primarily due to incorrect fit and sizing (Li et al. 2024). In garment designing, 3D body scanning is used to design products that best fit the body since people have different body shapes. In technical garments, poor fit is important as it can negatively affect garment protection. Once the body is scanned, the digital body model can be shared with other computer programs for patternmaking and visual garment fitting (Sitotaw et al. 2020; Maghrabi et al. 2015; Liu et al. 2010).

In 3D virtual garment fitting programs, measurements from the real body are used to create a virtual body and the customer can visualize the garment fit and drape and behavior of the product in preproduction to prevent material wastage compared to the conventional method, where a real garment prototype has to be worn by the wearer. Based on the evaluation of the virtual garment fitting, changes from the customer and designer can be made, leading to improved human-centered (personalized) fashion style, customer satisfaction, production times and waste reduction (Dik et al. 2023; Liu et al. 2017; Chen et al. 2015).

Historically, creating clothing patterns is a manual process where design specifications are used to manually draw the construction lines onto paper. The lines define the shape and dimensions of each component of the particular garment size (Liu et al. 2018; Tao et al. 2018). Once the drafting is complete, the paper templates are cut and used as guides for fabric cutting to ensure that the correct shapes and dimensions of fabric pieces are cut for garment assembly. This manual approach is considered not to be efficient as it is labor-intensive and time-consuming. Furthermore, it relies heavily on the skill of the patternmaker. This method is not well suited for

FIGURE 3.5 Avatars with arms stretched and raised (left) and sitting and squatting (right) postures (Yang and Baytar 2024).

mass garment production that accommodates individual garment fitting and style preferences. To overcome these challenges, 3D models are being explored, whereby computer programs are used to generate 3D body models that are used to make 3D patterns that are transformed into 2D patterns (Figure 3.5). A 3D body scanner can also be used to generate 3D body models (Liu et al. 2018). CAD systems with 3D animation technology are also being explored to generate a visual 3D prototype garment for preproduction, accommodating different garment fitting and style preferences. This allows the designer to incorporate customer input. Adjustments to the garments can be made in real time and visualized to accommodate the customer's body type and style, making the garment construction fast, efficient and save materials (Figure 3.6). However, with the conventional garment construction, it is labor-intensive and time-consuming, as it requires the construction of physical prototypes (Dik et al. 2023; Tao et al. 2018).

3D technology is also used in virtual fitting rooms or virtual try-on technology by fashion companies for customers to try a virtual garment. Customers can try different available products, thereby enhancing body fitting and appearance in real life, resulting in fewer returns of the garments (Batool and Mou 2024).

3.2.6 Biotechnology

Biotechnological treatments of textiles, particularly enzymes in the textile industry, have received attention due to their specificity and ability to operate under mild conditions; they can be used for targeted treatment of textiles. Treated textiles are

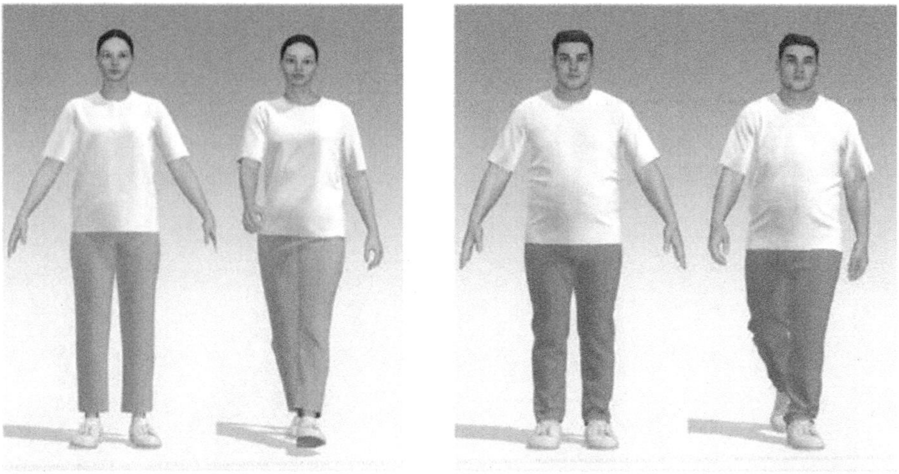

FIGURE 3.6 Female (left) and male (right) avatars in standing and walking postures (Li et al. 2024).

without damage. Conventional chemical methods are operated at extreme conditions and generate toxic waste. Enzymes such as amylases, lipases, cellulases, cutinases, pectinases, proteases, laccases and oxidoreductase are increasingly used in various preparatory wet processing techniques (Li et al. 2023; Rahman et al. 2020). To facilitate their effective migration into textile substrates, mechanical agitation is used when using conventional methods. However, ultrasonic treatment can serve as an alternative approach, enhancing enzyme penetration and improving overall process efficiency. Enzymes with high specificity offer an eco-friendly and less waste-generating method that can be applied in desizing, scouring, bleaching and wastewater detoxification (Harish et al. 2025; Srivastava and Bandhu 2022; Imran et al. 2015; Abdel-Halim and Al-Deyab 2013).

Microorganisms such as bacteria and fungi can be used for dyeing and dye production instead of using conventional chemicals. Genetically modified bacteria, *Escherichia coli* was used to produce Indigo and Tyrian Purple dyes from tryptophan, an amino acid which was used as a substrate (Alves et al. 2024). The fermentation of biodegradable materials by genetically modified microorganisms can produce bio-pigments that can be fed directly to dye the fabric. This process is reported to reduce water and energy consumption by 90% and 20%, respectively. Some of the challenges with bio-dyes are poor stability, yields and fixation when compared to conventional chemical dyes. Furthermore, complex equipment and highly skilled personnel are required (Rahman et al. 2020). Despite the challenges, biodegradable materials can serve as substitutes for hazardous substances. Furthermore, their moderate to low energy and water consumption contributes significantly to sustainable production practices (Alves et al. 2024; Rahman et al. 2020).

It is estimated that about 80% of the dyes are azo dyes, and these dyes pose a danger to the environment and animals since they are carcinogenic and not easily degraded. The treatment of azo dyes uses physical and chemical methods. Biological treatment does not produce toxic residues. Microorganisms such as fungi and bacteria can be used to degrade and decolorize azo dyes. *Halobacillus trueperi* bacteria produce protease enzymes with potential applications in effluent treatment. The ability to tolerate high salt content, heavy metals and oxyanion concentrations, commonly found in effluent water, makes these bacteria excellent candidates for treating wastewater (Harish et al. 2025).

Enzymes are used in retting processes to separate fibers from stems of plants. Fiber extraction from plant stems using *Aspergillus niger*, which produces cellulase, pectinase, chitinase and peroxidase enzymes, results in improved yield and enhanced fiber quality. Moreover, there is no significant difference in the mechanical properties between enzyme-treated and untreated fibers, indicating that enzymatic processing preserves the structural integrity of the fibers (Manjusha et al. 2023). For surface treatment, cellulase can be used as a pretreatment method for cellulose-containing materials. By selectively targeting and degrading exposed cellulose material on the surface, this method enhances surface properties, facilitating subsequent applications such as antimicrobial coating, where the antimicrobial agents can be effectively adsorbed onto the fibers. Cotton fibers treated with Tulsi extract retained a degree of antimicrobial activity after four washing cycles; however, this activity progressively decreased with each subsequent wash. It is noted that exposure of the treated material to mechanical forces can compromise its protective efficacy (Vajpayee et al. 2025; Vajpayee et al. 2021).

Cellulosic materials like cotton are treated with bleaching agents to remove coloring pigments and improve their whiteness. For example, H_2O_2 requires an alkaline pH of about 10–11 and temperatures of approximately 90°C–100°C. Higher levels of whitening require increasing concentration. A concentration balanced with an acid rinse to neutralize the alkaline pH. This process consumes a lot of energy, water and chemical waste. An alternative to H_2O_2 is titanium dioxide and single-walled carbon nanotubes, which enhanced the whiteness of cotton fabric at a pH of 7 and a lower temperature of about 50°C compared to other methods (Cigeroglu and Basyigit 2024).

Sizing agents such as starch and polyvinyl alcohol are applied to yarns to prevent yarn breakage. Before the dyeing or finishing processes are undertaken, the desizing agent is used to remove the sizing agent. Conventional methods rely on chemicals to remove sizing agents, which is not an environmentally friendly approach. Enzymes offer a sustainable alternative; for example, amylase effectively degrades starch-based sizing agents, while catalase breaks down hydrogen peroxide residues remaining after bleaching into water and oxygen (Imran et al. 2015; Chen et al. 2007).

3.2.7 Ozone

Ozone treatment offers an alternative eco-friendly method to treat fibers and contaminated water (Nadeem et al. 2024). Ozone has a higher oxidizing capability than oxygen, H_2O_2 and chlorine (Wei et al. 2026), and it has found applications in the treatment of toxic chemicals and the removal of color in water and fibrous textile materials (Sargunamani and Selvakumar 2006). Textile fibers can be treated with bleaching agents like ozone that do not contain hazardous elements like chlorine (Yang et al. 2025; Nadeem et al. 2024). Cotton jean fabric can be treated with ozone to degrade the indigo dye without causing significant damage to the fibers especially when used with ultrasonic, reducing the treatment time (Hamida et al. 2017). When ozone was used with hydrodynamic cavitation to activate sodium percarbonate and ozone, it enhanced the degradation of 1,4-dioxane, and there was also less fiber damage (Li et al. 2026).

Conventional color fading treatment methods on garments use stones or bleaching agents like sodium hypochlorite, which cause damage to the fabric's structural integrity. Increasing the stone load or concentration of the bleaching agent to achieve better color fading causes more fabric damage, and additionally, more dust and chemical waste are generated (Nadeem et al. 2024). During ozone treatment, one of the factors that must be controlled is time. The degree of whiteness of cotton slivers increased from 74 to 84 when the ozonation time was increased from 1 to 15 min (Erdem and Bahtiyari 2018). Because bulkier fabrics retain higher moisture levels, they tend to bleach more effectively. Areas of the fabric subjected to ozone treatment are noticeably brighter, resulting in a nonuniform brightness across the material (Hamida et al. 2017). Furthermore, as a strong oxidizing agent, it can attack textile fibers, causing degradation. The cost of ozone can be prohibitive, because it is unstable and decomposes quickly, making its storage challenging and necessitating on-site generation (Wei et al. 2026).

3.3 CHALLENGES

Cleaner production technologies have transformed production, enabling it to be optimized to improve production speed, reduce material and energy consumption, which in turn leads to less waste and pollution (Vadruccio et al. 2025; Wang et al. 2025; Luo et al. 2024; Ashima et al. 2022). Reducing resource consumption is important, as the price of the raw materials is the biggest component of the final product price (Nalajala et al. 2023).

Although cleaner production technologies are efficient and generate less pollution, they are expensive and complicated, which can be a hindrance for large-scale industrial applications. For technology like 3D, Ashima et al. (2022) indicated that it is limited by the size of the layer. The initial capital cost to acquire equipment and other technologies is high, particularly for small companies with limited financial resources. Furthermore, the lack of in-house expertise to fully take advantage of these technologies may necessitate seeking external support, which can, in turn, lead to resistance to change (Hadi 2024; Yusup et al. 2014).

Poor environmental regulations and enforcement are also important, as they do not incentivize companies to adopt eco-friendly technologies and practices, as other priorities are taken care of first (Yukseler et al. 2017; Yusup et al. 2014). The adoption of artificial intelligence (AI) in systems will initially face some challenges, because AI models require large amounts of data to be trained.

3.4 ADVANTAGES

Despite their challenges, the adoption of cleaner production technologies has several advantages, such as:

- **Specificity**: Targeted area treatment, such as when plasma, laser and enzymes are used to treat only the textile surface, results in limited damage to the overall structural integrity. Enzymes have the additional benefit of having the ability to operate under mild conditions (Atav et al. 2025; Cheng et al. 2024; Li et al. 2023; Rahman et al. 2020; Vesel et al. 2010).
- **Toxicity**: No toxic residues are generated, such as when $ScCO_2$ is used as a solvent for dyeing, unlike when conventional solvents are used. Furthermore, CO_2 can be recovered and reused (Allafi et al. 2021; Huang et al. 2026; Li et al. 2020; Wang et al. 2023). Ozone can be used to treat toxic chemicals and for the removal of color in effluent water and fibrous textile materials (Sargunamani and Selvakumar 2006).
- **Efficiency**: Augmenting conventional textile processes improves efficiency, resulting in reduced chemical consumption and shorter processing times, for example, by combining ultrasound technology with conventional methods (Ochca et al. 2019; Jajpura and Nayak 2020; Ma et al. 2014).
- **Customization**: Quick real-time customization of textiles, especially garments, results in fewer returns and less waste due to correct garment fit (Dik et al. 2023; Li et al. 2024; Liu et al. 2017; Tao et al. 2018).

3.5 CONCLUSION

The implementation of cleaner production technologies has significantly transformed textile manufacturing by reducing fiber damage, enhancing efficiency and reducing resource consumption, thereby contributing to lower emissions and waste. In contrast, conventional chemical and mechanical methods cause damage to fibrous materials and the environment. When cleaner production technologies are integrated with conventional ones, the efficiency of conventional technologies is enhanced. The large-scale adoption of cleaner production technologies continues to face challenges, primarily due to their higher cost compared to conventional ones.

REFERENCES

Abdel-Halim, E. S. and Al-Deyab, S. S. 2013. One-step bleaching process for cotton fabrics using activated hydrogen peroxide. *Carbohydrate Polymers* 92: 1844–1849.

Allafi, F. A., Hossain, M. O., Kadir, M. O., Shaah, M. A., Lalung, J. and Ahmad, M. I. 2021. Waterless processing of sheep wool fiber in textile industry with supercritical CO_2: Potential and challenges. *Journal of Cleaner Production* 285: 124819.

Alves, M. B., Jorge, A. M. and Pereira, J. F. 2024. The biotechnology revolution in textile dyeing. *Trends in Biotechnology* 42 (10): 1211–1214.

Angelova, Y. P. 2020. Factors influencing the laser treatment of textile materials: An overview. *Journal of Engineered Fibers and Fabrics* 15: 1–16.

Ashima, R., Haleem, A., Javaid, M. and Rab, S. 2022. Understanding the role and capabilities of Internet of Things-enabled additive manufacturing through its application areas. *Advanced Industrial and Engineering Polymer Research* 5: 137–142.

Atav, R., Gunduz, O., Yaz, S., Cakan, G. and Serin, M. 2025. Obtaining color effects in garment dyeing with laser technology. *Coloration Technology* 1–10. https://doi.org/10.1111/cote.70040

Atwee, T., Elshafei, N., Mostafa, H. Y. and Bassyouni, F. 2025. Application of laser technology to the surface treatments of various synthetic textile samples. *World Journal of Advances in Applied Physics and Mathematical Theories* 1 (102): 1–11.

Ayesh, M., Horrocks, A. R. and Kandola, B. K. 2024. The effect of combined atmospheric plasma/UV treatments on improving the durability of organophosphorus flame retardants applied to polyester fabrics. *Polymer Degradation and Stability* 225: 110789.

Bamforth, P., Williams, K. and Jackson, M. R. 2006. Edge quality optimization for CO_2 laser cutting of nylon textiles. *Applied Thermal Engineering* 26: 403–412.

Batool, R. and Mou, J. 2024. A systematic literature review and analysis of try-on technology: Virtual fitting rooms. *Data and Information Management* 8: 100060.

Bekele, T. and Alamnie, G. 2025. The photocatalytic degradation of organic pollutants: A comprehensive overview. *Results in Chemistry* 18: 102758.

Beyene, D. and Nalankilli, G. 2017. Anionic dyeability of polyester fabric by chemical surface modification. *International Journal of Modern Trends in Engineering and Science* 4 (9): 3–14.

Chen, J., Wang, Q., Hua, Z. and Du, G. 2007. Research and application of biotechnology in textile industries in China. *Enzyme and Microbial Technology* 40: 151–1655.

Chen, X., Tao, X., Zeng, X., Koehl, L. and Boulenguez-Phippen, J. 2015. Control and optimization of human perception on virtual garment products by learning from experimental data. *Knowledge-Based Systems* 87: 92–101.

Cheng, J., Tang, S., Lin, Z., Wang, Z., Wu, D., Wang, D., Liu, C. and Cao, Z. 2024. Wearable textile with laser-induced hydrophilic porous graphene oxide coating for rapid adsorption of volatile organic compounds. *Applied Surface Science* 648: 159073.

Choi, J., Kim, T.-H., Kim, H.-Y. and Kim, W. 2016. Ultrasonic washing of textiles. *Ultrasonics Sonochemistry* 29: 563–567.

Cigeroglu, Z. and Basyigit, Z. O. 2024. Ecofriendly bleaching method for cotton fabric: Development of titanium dioxide doped single wall carbon nanocomposites by photocatalytic treatment for sustainable textile pre-finishing applications. *Journal of Molecular Liquids* 412: 125895.

Dik, N. Y., Tsang, P. W., Chan, A. P., Lo, C. K. and Chu, W. C. 2023. A novel approach in predicting virtual garment fitting sizes with psychographic characteristics and 3D body measurements using artificial neural network and visualizing fitted bodies using generative adversarial network. *Heliyon* 9: e17916.

Erdem, A. and Bahtiyari, M. I. 2018. Ultrasonic-bioscouring and ozone-based bleaching of cotton slivers and coloration of them with natural dye sources. *Journal of Cleaner Production* 188: 670–677.

Gong, D., Jing, X., Zhao, Y., Zheng, H. and Zheng, L. 2021. One-step supercritical CO_2 color matching polyester with dye mixtures. *Journal of CO_2 Utilization* 44: 101396.

Gotoh, K. and Harayama, K. 2013. Application of ultrasound to textiles washing in aqueous solutions. *Ultrasonics Sonochemistry* 20: 747–753.

Gotoh, K., Harayama, K. and Handa, K. 2015. Combination effect of ultrasound and shake as a mechanical action for textile cleaning. *Ultrasonics Sonochemistry* 22: 412–421.

Hadi, G. A. 2024. The implementation of cleaner production throughout various industries. *Journal of Waste and Sustainable Consumption* 1 (2): 64–76.

Hamida, S. B., Srivastava, V., Sillanpaa, M., Shestakova, M., Tang, W. Z. and Ladhari, N. 2017. Eco-friendly bleaching of indigo dyed garment by advanced oxidation processes. *Journal of Cleaner Production* 158: 134–142.

Han, Y., Zheng, H., Jing, X. and Zheng, L. 2018. Swelling behavior of polyester in supercritical carbon dioxide. *Journal of CO_2 Utilization* 26: 45–51.

Harish, A., Sekar, R. R., Sreeja, S. J., Vijila, S. M., Vidhya, V. and Ponmurugaraj, N. 2025. Biodegradation and phytotoxicity evaluation of textile azo dye effluent using alkaline protease enzyme from halophilic bacterium *Halobacillus trueperi* RRJS1. *Journal of the Indian Chemical Society* 102: 102060.

Hassan, M. M. 2025. A comprehensive review of the advances in process engineering and greener solvents in dyeing to impart sustainable textile manufacturing. *Sustainable Materials and Technologies* 45: e01490.

Hitz, B., Ewing, J. J. and Hecht, J. 2001. *Introduction to Laser Technology*. 3rd ed. Wiley Publishers, Hoboken, NJ.

Huang, Y., Gong, D., Zheng, H., Chen, J., Zhu, S. and Zheng, L. 2026. Solubility of triazine reactive disperse dyes in supercritical CO_2: Experiments and modeling. *Dyes and Pigments* 244: 113128.

Hung, O. and Kan, C. 2017. A study of CO_2 laser treatment on color properties of cotton-based fabrics. *Coatings* 7: 131.

Imani, H., Gharanjig, K., Nahavandi, A. M. and Saadati, S. H. 2024. A cost-effective, efficient, and environmentally-safe method for deep shade dyeing of wool with extracted dye from *Reseda luteola* L.: Laboratory and industrial scales. *Sustainable Chemistry and Pharmacy* 40: 101636.

Imran, M. A., Hussain, T., Memon, M. H. and Rehman, M. M. 2015. Sustainable and economical one-step desizing, scouring and bleaching method for industrial scale pretreatment of woven fabrics. *Journal of Cleaner Production* 108: 494–502.

Jabbar, A., Bryant, M., Armitage, J. and Tausif, M. 2024. Oxygen plasma treatment to mitigate the shedding of fragmented fibers (microplastics) from polyester textiles. *Cleaner Engineering and Technology* 23: 100851.

Jabbar, A., Palacios-Marın, A. V., Ghanbarzadeh, A., Yang, D. and Tausif, M. 2023. Impact of conventional and modified ring-spun yarn structures on the generation and release of fragmented fibers (microfibers) during abrasive wear and laundering. *Textile Research Journal* 9: 1099–1112.

Jajpura, L. and Nayak, R. 2020. Ultrasound applications in textiles and apparels. In *Sustainable Technologies for Fashion and Textiles* (Ed. Nayak, R), Woodhead Publishing, Cambridge, UK, 143–161.

Jones, L. C., Wheeler, F., Williams, T., Robus, J. J., Roberts, E., Lloyd, D. and Tyrer, J. R. 2025. From steam power to laser power: Resource analysis of direct surface heating for textile dyeing. *Procedia CIRP* 135: 462–467.

Li, J., Luo, J., Wang, J., Yu, Y., Wang, Q. and Wang, P. 2023. An eco-friendly technique for shrink-resist finishing of dyed wool via combined use of hydrolase and cyclamate. *Journal of Cleaner Production* 427: 139276.

Li, J., Su, X., Liang, J., Mok, P. Y. and Fan, J. 2024. Tailoring garment fit for personalized body image enhancement: insights from digital fitting research. *Journal of Theoretical and Applied Electronic Commerce Research* 19: 942–957.

Li, Q., Xiao, X., Jiang, H. and Zhou, Y. 2025. Assembling plasma and ultraviolet radiation with cold-batch process towards a water and energy conservative cotton bleaching technology. *Sustainable Materials and Technologies* 46: e01644.

Li, S., Zheng, H., Su, Y., Zhao, Y., He, T. and Zheng, L. 2020. Effect of fluid field on the eco-friendly utilization and recycling of CO_2 and dyes in the waterless dyeing. *Journal of CO_2 Utilization* 42: 101311.

Li, X., Gao, C., Li, W., Shi, S., Zhu, E., Zhang, M. and Wang, S. 2026. Study on ball milling driven solvent extraction: Ozone oxidation decolorization of waste printed polyester fabrics. *Process Safety and Environmental Protection* 209: 108545.

Liu, J.-J., Zhang, D.-L. and Yuen, M. M.-F. 2010. A survey on CAD methods in 3D garment design. *Computers in Industry* 61: 576–593.

Liu, K., Zeng, X., Bruniaux, P., Tao, X., Yao, X., Li, V. and Wang, J. 2018. 3D interactive garment pattern-making technology. *Computer-Aided Design* 104: 113–124.

Liu, K., Zeng, X., Bruniaux, P., Wang, J., Kamalha, E. and Tao, X. 2017. Fit evaluation of virtual garment try-on by learning from digital pressure data. *Knowledge-Based Systems* 133: 174–182.

Long, J.-J., Cui, C.-L., Zhang, Y.-Q. and Huan, G.-H. 2015. Clean fixation of dye on cotton in supercritical carbon dioxide with a heterogeneous and phase transfer catalytic reaction. *Dyes and Pigments* 115: 88–95.

Luo, S., Xiong, Z. and Liu, J. 2024. How does supply chain digitization affect green innovation? Evidence from a quasi-natural experiment in China. *Energy Economics* 136: 107745.

Ma, M., You, L., Chen, L. and Zhou, W. 2014. Effects of ultrasonic laundering on the properties of silk fabrics. *Textile Research Journal* 84 (20): 2166–2174.

Maghrabi, H. A., Vijayan, A., Wang, L. and Deb, P. 2015. Design of seamless knitted radiation shielding garments with 3d body scanning technology. *Procedia Technology* 20: 123–132.

Manjusha, W. A., Josphine, J. S., Deepa, P. K., Sujatha, S. and Ahil Raj, S. 2023. Biosoftening of banana pseudostem fiber using cellulase and pectinase enzyme isolated from *Aspergillus niger* for textile industry. *Journal of Genetic Engineering and Biotechnology* 21: 170.

Millward-Hopkins, J., Purnell, P. and Baurley, S. 2023. Scenarios for reducing the environmental impacts of the UK clothing economy. *Journal of Cleaner Production* 420: 138352.

Mishra, V., Mukherjee, P., Bhattacharya, S. and Sharma, R. S. 2025. Innovative sustainable solutions for detoxifying textile industry effluents using advanced oxidation and biological methods. *Journal of Environmental Management* 380: 124804.

Mittal, K. I. and Lei, W.-S. 2018. *Laser Technology: Applications in Adhesion and Related Areas*. Scrivener Publishing, Beverly, MA.

Mousa, A. A., Mohamed, F. A., Abd El-Megied, S. A. and Youssef, Y. A. 2024. Dyeing of synthetic fiber-based wool blended fabrics in supercritical carbon dioxide. *Scientific Reports* 14: 30604.

Mukherjee, T. and Debroy, T. 2024. *Theory and Practice of Additive Manufacturing*. John Wiley & Sons, Hoboken, NJ.

Nadeem, N., Zubair, U., Javid, A., Raza, H S., Hussain, T. and Nawab, Y. 2024. Optimization of closed loop wet ozone process for controlled bleaching of Indigo colored apparels through central composite design. *Process Safety and Environmental Protection* 187: 749–761.

Nalajala, P., Gudikandhula, K., Shailaja, K., Tigadi, A., Rao, S. M. and Vijayan, D. S. 2023. Adopting internet of things for manufacturing firms business model development. *Journal of High Technology Management Research* 34: 100456.

Ochca, P., Villegas, I. F., Groves, R. M. and Benedictus, R. 2019. Diagnostic of manufacturing defects in ultrasonically welded thermoplastic composite joints using ultrasonic guided waves. *NDT and E International* 107: 102126.

Park, S. and Lee, S. 2025. Synergistic effect of ultrasonic cavitation and enzymatic activity on washing efficiency and dimensional stability of wool knitted fabrics. *Fashion and Textiles* 12: 18.

Rahman, M., Billah, M., Hack-Polay, D. and Alam, A. 2020. The use of biotechnologies in textile processing and environmental sustainability: An emerging market context. *Technological Forecasting & Social Change* 159: 120204.

Rathinamoorthy, S. and Balasaraswathi, S. R. 2024. Assessing the contribution of sewing threads to microfiber release during domestic laundering. *Environmental Pollution* 362: 124966.

Sargunamani, D. and Selvakumar, N. 2006. A study on the effects of ozone treatment on the properties of raw and degummed mulberry silk fabrics. *Polymer Degradation and Stability* 91: 2644–2653.

Serrano-Martínez, V. M., Ruzafa-Silvestre, C., Hernández-Fernández, C., Banon-Gil, E., Aran-Ais, F. and Orgiles-Calpena, E. 2024. Improving cotton fabric dyeability by oxygen plasma surface activation. *Surfaces* 7: 1079–1095.

Shahidi, S., Rashidian, M. and Dorranian, D. 2018. Preparation of antibacterial textile using laser ablation method. *Optics and Laser Technology* 99: 145–153.

Sitotaw, D. B., Ahrendt, D., Kyosev, Y. and Kabish, A. K. 2020. Additive manufacturing and textiles-state-of-the-art. *Applied Sciences* 10: 5033.

Srivastava, A. and Bandhu, S. 2022. Biotechnological advancements and challenges in textile effluents management for a sustainable bioeconomy: Indian case studies. *Case Studies in Chemical and Environmental Engineering* 5: 100186.

Srivastava, P., Goyal, S. and Patnala, P. K. 2014. Degradation of reactive, acid and basic textile dyes in the presence of ultrasound and rare earths [Lanthanum and Praseodymium]. *Ultrasonics Sonochemistry* 21: 1994–2009.

Stella, F., Garofalo, S. F., Cavallini, N., Fino, D. and Deorsola, F. A. 2024. Closing the loop: Analysis of biotechnological processes for sustainable valorization of textile waste from the fast fashion industry. *Sustainable Chemistry and Pharmacy* 38: 101481.

Tao, X., Chen, X., Zeng, X. and Koehl, L. 2018. A customized garment collaborative design process by using virtual reality and sensory evaluation on garment fit. *Computers & Industrial Engineering* 115: 683–695.

Thakore, K. A. and Abate, B. 2017. Application of ultrasound in the pretreatment of cotton. *Cellulose Chemistry and Technology* 51 (9–10): 983–992.

Vadruccio, R., Scarpino, C. and Tumino, A. 2025. Unveiling industrial Internet of Things potential for sustainable manufacturing: a triple bottom line approach. *Procedia Computer Science* 253: 1083–1092.

Vajpayee, M., Singh, M. and Ledwani, L. 2021. Non-thermal plasma treatment of cellulosic biopolymer to enhance its surface property for various applications: A review. *Materials Today: Proceedings* 43: 3250–3255.

Vajpayee, M., Singh, M., Soni, K. and Ledwani, L. 2025. Study of synergistic effects of pre-enzyme treatment and dielectric barrier discharge plasma for surface modification of cellulosic textile polymer. *International Journal of Biological Macromolecules* 307: 141879.

Venkatraman, P. D. and Liauw, C. M. 2019. Use of a carbon dioxide laser for environmentally beneficial generation of distressed/faded effects on indigo dyed denim fabric: Evaluation of color change, fiber morphology, degradation and textile properties. *Optics and Laser Technology* 111: 701–713.

Vesel, A., Mozetic, M., Strnad, S., Persln, Z., Stana-Kleinschek, K. and Hauptman, N. 2010. Plasma modification of viscose textile. *Vacuum* 84: 79–82.

Vijayakumar, A. A., Zahid, M., Corvaglia, S. G., Montagna, F., Lionetto, F. and Maffezzoli, A. 2025. Characterization of thermoplastic sandwich panels manufactured by continuous ultrasonic welding of skins to the core. *Composites Part B* 307: 112927.

Wang, C., Ma, Y. and Sui, W. 2023. The secondary flows in a cyclone separator: A review. *Processes* 11: 2935.

Wang, Y., Chai, Z. and Xinjie, X. 2025. The route towards green transformation: Mechanism of clean production enabled by enterprise digitization. *Energy Reports* 14: 386–400.

Wei, L., Xu, S., Sun, S. and Zhang, Y. 2026. Research progress in ozone generation technology and applications: A review. *Process Safety and Environmental Protection* 208: 108460.

Wu, C., Rong, Y., Li, M., Zhang, G., Huang, Y. and Zhang, T. 2022. High-quality cutting of polypropylene (PP) film by UV nanosecond laser based on thermal ablation. *Optics and Laser Technology* 147: 107600.

Xu, C., Yu, H., Zhang, S., Shen, C., Ma, C., Wang, J. and Li, F. 2024. Cleaner production evaluation system for textile industry: An empirical study from LCA perspectives. *Science of the Total Environment* 913: 169632.

Yang, S., Ding, J., Cao, Q., Bi, X., Li, J., Jia, X., Zhang, B., Zhang, P. and Yu, C. 2025. Sustainable and low oxidative damage bleaching strategy for degummed ramie cellulose fibers using NHPI selective catalytic oxidation system. *International Journal of Biological Macromolecules* 328: 147529.

Yang, Y. and Baytar, F. 2024. Fit models' roles in identifying fit issues in the apparel technical design process and implications for improving 3D virtual fitting. *Fashion and Textiles* 11: 32.

Yukseler, H., Uzal, N., Sahinkaya, E., Kitis, M., Dilek, F. B. and Yetis, U. 2017. Analysis of the best available techniques for wastewaters from a denim manufacturing textile mill. *Journal of Environmental Management* 203: 1118–1125.

Yusup, M. Z., Mahmood, W. H., Salleh, M. R. and Muhamad, M. R. 2014. The sustainability challenges in the adoption of cleaner production system: A review. *Jurnal Teknologi* 70 (1): 117–123.

Zhang, Y.-Q., Wei, X.-C. and Long, J. J. 2016. Ecofriendly synthesis and application of special disperse reactive dyes in waterless coloration of wool with supercritical carbon dioxide. *Journal of Cleaner Production* 113: 746–756.

Zhao, H., Wang, Y., Zhao, Q. and Xiong, X. 2025. A supercritical carbon dioxide dyeing strategy for polyester/cotton blended fabrics with excellent dyeing performance. *The Journal of Supercritical Fluids* 226: 106726.

Zhao, P., Ma, P., Guo, L., Jia, Y., Zhang, R., Chen, M., Wang, Z., Liu, D., Zhao, Y., Wang, X. and Rong, M. 2024. Inactivation of microorganisms on fabrics using plasma-activated nebulized mist driven by different plasma gases. *Journal of Hazardous Materials* 471: 134365.

Zhao, X.-Y., Jun, F., Wang, F., Cao, X.-X., Du, Z.-Q., Wang, Y. and Long, J. J. 2023. Investigation of the uptake and compatibility behaviors of special disperse dyes developed for sustainable color matching in supercritical carbon dioxide. *Journal of CO_2 Utilization* 72: 102478.

4 Current Sustainable Initiatives Transforming Clothing and Textile Sectors

Rushdeen Rose and Shamil Isaacs

4.1 INTRODUCTION

Global textile fiber production grew from 58 million tons (MT) in 2000 to 116 MT in 2022, estimated to reach 147 MT by 2030. The textile industry is a major source of pollutions due to their growing waste generation, use of a lot of resources (energy, water, chemicals), and their long-term harmful effects on earth's inhabitants (Niinimäki et al. 2020).

Estimates indicate that over the next 30 years, the textiles industry will emit 21 billion tons of carbon dioxide (CO_2), consume 180 million terajoules of electricity, and use 60 trillion L of water. End-of-life waste creates toxic fiber mountains, seeping chemicals into the earth and water indefinitely. Synthetic fibers last 20–50 years if exposed to sunlight but can take up to 500 years if buried (Textile Exchange 2023).

Efforts to reduce overproduction and pollution are complicated by the textiles, clothing, and footwear industry's significant employment footprint, which supports an estimated 75 million formal jobs worldwide, not including informal workers. The sector adds $2.4 trillion to global manufacturing and has a $1.12 trillion market. It spans sectors like agriculture, chemicals, logistics, and tourism, complicating sustainable change. Ten years ago, attempts to shift to sustainable methods faced resistance, as the industry prioritized mass production and profit, making system change seem financially illogical.

Recently, industry sentiments on sustainability shifted from defensive resistance to a strategic approach. Companies once saw sustainability reporting as wasteful, leading to greenwashing. Now, brands see this as an opportunity to connect with customers and stakeholders who are conscious about sustainability. Many companies pursue sustainability to genuinely reduce environmental and social impacts. While challenges remain, substantial growth in practical sustainability initiatives is reshaping the clothing and textile industries. This chapter highlights key organizations and maps their initiative ecosystem, focusing on current efforts from international policy to consumer participation. The United Nations offers global frameworks, regions develop policies, and organizations translate goals into on-site actions. To clarify these activities, the chapter reviews global, regional, and industrial efforts, ending with an evaluation of current progress and outlook.

DOI: 10.1201/9781003689225-4

4.2 GLOBAL INITIATIVES

The sustainability trend we are witnessing today is the result of seeds planted over half a century ago. The 1972 Stockholm Conference discussed the adverse environmental impacts of industrialization, including climate change, biodiversity loss, and pollution. The event was a landmark event for sustainability, which initiated the United Nations Environment Program (UNEP), an independent mechanism for changing resource consumption behavior (United Nations 1972). The conference developed into a convention/treaty to resolve issues related to toxic chemicals, health, and the environment.

After a prolonged period of inaction, major industrial disasters reinforced the urgency for global action. The Brundtland report of the World Commission on Environment and Development (WCED), titled "Our Common Future" (1987), called for global action and investment to address harmful phenomena such as nuclear disaster, ozone layer depletion, and multiple oil spills at sea. The Brundtland report coined the term "sustainable development". It provided the definition: "development that meets the needs of the present without compromising the ability of future generations to meet their own". The report demanded practical change—its overwhelming influence led to the 1994 Rio Earth Summit, which produced significant outcomes, including the Rio Declaration on Environment and Development, Agenda 21 (now Agenda 2030), the Convention on Biological Diversity (CBD), and the United Nations Framework Convention on Climate Change (WCED 1987).

This section explores the development of sustainability and the major factors influencing the clothing and textile sectors' industrial reforms. It begins with an analysis of the United Nations as a key organization directing its clothing and textile initiatives. Next, it reviews regional efforts to adopt sustainable clothing and textile initiatives, which frequently align with UN goals.

4.2.1 United Nations

The previous section highlighted key events such as the Stockholm Conference and the Rio Summit, which established frameworks to address global challenges. Within the UN framework, the process begins with the 193 member states, who meet to discuss and identify global sustainability concerns. The main forum for deciding on actions to meet these challenges is the Conference of the Parties (COP). The COP is a formal, treaty-focused forum in which governments negotiate sustainability commitments. It has led to various conventions, including Stockholm Convention, United Nations Framework Convention on Climate Change (UNFCCC), and the Basel Convention. In these conventions, countries' intentions are expressed through pledges or treaties. In terms of clothing and textile industries, commitments will relate to climate change, chemicals, waste, biodiversity, and social impact (United Nations 2015). The UN has multiple complex frameworks to govern the various spheres of sustainability. For example, the UNFCCC is the legal agreement that implements efforts to reduce climate change.

4.2.1.1 How the UN Sets Sustainability Priorities for Clothing and Textile

International agreements play a crucial role in regulating the fashion industry. These binding rules, established through forums such as the Stockholm COPs on Persistent Organic Pollutants (POPs) and the Basel COPs on Hazardous Waste, directly govern textile production by banning certain chemicals, restricting emissions, and setting criteria for waste and sustainability reporting. At the international level, the UN Alliance for Sustainable Fashion serves as a central platform to align the activities of various UN agencies. Within the industry, commitments like the UN Fashion Industry Charter for Climate Action drive progress, with signatories pledging to halve emissions by 2030 and achieve net-zero by 2050. Systemically, the UNEP promotes circularity by reframing waste and pollution as design flaws and advocating for systemic innovation in policies and business models. Collectively, these specialized programs support the broader goals of global frameworks, notably the UN's Agenda 2030 and the Paris Agreement.

4.2.1.2 UN Agenda 2030 Sustainable Development Goals (SDGs)

UN Agenda 2030 sets **broad guidelines** for achieving sustainability through 17 Sustainable Development Goals (SDGs). The SDGs, particularly Goal 12 (Responsible Consumption and Production), provide a universal benchmark and reporting framework against which national governments and corporations are aligning their clothing and textile sector strategies (United Nations 2015). Table 4.1 highlights the most relevant clothing and textile goals.

TABLE 4.1
SDGs Most Relevant to the Clothing and Textiles Value Chain

Value Chain Stage	Relevant SDGs	Why These SDGs Apply
Fiber production	SDG 6, 12, 13, 15, 1	Water use, pesticides, land impact, emissions, and rural livelihoods
Yarn and fabric manufacturing	SDG 7, 12, 13, 8	Energy use, clean energy transition, emissions, and worker conditions
Dyeing, printing, and finishing	SDG 6, 12, 3, 13	Water pollution, chemicals, worker health, and high energy use
Garment manufacturing	SDG 8, 5, 1, 3, 12	Labor rights, gender equality, wages, and safe working conditions
Logistics and distribution	SDG 13, 9	Transport emissions, infrastructure efficiency
Retail and marketing	SDG 12, 13	Sustainable consumption, transparency, climate impact
Consumer use phase	SDG 12, 13, 6	Washing, energy use, microplastics, product lifespan
End of life	SDG 12, 13, 14, 15	Waste, recycling, ocean plastics, and biodiversity impacts

4.2.2 The Paris Agreement

The Paris Agreement is an international treaty that establishes a legal framework to combat climate change, aiming to keep global warming below 1.5°C. It was signed by 193 Parties at COP21 in Paris on 12 December 2015 and is implemented by the UN through agencies, programs, and units. UNEP promotes chemical management, pollution reduction, and circular-economy practices relevant to clothing and textiles. United Nations Development Program (UNDP), via United Nations Industrial Development Organization (UNIDO), supports cleaner production, industrial upgrading, and capacity building in textile-producing countries. These platforms help turn commitments into action within CAT supply chains. Later, we will introduce policies around the world that are built on the Paris Agreement, the key initiative aimed at preventing catastrophic disasters from climate change (UNFCCC 2015).

Implementation into countries: Governments are responsible for implementing UNEP guidelines, integrating international agreements into domestic policies, restricting chemicals, setting emission targets, and reviewing clothing and textile waste strategies. High-level political forums report on progress and compliance, to be discussed at the next COP. The following subsections show organizations that started regionally and expanded into global entities (United Nations 2015).

4.2.2.1 Global Organic Textile Standard (GOTS)

In its essence, the Global Organic Textile Standard (GOTS) certifies organic and fair labor practices. It limits chemical use to an approved "Positive List," explicitly forbidding hazardous substances such as per- and polyfluoroalkyl chemicals (PFAS) and Genetically Modified Organisms (GMOs), and mandates rigorous wastewater treatment and resource monitoring. Socially, facilities must respect human rights in line with International Labor Organization (ILO) standards, banning forced and child labor, promoting occupational safety, and evaluating "living wage gaps" to ensure fair pay. These requirements are enforced through a mandatory six-step due diligence process and annual on-site audits to ensure compliance and product traceability (GOTS 2023).

4.2.2.2 OEKO-TEX

OEKO-TEX is a global organization that assesses products for harmful substances in the clothing and textile industry through rigorous scientific testing and certification. They collaborate with a worldwide network of independent institutes to offer standardized labels, ensuring products are environmentally friendly and safe for consumers. OEKO-TEX (2024) highlights their extensive ecosystem of strategic partnerships, including digital platforms that enhance supply-chain traceability and chemical management. Furthermore, the organization partners with various non-profits and government initiatives to advance larger global ecological goals and reduce industrial carbon footprints. The aim is that their services empower businesses and consumers to make responsible choices that help preserve natural resources.

4.2.2.3 ZDHC Manufacturing Restricted Substances List (MRSL)

The Zero Discharge of Hazardous Chemical Manufacturing Restricted Substance List (ZDHC MRSL) lists chemicals prohibited for intentional use in textile and leather processing facilities. It emphasizes controlling what enters the factory rather than focusing solely on residues on the final garment, thereby safeguarding workers and reducing the risk of toxic chemicals polluting local water systems (ZDHC 2015).

4.2.2.3.1 Summary

This section examined the historical development and current structure of global sustainability frameworks, especially those affecting the clothing and textile industries. It outlined how international cooperation has evolved from the 1972 Stockholm Conference to the creation of the SDGs and the Paris Agreement. In addition, the section highlighted the vital role of the United Nations in hosting COP summits, where binding treaties are negotiated to address climate change, hazardous chemicals, and resource management. It also explained how industry-specific initiatives such as GOTS, OEKO-TEX, and ZDHC implement these broad goals through stringent certification standards and chemical restrictions. By linking high-level policies with factory practices, the section demonstrated a coordinated effort to steer the fashion industry toward a circular economy and ethical production. Major priorities include reducing carbon emissions, conserving biodiversity, and promoting fair labor practices throughout the clothing and textile supply chain. The following section will investigate how global objectives are disseminated and implemented on a regional level.

4.3 REGIONAL INITIATIVES

As discussed earlier, each region or bloc is responsible for aligning its clothing and textile policies with those of the UN to ensure compliance. In this section, we will examine the sustainability initiatives related to clothing and textile undertaken by the regions of Europe, Southeast Asia, the Americas, Africa, and the Middle East.

4.3.1 Europe

The European Commission sets policies that evolve from United Nations or COP meetings, as well as from proposals by member countries of the European Union (EU). Key to this effort is the European Green Deal, the Green Claims Directive, the Ecodesign for Sustainable Products Regulation, and other policies that mandate durability and recyclability. In addition, the Digital Product Passport (DPP) policy aims to foster a system of supply-chain transparency. These initiatives aim to shift away from fast fashion, promoting repairable, eco-friendly textiles as the norm. Producers are made responsible through Extended Producer Responsibility (EPR) schemes, ensuring they oversee their products throughout their lifecycle.

4.3.2 EU Sustainable Policies

4.3.2.1 The European Green Deal

The European Green Deal (EGD) is the EU's strategy to make Europe the first continent to achieve carbon neutrality. Its policies focus on the energy, agriculture, transport, and industrial sectors to reduce greenhouse gas (GHG) emissions. The EGD aims to achieve a 50% reduction in emissions by 2030 and full carbon neutrality by 2050. Carbon neutrality occurs when emissions released into the atmosphere are balanced by the amount removed, such as through planting trees. To track progress, UN members report their GHG levels every 5 years. At the end of this chapter, we will evaluate the success of the EGD thus far by examining data and trends of GHG emissions.

4.3.2.2 Green Claims Directive

As mentioned in the introduction chapter, achieving the targets set by the EGD can be challenging from an economic, technological, legal, and social perspective. Some organizations have resorted to greenwashing, and some continue to do so. The European Commission introduced the Green Claims Directive (GCD), a set of criteria to prevent companies from making false environmental claims about their products and services. This provides consumers with certainty when buying sustainable and safe products. It also allows producers to make high-quality products without fear of being undercut by cheap, harmful products (European Commission 2024).

The GCD operates on the principle that the information provided to consumers must be reliable, comparable, and verifiable. Previously, manufacturers were left to their own devices in how they presented information, as there was neither a formal verification process nor any legal implications. However, the GCD now requires any organization to substantiate claims. For example, products claiming to be made from recycled plastic or manufactured with reduced water use must be verified by evidence.

4.3.2.3 Ecodesign for Sustainable Products Regulation

The Ecodesign for Sustainable Products Regulation (ESPR) is the EU policy aimed at improving product sustainability by promoting circularity, recyclability, durability, and energy efficiency (European Commission 2024). The ESPR mandates that companies integrate sustainability into product development by meeting specific ecodesign criteria. These include enhancing product durability, reusability, and upgradability; ensuring products are repairable; establishing maintenance and restoration systems; minimizing resource and energy use; eliminating harmful substances that impede circularity; increasing the use of recycled parts; boosting the recyclability or re-manufacturability of products; reducing waste; setting standards for carbon and environmental footprints; and increasing transparency by providing sustainability information.

Companies such as Patagonia, Nudie Jeans, Arc'teryx, and The North Face have offered to repair textile products, provide the necessary knowledge and repair kits, or buy back your old products and resell them as second-hand (Turunen and Gossen 2024; Patagonia 2022; The North Face 2024). Other institutions have developed mechanisms to aid ecodesign, such as the Textiles 2030 Circular Design toolkit by the Waste and Resources Action Program (WRAP). This toolkit provides a comprehensive guide on how to design products with circularity in mind (WRAP 2023).

4.3.2.4 Universal PFAS Restriction

The universal PFAS restriction aims to limit exposure to about 10,000 PFAS due to their high environmental persistence. Since PFAS are widely used in textiles for water and oil resistance, this regulation compels brands to seek safer alternatives to prevent human exposure and environmental harm (Doyle et al. 2024).

4.3.2.5 WRAP and the UK Textiles Pact

The UK Textiles Pact (UKTP) is another WRAP initiative that supports fashion and textiles organizations in transitioning to sustainable, circular practices by 2030. WRAP collaborates with stakeholders, including brands, retailers, manufacturers, reuse and recycling organizations, industry bodies, research and innovation organizations, and policymakers, to implement circular solutions across the fashion and textiles value chain (WRAP 2023, 2025). The UKTP aligns with the European Green Deal's decarbonization targets by matching the 50% and carbon-neutral goals for 2030 and 2050, respectively. UKTP also includes 30% reductions in water use and water pollution.

WRAP concentrates on four main areas to support organizations, industries, and countries in sustainable adaptation and compliance with the UKTP. These areas include designing for circularity, developing circular business models, closing the loop on materials, and decarbonizing supply chains. While these categories are well defined, WRAP's interventions are tailored to each specific context. Essentially, WRAP examines the system targeted for change to identify potential unintended effects, such as how a new EPR scheme might impact second-hand goods sellers.

Ireland has one of the highest clothing disposal rates per capita in Europe. To address this, WRAP conducted a detailed macro-environmental analysis, resulting in a customized national EPR roadmap for the Irish government. The Irish EPR program ensures that clothing producers take responsibility for the retrieval of clothing sold to customers. This is achieved through awareness campaigns to inform customers about the scheme and by establishing clothing collection services. The aim is to ensure that Ireland's fashion economy becomes circular and that waste is managed consciously and effectively (WRAP 2025).

4.3.2.6 Digital Product Passport (DPP)

The DPP essentially provides a view into a product's history. It details where and how raw materials like fibers were sourced, where they were processed into textiles, who assembled the product and under what conditions, where it was sold and purchased, how long the owner kept it before returning for repairs, whether it was sold second-hand, and other relevant details. This information empowers consumers to make well-informed purchasing choices. They can scan the product to assess the sustainability of its materials and production processes and verify if it was made ethically. Technologies such as QR codes, NFC tags, or RFID tags—attached to the product or label—facilitate access to the label (EU 2024). The following section will discuss the enabling technologies for the DPP policy.

4.3.3 EU Implementation

4.3.3.1 CISUTAC

Whereas WRAP is a connector between government and business, CISUTAC is a research and development organization that develops industrial technologies and infrastructure and collects data to assess the textiles value chain. The solutions developed by CISUTAC are prototyped and piloted for industrial applications, making

CISUTAC's contribution essential for practical transitioning to a circular economy (CISUTAC 2025).

Tracking the product's journey through its lifecycle requires new technologies such as information storage devices, scanning technologies and standardized data recording methods. In line with the ESPR, CISUTAC is developing the DPP's infrastructure. The DPP would be accessible throughout the supply chain for circular, recycling, second-hand and repair markets. In other words, technology would help record cradle-to-grave information (CISUTAC 2025).

4.3.3.2 Registration, Evaluation, Authorization and Restriction of Chemicals (REACH)

The EU's REACH regulation is a foundational global initiative that systematically restricts hazardous substances in textiles placed on the EU market, thereby protecting human health and the environment (ECHA 2023). This framework mandates supply-chain transparency by requiring communication of information on Substances of Very High Concern (SVHCs) throughout the supply chain, compelling brands to know and control their material inputs. The EU's Sustainable Products Initiative (SPI) and ESPR build upon this chemical safety foundation by establishing new, legally binding requirements for durability, repairability, and recycled content for textiles (European Commission 2024).

4.3.3.3 Ellen MacArthur Foundation

The Ellen MacArthur Foundation is a UK-based organization that aims to speed up the shift toward a circular economy, which is an industrial system designed to be restorative and regenerative instead of extractive and linear. According to the foundation, its mission rests on three core principles: eliminating waste and pollution through thoughtful design, maintaining product and material circulation at their highest value, and restoring natural ecosystems. The organization advocates a new framework for the fashion and textile sectors and drives systemic change through global education programs and extensive networks of businesses and knowledge-sharing (Ellen MacArthur Foundation 2023).

4.3.4 Southeast Asia (ASEAN)

The Southeast Asian bloc supplying Western regions, such as the EU, is the Association of Southeast Asian Nations (ASEAN), comprising Indonesia, Malaysia, the Philippines, Thailand, Brunei Darussalam, Singapore, Vietnam, Myanmar, Cambodia, Timor-Leste, and Lao PDR. ASEAN's environmental and climate frameworks consistently align with UN treaties and the SDGs. According to Berto (2022), Southeast Asia, a significant production hub for the West since the 1960s, has been experiencing adverse environmental impacts-specifically in Vietnam, Cambodia, and Indonesia.

A main initiative for the bloc is the ASEAN-UN Plan, which focuses on haze pollution, marine plastic pollution, and climate finance. The bloc positions itself as a regional platform that supports the UN agenda while respecting national sovereignty (ASEAN 2025). The region relies heavily on cooperation between member states rather than enforcement legislation.

4.3.4.1 ASEAN Sustainable Policies

There are three sustainable initiatives at different levels in the textiles sector, including national circular-economy partnerships, textile circularity consortia, and on-the-ground innovation in pineapple fiber fabrics.

4.3.4.2 Circular Fashion Partnership (Indonesia)

In 2024, Indonesia launched its Circular Fashion Partnership to address the major post-industrial waste problem currently facing the country. Being one of the largest manufacturing hubs, the country has launched this program to align its waste material flows with circular practices. The aim is to capture and reuse waste, reducing the need for virgin textiles. The program trains various industry stakeholders through train-the-trainer programs to enable circularity-focused knowledge dissemination across the industry. This initiative has brought on board global brands such as Adidas, H&M, and Lululemon (CFA 2025).

4.3.4.3 Textile Circularity Consortium (Singapore)

In 2025, the Singapore Fashion Council hosted the "Be the Change Summit" to discuss and deliberate on the path to a circular economy. Like Indonesia, Singapore generates 41,000 tons of textile waste each year. To combat this, the council established the Singapore Textile Circularity Consortium. They acknowledge the role that technology will play in the solution and reward innovators in the space by showcasing their solutions, including an AI-powered platform to assist designers in developing eco-friendly design recommendations, posture-correcting t-shirts, and the use of pomegranate waste to produce affordable natural dyes (Singapore Fashion Council 2025).

4.3.5 ASEAN Initiative

4.3.5.1 PINALINA (Vietnam)

In a world facing resource scarcity, humans need to maximize resource use and reduce waste. Once a by-product of agriculture, pineapple leaves are now transformed into fibers that can be blended with cotton and silk to enhance durability. Traditionally, pineapple fabrics were produced as artisanal products in the Philippines, but they have now become mainstream thanks to Faslink, a textiles R&D company, under the brand name Pinalina. Vietnam has actively promoted pineapple fabric products through its "Wear Pineapple" campaign. Pinalina offers numerous benefits. In Vietnam, around 3 MT of pineapple leaves are discarded each year after harvest. Previously, these leaves were burned, but Pinalina converts this agricultural waste into a valuable and sustainable resource. Using these leaves reduces Vietnam's reliance on imported raw materials. According to suppliers, the fabric is safe, comfortable, aesthetically pleasing, rustic, and highly desirable. Coteccons, a major construction firm, has produced all 8,000 of its uniform shirts from Pinalina (Duyen 2025; Tre 2025).

4.3.6 United States–Mexico–Canada (USMCA)

4.3.6.1 USMCA Sustainable Policies

Sustainability efforts in North America are reinforced by the United States–Mexico–Canada (USMCA) Agreement, as well as industry-led initiatives including the

Sustainable Apparel Coalition (SAC), cleaner production strategies, and circularity programs implemented throughout Canada and the US. SAC has developed a benchmarking tool known as the Higg Index, which measures the environmental, social, and labor impacts of making and selling its products and services (SAC 2021). Under USMCA, environmental cooperation extends to chemicals, waste, and supply-chain transparency, which influences textile-related industries. This region's approach tends to focus on measurement, reporting, and innovation rather than regulatory enforcement.

4.3.6.2 The Commission for Environmental Cooperation (CEC)

The CEC develops and implements sustainability standards for the USMCA and has its own strategic sustainability goals, known as the Pillars for 2025. These pillars include clean air, land, and water; preventing and reducing pollution in the marine environment; circular economy and sustainable materials management; shared ecosystems and species; resilient economies and communities; and effective enforcement and environmental laws (CEC 2021). The CEC allows citizens to report countries that have failed to implement its environmental laws. When stakeholders submit a report, the non-compliance is made public for discussion. This approach uses public discourse to address non-compliance at the national level. An example would be a government's inaction in resolving water pollution caused by textile effluent. Citizens can report the government to the CEC, which will raise it as a finding at the USMCA level.

4.3.6.3 USMCA Implementation

4.3.6.3.1 US Cotton Trust Protocol

This voluntary sustainability program for US growers utilizes the Protocol Consumption Management Solution (PCMS) to monitor cotton from farms to final products. It offers brands audit trails and transparency maps, crucial for demonstrating compliance with Uyghur Forced Labor Prevention Act (UFLPA) requirements related to forced labor. The protocol also enforces strict standards for worker rights and well being, excluding members who do not meet its safety and labor standards (Trust US Cotton Protocol 2025).

4.3.6.3.2 Higgs Index

The Facility Environmental Module (FEM) provides a standardized method used by global manufacturing facilities to assess and improve their environmental performance. The Higg FEM is a comprehensive tool used by manufacturers and retailers that utilizes Life Cycle Assessment (LCA) data evaluate sustainability across seven key impact areas, including environment, emissions, effluent, chemical, energy, water use, and waste. Developed by Cascale, this module requires facilities to submit operational data annually, typically from January to April, to establish a baseline for progress. Its primary goal is to encourage systematic change, reduce redundant reporting, and create a shared language for communicating impacts to stakeholders, including brands, retailers, and consumers. Facilities must perform self-assessments in compliance with various requirements, including obtaining site permits and managing chemicals, and these are also subject to third-party verification. Famous brands and manufacturers using the tool include H&M, Zara, Nike, Adidas, Patagonia, Gap Inc., Puma, SHEIN, Walmart, United Colours of Benetton, TAL Apparel, and Yunus Textile Mills (Cascale 2026).

Greenpeace researchers, in their report "Greenwash Danger Zone", challenge the claims made by the Higg Index. They argue the tool is inherently flawed, citing issues such as the promotion of synthetic fibers, incomplete lifecycle data, microplastic fiber release into air, water, and soil, many garments being non-recyclable, reliance on landfill incineration, waste exports, and conflicts of interest where members set and write their own rules (Cobbing et al. 2023).

4.3.6.3.3 Nuuly

The Higg Index (standardized tools for measuring environmental and social sustainability across the apparel footwear and textile industries), formerly the SAC, supports brands in monitoring and improving their sustainability goals, but there are also other initiatives on the ground that are changing how we interact with, buy, sell, and wear clothing. Rental platforms like Nuuly have moved from niche to mainstream, offering consumers access rather than ownership across a wide range of apparel categories and increasing garment utilization (Lang and Zhang 2023; Nuuly 2025).

4.3.7 Africa

4.3.7.1 African Union's (AU) Sustainability Policies

4.3.7.1.1 AU Agenda 2063

Agenda 2063 is the African Union's (AU) 50-year plan for the continent's sustainability. In essence, the seven aspirations, namely inclusive growth, the elimination of poverty, modern agriculture, strong education and skills, and climate-resilient economies, align well with the SDGs. The African Circular Economy Alliance collaborates directly with UNEP and UNIDO, anchoring African sustainability transitions within UN structures. Africa also negotiates collectively on issues of equity, adaptation, and climate finance under the UNFCCC (African Union 2024).

4.3.7.1.2 Continental Circular-Economy Action Plan for Africa

Inspired by the lessons from industrial revolutions in developed countries, the continent aims to leapfrog into sustainable development. The Continental Circular-Economy Action Plan for Africa (CCEAPA) is a 10-year initiative aimed at preventing harmful economic practices, specifically targeting challenges such as waste imports, climate change, biodiversity loss, and pollution. The plan's objective is to replace reliance on environmentally damaging commodity exports with a circular-economy approach that encourages sustainable growth across Africa (African Union 2024). In addition, the African Circular-Economy Alliance (ACEA) Fashion and Textiles Group is a government-led initiative of African countries, including South Africa, Nigeria, and Rwanda, that aims to catalyze Africa's transition to a circular economy. The coalition has gathered a specialized team for the clothing and textile sector, recognizing the circular economy as key to advancing economic growth.

4.3.8 AU Implementation

4.3.8.1 Localization

Another key strategy is localizing supply chains to promote growth, reduce unemployment, and create jobs for marginalized communities. It emphasizes leveraging African resources such as indigenous knowledge and unique local resources to produce high-value products.

Furthermore, Africa must shift from a consumer-based import economy to a supplier-based export economy. Cotton is one area where this opportunity exists. Africa accounts for 9% of the world's cotton production, worth approximately US$ 31 billion. With a growing, youthful population, the goal is to boost cotton sales and enhance local value-added processing to create significant employment opportunities. The African Export-Import Bank invested US$ 475 million in Africotin, a Pan-African project focused on the cotton supply chain. Additionally, Afreximbank allocated US$ 195 million to textile and cotton parks in Burkina Faso and Nigeria.

4.3.8.2 Africa Fashion Reception

Pan-Africanism, a movement emphasizing solidarity, self-determination, and harmony in Africa, is the guiding perspective for Africa's development. The continent aims to boost internal trade among member nations and enhance its product development and supply chains. Currently, roughly 16% of Africa's trade is intra-African, compared to about 60% in Europe (Eurostat 2024; Mathekga 2025). The Africa Fashion Reception initiative provides a platform for connecting African fashion organizations. It advocates for free trade agreements within the African Continental Free Trade Area (AfCFTA), a key project of Agenda 2063.

4.3.9 Middle East and North Africa (MENA)

The Middle East and North Africa (MENA) regions are key players because of their well-established clothing and textile supply chain. Countries such as Egypt, Morocco, Jordan, and Tunisia are closely connected to European markets through their geographic proximity and trade agreements. Although MENA is a low-margin area, its closeness to the EU means it faces pressure to fully absorb the costs of adapting systems to meet sustainability standards. The region also faces severe environmental challenges, including water shortages, land degradation, and reliance on energy. Therefore, sustainability isn't just about meeting export regulations; it's critical for addressing local issues. For example, the UAE alone uses 500 million textile items annually, discarding 210,000 tons—90% of which end up in landfills (Keerthana 2024).

4.3.9.1 MENA Sustainability Policies

4.3.9.1.1 MENA Vision 2030

In Saudi Arabia, under its decadal strategic plan, Vision 2030, and its sub-project, the Saudi Green Initiative, the country has made great strides toward transitioning away from oil, improving waste management efficiency, and expanding renewable energy production.

4.3.9.1.2 MENATEX

MENATEX is an extension of the Global Textiles and Clothing Program (GTEX). Funded by the Swiss State Secretariat for Economic Affairs (SECO) and the Swedish International Development Cooperation Agency (SIDA), it is governed by the International Trade Centre (ITC). MENATEX aims to enhance supply chains for Swiss and Swedish companies, ensuring compliance with EGD and the UNEP goals. The program helps supplier countries such as Egypt, Jordan, Morocco, and Tunisia transition to digital and sustainable operations. Its goals include reducing waste, establishing sustainable systems, and boosting e-commerce for small and medium enterprises (SMEs) in these regions. Focused on strengthening connections within the MENA region to promote circularity, the initiative created 3,900 jobs and generated US$ 75 million in exports from 2018 to 2022. It also reduced carbon emissions by 81,241 tons and conserved 150,000 m^3 of water. GTEX Phase II has secured an additional US$ 13 million for the MENATEX project (Duyen 2025).

4.3.10 MENA Implementation

4.3.10.1 Kering Generation Award X MENA

To drive change at the grassroots level, the Saudi Fashion Commission, in partnership with Kering, launched the Kering Generation Award X MENA, recognizing startups with innovative concepts in clean production, regenerative materials, and circular design. Winners receive substantial resources, including a trip to the Change Now global innovation summit in Paris, expert mentorship, and incubation at "The Garage" in Riyadh (Kering 2026).

4.3.11 Summary

In the UK, the notable Textiles 2030 initiative, spearheaded by WRAP, has brought together major retailers, brands, and charities to set ambitious, measurable goals for reducing carbon emissions and water use, and promoting circular textile practices. This initiative showcases a successful national implementation model (WRAP 2023). Industry-led groups such as the SAC's Higg Index offer standardized tools and metrics that hundreds of influential brands currently use to monitor and report environmental impacts, thereby enhancing regulatory data compliance (SAC 2021). The Ellen MacArthur Foundation's Make Fashion Circular campaign continues to gather global brands around practical, pre-competitive roadmaps for circularity, shaping policy development (Ellen MacArthur Foundation 2023). EU-funded projects like CISUTAC demonstrate efforts to develop systemic, scalable solutions by tackling technical and logistical challenges across the entire textiles sector (CISUTAC 2025). In regions like Bangladesh, the clothing and textile industry is a major economic sector, but it also faces challenges related to sustainable practices and labor conditions.

The transition toward a circular economy in Africa and the MENA region provides a strategic framework for replacing environmentally damaging commodity exports with sustainable, value-added growth. In Africa, the Continental Circular-Economy Action Plan (CCEAPA) and Agenda 2063 aim to reform traditional industrial stages by localizing supply chains and shifting from a consumer-based import economy to a supplier-based export powerhouse, particularly in the multi-billion-dollar cotton industry. At the same time, the MENA region is responding to both European market pressures and local environmental crises—such as severe water shortages and massive textile waste—through initiatives like MENATEX and Saudi Arabia's Vision 2030, which have already reduced carbon emissions and created thousands of jobs. By fostering grassroots innovation through platforms like the Kering Generation Award and expanding intra-continental trade via the AfCFTA, these regions are building economic resilience while addressing urgent ecological needs.

4.4 EVALUATING CURRENT PROGRESS

In this section, we will look at EU progress toward achieving the EGD emission targets. The Intergovernmental Panel on Climate Change (IPCC) oversees the tracking of carbon emissions and sequestration in land and oceans under Land Use, Land-Use Change, and Forestry (LULUCF) policies. This record allows one to analyze the trajectory and achievability of EGD objectives, which are 50% by 2030 and full net-zero emissions by 2050. There was a consistent decline in total GHGs with LULUCF, with a sharp drop in 2021 due to COVID-19. From 1990 to 2021, emissions decreased by about 0.045 units per year. At this rate, achieving carbon neutrality would take around 71 more years, approximately by 2092 (as calculated). The ambition for Europe to become carbon-neutral by 2050 is challenging; based on current trends, it's nearly impossible. Carbon emissions have fallen gradually from 1990 to 2018, driven by the transition from fossil fuels to renewable energy, a structural shift from heavy industry to service-based sectors, higher carbon taxes and stricter carbon policies. The change required to meet the goals demands a steep, rapid decline in emissions, with reductions far greater than ever before. For the clothing and textile industry, there are a few key developments that could have a significant, exponential impact on how we reduce GHG emissions. These changes are linked to overhauling and changing the supply chain, which is the root of the problem.

4.5 THE SUPPLY-CHAIN TRANSFORMATION

4.5.1 Digitization

A key focus on sustainability in this chapter is waste reduction. Unnecessary waste contributes to GHG emissions because it results from production and energy use. The most effective way to save energy and cut GHG emissions is to avoid producing items that aren't needed. Furthermore, making fewer items that last longer can greatly decrease GHG emissions.

Digital fashion tools can reduce waste and significantly support GHG reduction and other sustainable objectives. Digitization helps avoid unnecessary sample production and disposal, ensures products meet consumer preferences, and guarantees better-fitting garments, leading to items that are worn and enjoyed for extended periods.

Fast fashion's current trend involves producing numerous samples because variety and speed are key to selling large quantities quickly. Selecting designs for market release requires multiple phases of sampling, often across various fabrics and colors. As a result, a single garment style can have countless variations, and some samples are repeated to evaluate color, fit, and fabric suitability. This process not only incurs high costs in terms of time, transportation, energy, and resources globally but also generates significant waste, as many samples go unused. In production, once samples are chosen for bulk manufacturing, it's not guaranteed that the product will sell successfully. Unsold items lead to wasted resources. There have been instances where entire clothing ranges were discarded because the styles did not sell.

Styles that do not sell well can stem from various issues, such as inappropriate colors, unsuitable fabrics, or poor fit. Modern digital design technologies—like 3D scanning, 2D and 3D Computer-Aided Design (CAD) software, Internet of Things (IoTs), and artificial intelligence (AI)—can greatly minimize resource waste. These technologies will play a crucial role in lowering our carbon emissions enough to meet the 2050 target of carbon neutrality.

4.5.1.1 3D Computer-Aided Design (CAD)

3D CAD systems have advanced significantly, enabling designers to upload hundreds of small fabric swatches from suppliers or fabric designers to 3D CAD software programs such as Clo3D, Tukatech, and Style3D. Designers can create patterns, evaluate fit, choose fabrics and trims, and simulate a range of designs on a 3D model. Multiple colorways can be applied, allowing designers to review different versions before selecting a sample for production (Figure 4.1). This can effectively eliminate most of the waste in the design phase.

FIGURE 4.1 3D virtual design.

4.5.1.2 3D Scanning

3D scanning technology, besides reducing the time and effort required to take measurements, also enables proper sizing assessments by allowing us to capture large amounts of data and understand the body shapes we need to design clothes for. Clothing that doesn't fit properly is likely not be worn and would be perceived as poor quality. Scanning enables comprehensive sizing studies, allowing us to assess not only the shape but also the number of sizes within each shape (Figure 4.2). Essentially, revealing how exactly to size and shape your garments and therefore improve wearability and customer satisfaction. 3D scanning, when performed thoroughly, can reveal various market segments. Brands can leverage this to determine their target audience and estimate market size. This helps produce the appropriate quantity of products for each customer segment.

4.5.1.3 Artificial Intelligence

With the advent of social media and AI, companies can now take 3D designs and showcase them as products on social media, in their stores, or as adverts. Products can be rendered to look realistic, and data on potential customers can be analyzed to assess the likelihood of the product selling well. Only after this assessment will the company produce in bulk. This allows businesses to reduce the risk of producing unwanted products.

Try-on software is another clothing technology that allows customers to virtually try on products before purchase, reducing the chance of returns and further waste.

4.5.2 Recycled Polyester (rPET)

Extracting crude oil from beneath the earth's surface inevitably results in carbon emissions, which conflict with the Paris Agreement commitments and the EU GHG

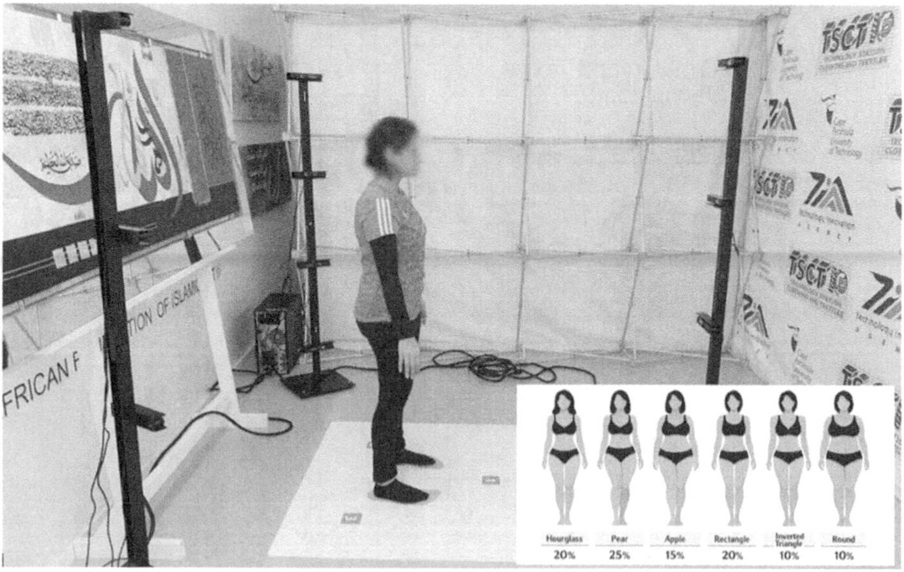

FIGURE 4.2 3D scanning with shape analysis.

targets. This natural resource contributes to various forms of pollution because it can be transformed into products such as polyester, which can pollute the environment for extremely long periods. The oil extraction process and the steps required to transform oil into usable products disturb the biosphere, causing harm at every stage of use (Luu and Baker 2021). Therefore, exploring alternative ways to sustain our consumption patterns and the versatility that polyester provides for textiles is essential. Recycled polyester (rPET) is made by reprocessing plastic waste—mainly post-consumer bottles—into new textile fibers. It is often favored in fashion because it avoids the resource-intensive, pollution-generating process of extracting raw petroleum for virgin polyester. To make recycled polyester a viable alternative to virgin polyester, one must develop products for its recycling. In other words, they must be designed for circularity. One must consider all parts of the product, focusing on mono-material components (all made from polyester (PET)) and easy disassembly. In an ideal world, it would be easy to recycle polyester textile products and prevent them from being downcycled or sent to landfills.

The goal must be to transition from using polyester derived from plastic bottles to recycled textile-based polyester. By designing circularity and leveraging technological and process advancements, we can significantly enhance our ability to recycle polyester from textiles. This progress is crucial in reducing the microplastics released into the environment beyond current levels and in further decreasing harmful emissions, making a compelling case for immediate action. There are some concerns for rPET process as it contributes to the open recycling routes rather than closed loop circular process. Furthermore, feedstock for this process is from the packaging industry, not the discarded textile waste, which should be the feeder material for rPET recycling process.

4.5.3 Transparency, Ethics, and Supply-Chain Reinvention

Regulations like REACH and DPPs are prompting companies such as TrusTrace and Fibertrace to adopt digital traceability platforms. This shift moves brands toward verifying materials and chemical information through physical-to-digital methods. Social auditing programs by groups like the Fair Labor Association (FLA) remain key on-site tools for overseeing factory conditions, guided by ILO conventions. Worker-led social responsibility initiatives, like the Fair-Trade Certified apparel program, provide direct financial benefits to factory workers, supporting the SDGs. Additionally, the increased use of renewable energy in manufacturing—fueled by initiatives such as the RE100 coalition—sees leading suppliers investing in solar and wind power.

4.6 Brand and Consumer Action

4.6.1 Product-Led Strategies for Longevity

Brands such as Patagonia, through its Worn Wear program, and Nudie Jeans have built their systems and products around dedicated repair services that are core to their brand identity and aid waste reduction. Incorporating recycled materials, conducting durability testing, and providing clear care instructions have become standard practices in designing sustainable collections. Additionally, limited-edition "re-crafted" or upcycled collections serve as concrete initiatives that directly address waste and promote circularity, encouraging consumer engagement.

4.6.2 Scaling New Business Models

Brand-integrated resale, powered by platforms like ThredUP and Trove, is now a standard offering from retailers like Recreational Equipment Inc (REI) and Lululemon, creating official channels for product recirculation. The "pre-order" or "made-to-order" model is successfully employed by smaller brands to precisely match production with demand and minimize deadstock.

4.6.3 The Empowered Consumer Ecosystem

The rise of the "conscious consumer" is evidenced by the growth of the second-hand apparel market, now projected to outpace fast fashion, demonstrating a behavioral shift toward circularity. Public-facing campaigns like WRAP's "Love Your Clothes" provide practical, evidence-based resources to citizens on extending garment life, care, and repair, directly influencing consumer behavior at scale. Online communities and influencers dedicated to "slow fashion" visibly promote practices such as mending and capsule wardrobes, thereby shifting cultural norms. Community-based initiatives like clothing swaps and repair cafés represent grassroots, decentralized efforts extending the life of clothing.

4.7 FUTURE OUTLOOK AND RESEARCH AGENDA

The expected growth of EPR schemes and the emergence of a global, ubiquitous DPP system will significantly change product economics and enhance supply-chain transparency, building on the data infrastructure established by chemical regulations. The idea of "sufficiency"—cutting down total consumption—is gaining popularity in policy and academic circles, challenging the growth-driven mindset and suggesting a shift toward mindful wardrobe management. This represents a vital area for behavioral and economic research. A crucial research goal is to measure the systemic effects of these converging initiatives and develop comprehensive metrics that reflect environmental, social, and economic impacts throughout the entire product lifecycle.

4.8 CONCLUSION: THE STATE OF THE SHIFT AND THE PATH AHEAD

The initiatives outlined here, from UN frameworks and key regulations like REACH to collaborative agreements such as Textiles 2030 and EU projects like CISUTAC, showcase a multi-level operational landscape actively reshaping the textile industry. The future requires faster integration and scaling of these efforts, ensuring that macro-level policies and regulations are backed by mid-level innovations in supply chains and micro-level adoption by brands and consumers. As the outlook indicates, real transformation hinges on the continued merging of collaborative policy, disruptive technology, and a cultural reassessment of values—shifting the industry from volume-driven to one focused on longevity, safety, and care. Although 52 of 55 countries have policies supporting a circular economy, actual implementation remains limited. The textiles sector is especially critical due to its economic significance and environmental footprint.

The move toward sustainability is well intentioned but must be transparent throughout to ensure it benefits vulnerable communities, promotes equity for exploited groups, and prevents sustainability from becoming another form of concentrated power masked by green rhetoric. It is essential to question who benefits financially from sustainability efforts, who owns the technology and data, who controls and constrains the flow of goods under sustainable policies, and why there is little emphasis on sufficiency or degrowth. These considerations are vital for a fair, empathetic, and honest transition.

Critically, some argue that the EU's current reduction in carbon emissions might not lower overall emissions but merely shift them elsewhere—such as Africa, where raw materials for renewable energy are mined, and Asia, where technologies like batteries are manufactured. Further research into the export of carbon emissions is necessary.

Organizations must shift their thinking from exploiting Earth's finite resources to adopting sustainable practices that support resource reuse and circularity. Production should decrease at all stages. Such a change must be rooted in a fundamental shift in thinking. The biggest obstacle is accepting that true decarbonization may require sacrificing wealth.

REFERENCES

African Union. 2024. Continental circular economy action plan for Africa. https://au.int/sites/default/files/documents/45336-doc-GIZ-AU_Continental_Circular_Action_V11.pdf [Accessed 10 February 2026].

ASEAN. 2025. Overview of ASEAN-United Nations relations. https://asean.org/ [Accessed 10 February 2026].

Berto, I. 2022. The phenomenon of environmental justice in Southeast Asia within the textile and apparel industry. Master's Thesis, Ca' Foscari University of Venice, Italy.

Cascale. 2026. Our members. Cascale. https://cascale.org/our-members/ [Accessed 15 February 2026].

CEC. 2021. Commission for Environmental Cooperation (CEC) Strategic Plan 2021–2025. https://www.cec.org/files/documents/strategic_plans/cec-strategic-plan-2021-2025.pdf [Accessed 15 February 2026].

CFA. 2025. A year of circular progress in Indonesia. Global Fashion Agenda. https://globalfashionagenda.org/news-article/a-year-of-circular-progress-in-indonesia/ [Accessed 14 February 2026].

CISUTAC. 2025. CISUTAC. https://www.cisutac.eu/ [Accessed 18 February 2026].

Cobbing, M., Wohlgemuth, V. and Vicaire, Y. 2023. Greenpeace_Report_Greenwash_Danger_Zone. https://www.greenpeace.de/publikationen/Greenpeace_Report_Greenwash_Danger_Zone.pdf [Accessed 18 February 2026].

Doyle, H., Flexman, K., Keyte, I., Reynolds, K., Tyers, J., Serrano Valverde, B., Whiting, R., Løfstedt, M. and Fogh Mortensen, L. 2024. An assessment on PFAS in textiles in Europe's circular economy. https://www.eea.europa.eu/en/analysis/publications/pfas-in-textiles-in-europes-circular-economy/an-assessment-on-pfas-in-textiles-in-europes-circular-economy/@@download/file [Accessed 18 February 2026].

Duyen, K. 2025. Is Pinalina™ the solution to fashion's sustainable raw material challenge? Tatler Asia. https://www.tatlerasia.com/power-purpose/impact/pinalina-the-solution-for-sustainable-materials [Accessed 14 February 2026].

Ellen MacArthur Foundation. 2023. About us: Ellen's story, Ellen MacArthur Foundation. https://www.ellenmacarthurfoundation.org/about-us/ellens-story [Accessed 16 February 2026].

European Chemicals Agency (ECHA). 2023. https://euratex.eu/the-european-chemicals-agency-echa-regulation/ [Accessed 18 February 2026].

European Commission (EC). 2024. Green claims. https://environment.ec.europa.eu/topics/circular-economy-topics/green-claims_en [Accessed 18 February 2026].

European Union (EU). 2024. EU's digital product passport: Advancing transparency and sustainability. https://data.europa.eu/en/news-events/news/eus-digital-product-passport-advancing-transparency-and-sustainability [Accessed 16 February 2026].

Eurostat. 2024. Intra-EU trade in goods-main features. https://ec.europa.eu/eurostat/statistics-explained/index.php?title=Intra-EU_trade_in_goods_-_main_features [Accessed 18 February 2026].

Global Organic Textile Standard Global Standard. 2023. Global Organic Textile Standard (GOTS): Version 7.0. 1–53.

Keerthana, R. 2024. Navigating the complexities of textile waste management. https://www.wasterecyclingmea.com/news/textile/navigating-the-complexities-of-textile-waste-management [Accessed 24 March 2026].

Kering. 2026. Kering and The Saudi Fashion Commission launch Second Edition of the "Kering Generation Award X MENA". https://www.kering.com/en/news/kering-and-the-saudi-fashion-commission-launch-second-edition-of-the-kering-generation-award-x-mena/ [Accessed 15 February 2026].

Lang, C. and Zhang, R. 2024. Motivators of circular fashion: The antecedents of Chinese consumers' fashion renting intentions. *Sustainability* 16(5): 2184.

Luu, T. T. A. and Baker, J. R. 2021. Exploring consumers' purchase intention of RPET bottle-based apparel in an emerging economy. *Journal of Open Innovation: Technology, Market, and Complexity* 7(1): 1–12.

Mathekga, R. 2025. The future of African trade: GIS reports. https://www.gisreportsonline.com/r/africa-trade-afcfta/ [Accessed 15 February 2026].

Niinimäki, K., Peters, G., Dahlbo, H., Perry, P., Rissanen, T. and Gwilt, A. 2020. The environmental price of fast fashion. *Nature Reviews Earth and Environment* 1(4): 189–200.

Nuuly Clothing Rental. 2025. https://www.nuuly.com/?srsltid=AfmBOooFoiG15L9etSq2jbljtU25ZWbsASFnW7NFXTRJ49h_fpEx_opk [Accessed 20 February 2026].

OEKO-TEX. 2024. OEKO-TEX®: Tailor-made services for the textile industry. https://www.oeko-tex.com/en/about-us/ [Accessed 18 February 2026].

Patagonia. 2022. Repair process. https://help.patagonia.com/s/article/Repair-Process [Accessed 20 February 2026].

SAC. 2021. Sustainable apparel coalition - Higg product tools. https://product.higg.org/page/sustainable-apparel-coalition 10 December 2025 [Accessed 18 February 2026].

Singapore Fashion Council. 2025. Be the Change Summit 2025- Singapore Fashion Council. https://www.sgfashioncouncil.org.sg/be-the-change-summit-2025/ [Accessed 20 February 2026].

Textile Exchange. 2023. Materials Market Report 2023. https://textileexchange.org/knowledge-center/reports/materials-market-report-2023/ [Accessed 18 June 2026].

The North Face. 2024. Renewed. https://www.thenorthfacerenewed.com/ [Accessed 20 February 2026].

Tre, B. T. 2025. 8,000 T-shirts made from pandan leaf fabric have been successfully produced in Vietnam. https://www.vietnam.vn/en/8-000-ao-thun-tu-vai-la-dua-san-xuat-thanh-cong-tai-viet-nam [Accessed 20 February 2026].

Trust US Cotton Protocol. 2025. https://trustuscotton.org [Accessed 10 March 2026].
Turunen, L. L. M. and Gossen, M. 2024. From preloved to reloved: How second-hand clothing companies facilitate the transaction of used garments. *Journal of Sustainability Research* 6(1): 1–22.
UNFCCC. 2015. The Paris Agreement. United Nations Framework Convention on Climate Change (UNFCCC). https://unfccc.int/process-and-meetings/the-paris-agreement [Accessed 10 January 2026].
United Nations. 1972. UN Conference on the Human Environment.
United Nations. 2015. THE 17 GOALS. Sustainable development. https://sdgs.un.org/goals [Accessed 10 January 2026].
WCED. 1987. Report of the World Commission on Environment and Development: Our Common Future Acronyms and Note on Terminology Chairman' s Foreword.
WRAP. 2023. Textiles 2030 circular design toolkit. https://wrap.org.uk/sites/default/files/2023-10/WRAP_Design_for_Circularity_Toolkit.pdf [Accessed 10 January 2026].
WRAP. 2025. Ireland: The beginnings of a textiles EPR scheme. https://www.wrap.ngo/resources/case-study/ireland-beginnings-textiles-epr-scheme [Accessed 10 January 2026].
ZDHC. 2015. Chapter 1: ZDHC MRSL. https://mrsl-30.roadmaptozero.com/ [Accessed 20 January 2026].

5 Sustainable Raw Material Manufacturing Processes

Cyrus Tshifularo

5.1 INTRODUCTION

The demand for sustainable and renewable materials increased significantly after the turn of the millennium, driven by growing environmental concerns, rising waste pollution, the extensive use of fossil fuels, energy security challenges, and the rapid depletion of natural resources (Ramos et al. 2019). Environmental concerns and pollution threaten food security, standard of living, and humans' life and health (Ahmed et al. 2022). In addition, environmental concerns globally include higher temperatures, worldwide depletion of petroleum resources, decrease of polar ice caps, and sea level increase (Gedik and Avinc 2020). Climate change influenced worldwide temperatures to increase by almost 1°C (Silva and Rosenbach 2021). Climate change alters precipitation patterns, contributes to rising sea levels and melting ice caps, increases the frequency of disease outbreaks, promotes the spread of fungal diseases (Aryal et al. 2020), and negatively affects livestock health and productivity (Godde et al. 2021). Continuously neglecting changing global temperature and the current rate of CO_2 will intensify the current problems encountered. Taib et al. (2023) forecast that atmospheric CO_2 concentrations will surpass 450 ppm within the next 30 years and global temperature will be above 1.5°C (Taib et al. 2023).

The fashion industry is expanding rapidly, with trends changing seasonally and driving increased production and consumption. This growth contributes significantly to waste generation during both the manufacturing process and the post-consumer phase, thereby adversely affecting the environment (Suen et al. 2023). Chen and colleagues predicted that textile waste created worldwide will be above 92 million tons by the year 2030. The swift growth of the fashion industry is influenced by the demand from consumers that is above 400% currently in comparison to 20 years back (Chen et al. 2021b). In addition, consumer demand is further driven by fast fashion trends and the perception of clothing as a disposable, single-use product purchased for specific occasions (Suen et al. 2023).

Waste from textile materials is influenced by rising living standards, decreased life cycle of textile products, enlarging global population, and availability of various textile materials (Xu et al. 2019). Okafor and colleagues reported that the second polluting industry globally is the textile industry (Okafor et al. 2021). This is attributed to the amount of textile waste created, higher amount of water and chemicals used, and pollution originating from the textile industry. It was also corroborated that the textile industry is the one mostly polluting our environment based on

DOI: 10.1201/9781003689225-5

the volume of liquid waste discharged into rivers. Effluent into rivers changes the quality of water drastically and decreases groundwater reservoir levels. Records from Turkish statistical institute show that around 15% of water is consumed in the textile and garment industry (Alkaya and Demirer 2014). Hazardous chemical use in the textile industry can be reduced by using enzymes for desizing instead of mineral acids, applying solvent-based scouring, substituting hydrogen peroxide for chlorine in bleaching, and avoiding preservatives in finishing processes (Alkaya and Demirer 2014).

Cotton is a key fiber in garment production, and depending on the intended application, either fine or medium to high-quality cotton fibers are utilized. Open-end yarns are produced using coarse cotton fibers. Cotton market share globally is around 40%, closer to polyester at 42%. The increase in the global market for cotton is due to price advantages in comparison to polyester. However, polyester competes significantly with cotton as it can be blended with it to produce novel products with excellent properties (Radhakrishnan 2017). Cotton production is a complex process that involves multiple stages, including fiber production, yarn spinning, and fabric manufacturing, ultimately leading to applications in apparel, home furnishings, and industrial goods. Production of cotton is intensive as it includes agrochemicals, water, and energy. During the production of a ton of cotton fiber, 2 tons of CO_2 are released (Hustvedt and Dickson 2009). In all production stages, higher amounts of toxic chemicals, energy, raw materials and water are utilized. In addition, the manufacturing process of cotton releases greenhouse gases (GHG) emissions (nitrous oxide, CO_2, and methane), and chemicals utilized during cultivation (cotton farms) affect environments and human health and contribute to higher temperature and global warming. Global warming is also caused by excessive consumption of fossil fuels, and it has resulted in researchers exploring alternative materials that are economical, biodegradable, harmless to the environment, and renewable (Sakthivel et al. 2018) and minimizing dependence on one source of energy. This was also supported and accepted by many international treaties such as Paris Agreement (COP 21), with the aim of lowering carbon emissions by 40% in 2030 and targeting over 70% in 2050 (Atilhan et al. 2021).

5.2 SUSTAINABLE RAW MATERIALS

Sustainability must be expressed in consideration of the environment, economy, and earthling (Islam 2020). Sustainable raw materials include synthetic, natural, and regenerating fibers (acetate, modal, and viscose). Synthetic fibers include polyethylene terephthalate (PET), polypropylene (PP), acrylic, nylon, polylactic acid (PLA), and polyhydroxyalkanoates (PHA), whereas lyocell, alginate, hemp, flax jute, kenaf, banana, and wool are some examples of natural fibers (Azam et al. 2023; Filimonova et al. 2022; Qin et al. 2019).

Report from Textile Exchange in 2021 shows a decline in fiber production worldwide from 111 to 109 million tons between 2019 and 2020. It was attributed to the COVID-19 pandemic (Textile Exchange 2021). Despite this, there were increases in utilization of regenerated cellulose fibers (RCFs), 3.9 million tons, cotton fibers, 7.8 million tons and recycled polyester, 8.4 million tons. The increase in the utilization of RCFs is attributed to their processing with ease where synthetic fibers are produced from bio-based feedstocks via fiber spinning (Sayyed et al. 2019). Industries, namely, textile and fashion, have noted feasible market opportunities from sustainable products (Yang and Dong 2017). Some biggest fashion brands, such as Adidas and H&M, utilize RCFs from forests (agricultural remains and recycled products). For example, H&M invested in Renewcell and Infinite fiber company that will focus on regenerated fibers only (H&M 2021). The following section highlights some of the widely used raw materials.

5.2.1 Cotton

Cotton is the most utilized natural fiber globally, and it is grown commercially in South and North America, Asia, Australia, and Africa. The global production of cotton fibers in 2024–2025 shows that China is the leading producer (7 million metric tons), followed by India (5 million metric tons), Brazil (3.7 metric tons), and the USA (3.1 metric tons) (Statista 2026). Cultivation of cotton differs depending on agro-ecological, environment, and climatic conditions. Cotton supply worldwide is categorized into six categories in terms of fiber quality, namely, fine, extra-fine, medium, high-medium, coarse count, and padding (Hustvedt and Dickson 2009). The performance of cotton fibers and their influence on the quality of products make them attractive to both fashion stores and farmers (Radhakrishnan 2017). Cotton fiber cultivation uses agricultural chemicals such as pesticides, herbicides, fertilizers. These chemicals are utilized to control defoliants, weeds and pests to obtain better fibers. Amount of glyphosate used when cultivating conventional cotton globally has increased in between 2011 and 2019 (Delate et al. 2021; Das et al. 2023). Utilization of pesticides and fertilizers increases agricultural productivity but negatively affects the environment and results in the loss of ecosystem and biodiversity, pollution (water, air), and soil erosion (Bonou-Zin et al. 2019). Organic cotton can be used as an alternative to traditional cotton (Alkaya and Demirer 2014).

5.2.2 Hemp

Hemp fibers are utilized in textile, construction, automotive, medicine, and agriculture applications. It can be utilized as an alternative to synthetic and cotton fibers (Islam 2020; Eyupoglu 2020; Addo et al. 2021; Li et al. 2021). Utilization of hemp fibers in the medical industry has resulted in a shift as farmers are cultivating it with the main purpose of extracting cannabidiol (CBD). In a nutshell, CBD is a chemical oil originating from hemp fibers used for health benefits. Advantages of using hemp fibers include multipurpose usability, uncomplicated carbon sequestering material, carbon-negative transformation possibility, lower production cost, and smaller production cycle (Ahmed et al. 2022; Gedik and Avinc 2020).

Zimniewska reported that hemp absorbs around 10 tons of CO_2 during vegetation time from the atmosphere, and therefore, results in thermal balance, healthy environment, and quality air (Zimniewska 2022). In addition, cultivating hemp results in squashing weed growth, reclamation properties, prevention of soil erosion, and draining heavy metals and hazardous substances from the soil (Stanwix and Sparrow 2014). The tall growth habit, dense foliage, and ability of hemp to be cultivated at high planting densities enable it to suppress weed growth effectively (Burns 2020). Furthermore, hemp is adaptable to different climatic conditions, less or no irrigation water is needed for its growth, and there is no use of chemical pesticides during its cultivation time (Islam 2020).

5.2.3 Flax

Flax fibers are grown either to produce seed oil from linseed or for fibers, and they can be cultivated in various parts of the world (Yan et al. 2014; Debnath 2021). Flax fibers are used in applications such as medicine (Debnath 2021), linen manufacture (Yan et al. 2014), and composite applications due to their unique properties, namely, elongation at break (≈2.4%), stiffness (50–70 GPa), fineness (0.25–0.33 g/km), length (40–60 mm), and density (1.4–1.5 g/cm^3). The amount of cellulose present in flax is above 65%, which contributes to fiber stiffness (Chowdhury et al. 2025).

5.2.4 Polylactic Acid

PLA belongs to the family of synthetic fibers and originated from renewable resources, such as sugarcane and corn. It can be used as an alternative to conventional plastics, which are currently used for economic benefit and sustainability. PLA possesses better physical and mechanical properties like that of PET and PP fibers. Pros of using PLA are environmentally friendliness, compostability, biocompatibility, easy processing, and biodegradability (Jem and Tan 2020; Taib et al. 2023). Properties of PLA allows it to be used in industrial applications such as packaging, automotive, and agriculture (Taib et al. 2023). In addition, PLA is utilized in different applications like manufacturing of medical devices and nonwoven materials. Utilization of PET was attributed to many factors such as lightweight, thermal stability, good mechanical properties, chemical resistance, and easy disposability. The high consumption and non-biodegradable nature of PET contribute significantly to environmental pollution (Qin et al. 2019).

5.2.5 Regenerated Fibers

Regenerated fibers are produced by processing low-value cotton linters and wood pulp, dissolving these raw materials in solvents to create polymer solutions. Both wet and dry spinning methods are utilized to regenerate polymer solutions to form solid fibers. However, fibers are solidified using chemical bath in numerous fiber spinning techniques. RCFs are commercially available in comparison to regenerated protein fibers (RPFs). This is attributed to the abundance of cellulosic materials and their excellent mechanical properties (EL Seoud et al. 2020; Sayyed et al. 2019). Examples of RPFs are wool waste (Cao et al. 2020) and milk protein (Thill et al. 2021). However, lower mechanical properties of RPFs hinder their utilization in different applications. Protein fibers are anticipated to be tougher than steel and stronger than spider silk (Yarger et al. 2018).

The fibers most regenerated in 2020 were viscose rayon at 5.2 million tons, followed by cellulose acetate at 0.9 million tons and lyocell at 0.3 million tons (Textile Exchange 2021). One of the main disadvantages of producing RCFs, such as viscose rayon, is the use of carbon disulfide (CS_2) and sulfuric acid, which can negatively impact the environment. Unlike in viscose rayon production, mild chemicals are utilized in the production of lyocell. Cellulose is dissolved using *N*-methylmorpholine-*N*-oxide (NMMO) in lyocell production. The viscous cellulose solution is spun and solidified by exposing it to an aqueous bath. Solvent recovery necessitates excessive electricity consumption, even though 99% of the solvents are recovered. Production of RCFs from wood materials results in deforestation (Zhang et al. 2018). The problem of deforestation can be minimized by utilizing textile waste, microbial mass (Dong et al. 2018), and alternative carbohydrate polymers (curdlan) to produce RCFs instead of wood (Suzuki et al. 2021). However, genuine techniques need to be developed so that regenerated fibers, excluding wood, can be utilized in textile and fashion industries to promote circular economy.

5.2.6 Waste Utilization

Global threat caused by waste from the textile industry opens doors for reuse and recycle of waste, which will lead to manufacturing novel products and open doors for new markets (Suen et al. 2023). Textile recycling is the reprocessing of textile waste from consumers to manufacture new products such as fabrics, fibers, and yarns. Waste can be recycled using different methods such as chemical, mechanical, quaternary, and primary recycling. However, recycling methods have their own advantages and disadvantages as some of them release toxic gases (Tshifularo and Patnaik 2020b). Universally, there is an existing challenge as only 15% of textile waste is recycled, but around 75–85% end up in either landfill sites or burned, even though standards and investments have been created for recycling programs.

Waste generated from cotton increases each year as textile products manufactured from these fibers are sold everywhere. This waste is categorized into pre-consumer (garment manufacturing, fabric, and yarn) and post-consumer waste (Tshifularo and Patnaik 2020b). Post-consumer waste is generated by discarding products after their service life.

5.2.7 Organic Farming

In addition, impacts of cotton cultivation on environment can be minimized by promoting organic farming where chemicals are excluded. A major benefit of organic farming is that its production methods do not adversely affect the environment, thereby supporting and preserving ecosystem services. Organic farming is part of the green economy, and it addresses social, economic, and environmental issues simultaneously (Reganold and Wachter 2016; Bonou-Zin et al. 2019). It produces less yields that are profitable and environmentally friendly in comparison to the traditional agriculture (Reganold and Wachter 2016).

5.3 RECENT DEVELOPMENTS IN SUSTAINABLE RAW MATERIALS AND THEIR MANUFACTURING

5.3.1 Use of Eco-friendly Cultivation and Manufacturing Processes

Sustainable production is expressed as manufacturing environmentally friendly products without transforming nature (Gedik and Avinc 2020). Radhakrishnan reported that agro-ecological, environmental, and climate concerns related to cotton cultivation can be minimized using biotechnology (Radhakrishnan 2017). Biotechnology is utilized to produce crops that are intransigence to diseases and pests. It enhances sustainable agriculture that is prone to insufficient soil nutrients as it can produce nutritious and healthy crops (Das et al. 2023). Cultivation of organic cotton involves resistant crops (abiotic and biotic), organic manures, vermicompost, rotation of seeds, and bio-fertilizers. Organic cultivation lowers persistent and poisonous pesticides, while the soil remains productive (Harshitha et al. 2021). The difference between organic cotton and cotton is that the former is cultivated without pesticides and manures, whereas the latter needs them. As mentioned earlier, both pesticides and manure affect environment negatively. Either organic cotton or cotton, they all need vitality and water during their production stages, but lower amount of water is utilized for the former. Nonetheless, their assembling process is the same. Currently, organic cotton is mostly produced in India, Turkey, and China. The following section discusses some of the important fibers.

5.3.1.1 Flax

The growing cycle of flax fibers is shorter. There is a growing need to reuse and recycle natural fibers, such as flax, due to the challenges associated with increasing natural fiber production, limited land availability for cultivation, low yield rates, and the rising generation of waste. It has been reported that above 39% of flax become solid waste in the process of spinning fibers into either thread or yarn (hackling process). Hackling process is a mechanical stage where bast fibers are converted into continuous and strong yarn using spinning process. The presence of low cohesive strength, along with residual lignin and pectin in flax fiber waste, contributes to the production of irregular and lower-quality yarns. Irregular yarns are also influenced by flax structure that is non-crimped, non-kinked, and straight. Valuable yarns may be produced by minimizing problems that occur during spinning process, and thus, an eco-friendly and sustainable approach is achieved.

Irrespective of flax's limitations, their morphological and structural properties can be improved via mechanical and chemical modifications. Bio-enzyme-based and chemical treatments can improve properties of flax fibers; however, the latter may negatively affect environment. Bio-enzymes can modify the structure and chemistry of flax fibers by catalyzing biochemical reactions. They are natural and derived from plant fermentation and composed of non-toxic and non-corrosive liquid enzyme (Li et al. 2024). Pros of using bio-enzyme treatments are excellent fiber quality, manifold nature, lower chemical oxygen demand and BOD, and less material loss and process duration. Majority of enzyme treatments decrease CO_2 emissions and water usage significantly and therefore are eco-friendly. Chowdhury and colleagues reported that recycling waste from flax fibers results in minimum carbon footprint unit in $kgCO_2e$/kg fiber production in comparison to that of synthetic fibers (Chowdhury et al. 2025).

Regular and fine long flax fibers are converted into yarns via spun process and utilized in linen production. Linen fabrics are utilized in household textiles, which include furnishing fabrics, bed linen, and interior decoration. Staple flax fibers are used in the manufacturing of heavier yarns and utilized in tents, sails, kitchen towels, etc. Continuous demand in linen utilization results in flax fibers contributing to the global economy, especially in those countries with favorable climatic conditions to grow them. The growth cycle of flax fibers is shorter as they can be sown in March and harvested at the end of July (Yan et al. 2014).

5.3.1.2 Hemp

Hemp is produced by cultivating hurds and seeds, and its stem can reach a height of around 4 m (Islam 2020), harvested at the end of blooming season. Its fiber quality depends on it harvesting time, growing conditions, and the method utilized to extract fibers. The number of herbicides needed when cultivating hemp can be minimized using crop rotation. Fibers are extracted by retting process, in which non-cellulosic materials attached to the fibers are removed. Dew, water, enzymatic, mechanical, and chemical are different types of retting processes. In addition, steam explosion and osmotic degumming are methods also used to extract fibers and to simultaneously remove hemicellulose, lignin, and pectin from the stem (Zimniewska 2022). In the chemical retting process, sodium hydroxide (NaOH), sodium benzoate ($NaC7H_5O_2$), and hydrogen peroxide (H_2O_2) solutions are heated to boiling temperature and applied to the stems to separate fibers. The limitation of chemical retting process includes degradation of fibers with resultant loss in tensile strength and change in the color of fibers due to high concentration of chemicals (Tshifularo et al. 2020a). Chemicals used to separate and improve fiber properties affect the environment negatively.

5.3.2 Recycling from Waste

Textile waste is grouped into pre- and post-consumer waste. Over 75% of pre-consumer cotton waste is reprocessed and utilized in the manufacture of insulation products, furnishings, automotive nonwovens, and lower-grade yarns. Yarns are spun using fibers extracted from cotton waste. Post-consumer waste includes yarns from cotton fibers that can be blended with synthetic fibers (polyester, acrylic, and nylon) and utilized to produce either knitted or woven fabrics (Chen et al. 2021b). It is difficult to reclaim post-consumer waste due to the impurities present on the discarded product, and wear and tear fluctuation. Lie et al. (2019) reported that above 90% of post-consumer waste from cotton end up being incinerated or in landfill (Liu et al. 2019). Incineration and cellulose decomposition release foul odors, hazardous chemicals, and greenhouse gases. The rate of recycling post-consumer synthetic fibers (nylon and polyester) is higher in comparison to that of cotton.

Collaboration between government, industries, and academia can assist in recycling cotton waste and decreasing the amount of energy and water utilized during their cultivation and disposal stages. Cotton waste can be recycled through

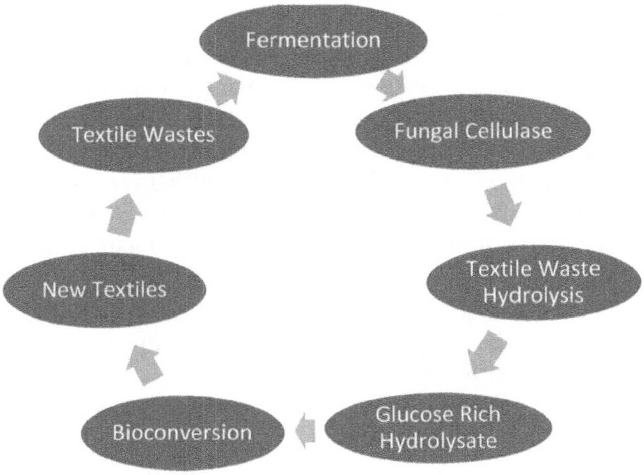

FIGURE 5.1 Lifecycle illustration of textile waste from cotton (Suen et al. 2023).

chemical recycling processes, which are capable of handling materials with varying degrees of wear, deterioration, and quality. It is not easy to dissolve cotton fibers using common solvents due to hydrogen bonding (intra- and inter-molecular) present in cellulose macromolecules (Medronho and Lindman 2015). Urea/alkali aqueous systems and ionic liquids are utilized in environmentally friendly cellulose dissolution. Alkali/urea aqueous system is not toxic, is available at lower cost, and can be adopted by industries. It is used to recycle post-consumer waste from cotton, and wet spinning is used to regenerate cellulose fibers. Aqueous systems use 1-allyl-3-methylimidazolium chloride (AMIMCI) as a solvent to reclaim cotton waste; therefore, regenerated fibers are produced. Pretreatment processes include hydrolysis, which is used to remove impurities and reduce the molecular weight of cotton fibers (Liu et al. 2019).

Figure 5.1 shows that textile waste can be used as a source of cellulose while contributing to the global economy. Textile waste goes through different pretreatments followed by enzymatic treatment. Majority of cellulose available is from rotten fungi, produced by solid-state fermentation and immersed fermentation (SmF) (De Jong and Jungmeier 2015). Nonetheless, SmF takes place when suspension flows with ease and produces end products with lower concentrations, and therefore, additional downstream processing stages are required with exorbitant production cost. Solid-state fermentation is a cost effective method whereby hard materials are utilized as substrates to produce cellulose from natural polymers originating from agro-industrial remnants. Almost 40% of bioprocesses depend on cellulose that is costly, and it needs to be minimized. In previous years, cellulose production has utilized cost-effective substrates derived from agricultural products and horticulture wastes, namely, rice straw, pruning's and grass clippings, however, cotton textile waste can be used alternatively as a substrate (Suen et al. 2023).

5.3.2.1 Cotton Recycling

Cotton can be reclaimed from its waste (pre-consumers and post-consumer). Post-consumer waste is from textile products (home textiles or apparel) that are no longer in use and discarded. Cotton waste is sorted based on its color and type before the recycling process. A stripping machine is used to break fabric and yarns into tiny pieces before being converted into fibers via the pulling process. Mostly, recycled cotton is blended with virgin cotton to improve the tensile strength of the yarn. Only 30% of recycled cotton is utilized in the finished fabric or yard due to the waste cotton that has already been dyed. Recycled cotton can be utilized in the production of high-quality paper as well as in industrial applications such as wiping and polishing materials. In addition, recycled cotton can be blended with plastic bottles to produce clothes and textile products (Ahmed et al. 2021).

The advantages of recycling cotton waste include reduced water and energy consumption, as well as increased landfill capacity due to reduced waste disposal (Senthil Kumar and Saravanan 2018). In addition, pollution is also caused by poor cotton spinning waste management, where waste ends up in the landfill. Higher amount of spinning waste from cotton is produced in Bangladesh because of the country's massive textile industries. Waste from cotton contains 50%–80% of micro dust (husk, leaf, and fiber fragments) and 10%–25% of sand. The higher number of fiber fragments shows that larger amount of micro dust occurs during processing (Ahmed et al. 2021).

5.3.2.2 Wool Recycling

Waste from wool is categorized as pre-consumer (post-industrial) and post-consumer. Post-consumer waste is defined as the type of clothes that are out of fashion, worn out, and damaged and ends up discarded. Waste created during garment manufacturing (including yarn, fibers, and smaller pieces of fabric) is called post-industrial waste. Majority of worn-out garments from wool cannot be reused but are recyclable. Wool fibers recycled from waste can be utilized as an alternative to pure fibers and in blends. Either closed- or open-loop mechanical recycling processes can be used to reclaim post-consumer wool waste; however, this method relies on the ability to reduce fiber breakages while increasing fiber length. Fiber breakage is influenced by yarn twists and fabric construction.

5.3.2.3 Open-Loop Recycling Process

Garments are mechanically pulled and return to fibrous state to be used in the production of novel products. Fibers produced are converted into fibrous web via garneting and air laying/carding. Mechanical and thermal bonding methods are used to produce nonwoven fabrics. In addition, the fibers produced by recycling wool waste can be used in the manufacturing of yarn and insulator pads (for furniture and automotive products). However, yarn spinning is difficult due to shorter fibers obtained during the mechanical recycling of woven fabrics (worsteds suits and coats). Recycled fibers from wool waste improve fire retardancy when utilized as mattress insular pads. The USA and countries of Europe and Asia buy bags of clothes (waste) and sort them into various grades (based on color shade and fiber composition). The commercial value of wool waste will be influenced by its homogeneity, fiber composition, shade/color, and fabric type (woven and knitwear) and length. Only selected manufactures pull waste and sell the fibers from waste (Russell et al. 2016).

5.3.2.4 Repurposing Waste Hemp

The extraction process of hemp fibers results in almost 90% of hemp biomass and residual waste. During 2019–2021, the waste generated from extracting hemp fibers was above 82,000 tons, which ended up composting, landfilling, and incinerating. The CBD hemp industry is expected to scale up, and this waste will affect our environment negatively (Griggs et al. 2025).

Waste from hemp is mostly dumped using various ways, namely, composting, landfill, incineration, and anaerobic digestion (AD). Discarding waste in landfills and incineration is a waste management strategy. This is due to the exorbitant market value of CBD industry that results in lower profit and demoralizes producers to reprocess biomass from hemp. Unused hemp biomass can be employed in various applications, namely, hempcrete, animal feed, bedding, and packaging. Nonetheless, hemp biomass contains many chemical solvents depending on the extraction method used. This may hinder the utilization of recycled hemp waste in food (animal and human). In AD, biomethane derived from hemp is used to minimize the utilization of fossil fuels (such as propane and natural gas) in the CBD industry. O'Connor and colleagues reported that AD is the preferred method for farmers to generate energy from discarded animal manure (O'Connor et al. 2021). Recycling hemp waste and minimizing greenhouse gases will play a major role in bio-based industries, to enhance sustainability and circularity.

Waste from hemp fabrics can also be minimized by recycling them via chemical process. The molar mass of unused hemp fabric was higher, and it was impossible to recycle using the dry-jet-wet spinning process. Therefore, an aqueous sulfuric acid was utilized to treat the fabric so that the viscosity level must be in between 400 and 500 mL/g. Samples hydrolyzed using acids were disintegrated in 1,5-diazabicyclonon-5-enium acetate, and it was possible to regenerate cellulose fibers via the dry-jet-wet spinning process. Regenerated fibers produced possess an excellent modulus of toughness around 55.9 MPa in comparison to that of virgin hemp fibers' 28.7 MPa. They were converted into yarn via spinning process and used in the manufacturing of knitted fabrics. Higher abrasion resistance was noticed in the fabric produced from regenerated fibers than virgin hemp fibers. This was attributed to even the fabric structure produced from regenerated fibers (Rissanen et al. 2023).

5.3.2.5 PET Recycling

As mentioned earlier, drawbacks of PET can be addressed by recycling its waste and minimizing greenhouse gas emissions, fossil fuel use, and energy consumption. Recycling plastic waste from PET is cheaper compared to the production of virgin plastic, and it is an advantage to global economy (Choudhary et al. 2019). Various modes such as energy recovery, and mechanical and chemical recycling are utilized to reclaim PET bottles. A chemical recycling method is the depolymerization of waste from PET into monomer units. The depolymerization process is associated with the formation of gaseous by-products. The monomers obtained are decontaminated using the distillation method followed by drying process to produce new polymers; however, the chemical recycling method is expensive.

In mechanical recycling, also known as physical recycling, discarded PET bottles are collected by people (in countries like South Africa and India) daily and sold to recycling companies as raw materials. Recycling companies sort and separate bottles into different groups based on contamination, labels, and neck rings, and the bottles are washed before being shredded (NCL Innovations 2018). Crushed PET bottles go through extrusion molding to produce PET granules. Disadvantages of physical recycling methods are degradation of properties, decrease in molar mass, and discoloration due to acidic pollution and hydrolysis because of water existence (Schneiderman and Hillmyer 2017). Meran and colleagues reported that the mechanical properties of the recycled products are lower in comparison to the original material (Meran et al. 2008). This results in limited quantity of post-consumer PET bottles recycled to manufacture novel products. Majority of recycled waste from PET bottles is employed in fiber and garment manufacturing (Schneiderman and Hillmyer 2017), manufacturing of pipes, contractile films, and containers for industrial purposes. Choudhary and colleagues reported that reclaimed PET bottles are used to manufacture monofilaments, bottles, and fabrics (Choudhary et al. 2019). However, the waste recycled from polyester, polypropylene, and polyethylene cannot be utilized in food and medicine sectors (Meran et al. 2008).

Almost 32% of PP is used in the food sector as lids, jars, tupperwares, bottles, and packaging. PP is used as an alternative to asbestos and has a slice of 3% in the construction industry. It is also used as door panels, dashboards, grilles, and bumpers in the automotive industries, and it holds above 9% of the market. Mannheim and Simenfalvi (2020) reported that majority of products from PP are not biodegradable after reaching their end-of-life stage and the resulted waste can be incinerated, recycled, and disposed in landfills (Mannheim and Simenfalvi 2020). Between 22% and 43% of waste from PP ends up in landfills (Alfarisi and Sutopo 2019). The waste from PP is mostly reclaimed via mechanical process, and the material produced relies on the processing conditions, physico-chemical properties, and the number of impurities from PP waste (Ragaert et al. 2017). The post-consumer waste from PP is recycled in Brazil using mechanical process where they are divided into different groups and crushed into pellets. Breaking machines incorporated with waltz that are 2 mm apart are utilized during separate processes. Crushed materials' dimensions vary irrespective of the constant intervals between waltzes. This problem is overcome by using filters to separate the crushed materials before blending them with raw materials. Filtering is done to obtain particle sizes between 650 and 1,000 μm and are injected with ease. Recycled PP waste is utilized in the manufacturing of automobile components and packaging. The disadvantage of products produced from recycled PP waste is the possibility of either breaking or cracking under stress, and therefore, they are brittle (Barbosa et al. 2017).

5.3.2.6 Acrylic Recycling

Acrylic fibers originated from fossil fuels and their manufacturing process influence climate changes due to GHG emissions . Almost 8 million tons of textiles produced using acrylic fibers are discarded in landfills and pollute environment due to poor biodegradability (Mu et al. 2024). This may result in toxic substance leaching into groundwater while reducing landfill capacity. Drawbacks of acrylic fibers can be minimized by recycling its waste. Acrylic fibers are utilized commercially as precursors in the production of carbon fibers (Jamaludin et al. 2024). However, the production of carbon fibers is lower than that of acrylic fibers (Sauer and Kuhnel 2019). With all the various recycling methods available, mechanical, chemical, and direct extrusion recycling cannot be used to recycle acrylic waste because of the complexity of the fiber composition. In addition, mechanical recycling reduces the length of fibers, and therefore, materials are degraded.

Mostly, textiles produced from acrylic fibers are blended with other fibers, namely, wool, cotton, and polyester. This may result in variation in density, small filament diameter, and interweaving and at the end increases cost. In direct wet extrusion methods, dyes present in acrylic fibers influence color consistency, acrylic dissolution, and dyeability of regenerated fibers. However, it is minimized by using the closed-loop recycling method. Dimethyl sulfoxide and butanediol are utilized to increase the swelling and dissolution of acrylic and enhance the separation of dyes and non-acrylic fibers from acrylic fibers with ease. The advantage of using the closed-loop recycling method is that there are no discharges released. Excellent mechanical properties and dyeability were achieved in regenerated acrylic fibers compared to pure fibers, but their properties can be improved by adding pure acrylic fibers. The average cost of energy and materials used to produce acrylic fibers from waste was 40% lower than that from fossil fuels (Mu et al. 2024).

5.3.2.7 Aramid Recycling

Aramid fibers are classified into staples, filaments, and pulp and can be utilized in the production of woven fabrics, paper, and yarns. In addition, aramid fibers are utilized to reinforce cables, sport fabrics, bullet proof vests ropes, and anti-cutting gloves (Chen et al. 2021b). Utilization of aramid fibers is attributed to their excellent properties, but their production is costly. Aramid fiber waste can also arise during manufacturing processes, such as yarn production and weaving. Aramid fibers maintain the original structure after reaching their lifespan, and there are only small damages on their mechanical properties noticed.

Both energy recovery through incineration and chemical recycling methods are not suitable for the reclamation of aramid waste. This is attributed to aramid fibers that do not dissolve in concentrated acids due to their higher chemical resistance. Properties of aramid fibers hinder reclaiming their waste, and they end up disposed of in landfill. Besides, lower amounts of aramid waste are recycled due to higher energy consumption and therefore are costly. Aramid waste from sources such as yarns, woven fabrics, gloves, and blends is recycled into aramid nanofibers (ANFs) via a deprotonation process assisted by a proton donor. Jia and colleagues reported that mechanical, pyrolysis, and depolymerization methods can also be utilized to recycle aramid fibers (Jia et al. 2026). Deprotonation method is economical, and it lowers the cost associated with virgin aramid fiber production almost seven times (Jia et al. 2026). ANFs recycled from aramid fiber waste possess chemical properties and molecular weights almost similar to pure aramid fibers (Yang et al. 2021).

5.3.2.8 Elastane Recycling

Elastane (EL), alternately known as spandex, contains almost 85% segment of polyurethane (PU), which increases the original length from 100% to 600%. It is used in textiles with lower concentrations to increase comfort, fitness, and flexibility. Higher amounts of EL are also utilized in medical and sports dresses due to its higher flexibility. However, EL, like other textile products, results in waste after its lifespan. Recycling waste from EL is associated with profound challenges, especially separating it from other textile materials. In addition, these challenges are also caused by advantages of EL during wear stages. Fabrics produced from polyamide and polyester contain smaller amounts of EL (20% by weight) and fabric from either wool or cotton, around 1%–5% by the weight of EL. Recycling textile waste containing EL results in clogged filter during melt-extrusion process, and the recycled fabrics will have small crumbs. EL needs to be removed from the textile waste before any recycling process is taken. Mostly textile materials blended will deteriorate EL when placed for 2 hours in an oven with a temperature above 230°C, ethanol is used to wash the residues, and it was observed that remnants can no longer be recycled. This is attributed to the degradation of PU (Boschmeier et al. 2023). Textile waste composed of a blend of EL and cotton was treated with various solvents, namely, LiCL/DMAc, DMSO, aqueous NaOH/urea, and H_2SO_4, to remove cotton fibers present, and therefore, only EL remains. However, there is higher possibility of LiCL/DMAc to damage EL fibers (Wang et al. 2022).

In addition, waste from EL and other fibers can be treated using tetrahydrofurfuryl alcohol (THFA) for four hours at 100°C for fiber separation. The process will be difficult to be converted to industrial production as it requires longer time, and it may not be profitable (Phan et al. 2023). DMSO solvent can successfully dissolute EL within ten minutes at a temperature of 120°C. EL fibers recovered after washing and drying was of smaller quantity. Sustainable process will only be achieved if filtration, centrifugation, and distillation operations are utilized to feed the solvent back. There was no structural change in the appearance of recycled EL fibers in comparison to that from virgin fibers (Boschmeier et al. 2023).

5.3.2.9 Nylon Recycling

Nylons are mostly utilized in textile, automotive, and food packaging industries due to their excellent tensile strength, durability, and thermal and mechanical properties. However, the products produced from nylon are non-biodegradable and significantly pollute the aquatic environments, for example, fishing nets (Tonsi et al. 2023; Lozano-González et al. 2000). The waste from nylon can be recycled using thermomechanical, chemical, and solvent dissolution recycling processes. In the total amount of polyamide produced globally, only 2% is produced via recycling process, and this value is lower as compared to PET which is at 15%. Thermomechanical recycling includes grinding and shredding nylon waste into smaller different pieces. Smaller pieces (pellets) are melted and molded into new products. Properties of the products produced from nylon waste are inferior to those of virgin polymer, and therefore, the product produced is utilized in applications where demand is lower (Tonsi et al. 2023).

Thermomechanical recycling process can be used in post-consumer waste from nylon and post-industrial waste. It is advised that the smaller pieces from nylon waste must be dried before performing mechanical recycling process. This is attributed to the possibility of thermal hydrolysis reaction that may decrease polymer molecular weight and affect mechanical properties of new products. Besides, thermal hydrolysis reaction can also be limited by adding stabilizer. Both trans-reactions and condensation are able to enhance molecular weight; however, nylon may degrade when the melting process is repeated many times (Lozano-Gonzalez et al. 2000). In the post-consumer waste from nylon, nylon is mostly blended with fibers, such as PET, PU, and glass fibers to obtain excellent mechanical properties. However, it becomes a problem in thermomechanical recycling as nylon has higher melting point and at least above 260°C re-extrusion can be done, and it may degrade polymers with lower chemical and thermal stability, for example, PU. This may result in formation of unwanted impurities that can significantly affect the properties of recycled products and hinder their applications (Tonsi et al. 2023).

5.4 CASE STUDIES

The sustainability of fibers, their environmental impact, and their effects on organisms and surrounding ecosystems are critical considerations in discussions of climate change driven by GHG emissions. Climate change is a result of human activities including fiber manufacturing and biodegradability, and utilization of pesticides and fertilizers. A scoring system has been developed to assess GHG emissions, water consumption, and energy use. The scoring system shows that natural fibers perform better in comparison to man-made fibers.

The available data on the ecological sustainability index and environmental impact index exclude the number of materials needed to meet the continuous growth of population. Higher population will prefer to utilize land to grow food rather than cultivating natural fibers. In addition, the data available did not mention the drawback of natural fibers and how they will be minimized. Furthermore, the impact on the environment caused by the methods used to improve the properties of natural fibers is not included (Islam 2020).

Industries and manufacturers are encouraged to be transparent to avoid illegal practices, weak supply chains, and unlawful trade which result in unsatisfied customers and hinder sustainable production (Agrawal et al. 2021). In addition, clothing and textile manufacturers need to be tracked to address various problems such as counterfeits, consumer demands that are volatile, and increase in competitors (Kumar et al. 2017). Furthermore, utilization of hazardous materials and unfair labor practices is common in the clothing and textile industries. Tracking has become the primary requirement in multi-site and multi-tier production to ensure quality, transparency, and transparency in clothing and textile manufacturers. This will be useful in enabling customers to access authentic product data more easily and in promoting ethical purchasing practices. In addition, tracking will minimize fear and challenges encountered by stakeholders, such as sharing information and manipulation of data due to advanced technology. Blockchain-based detectors are utilized to track garment supply chains and muti-tier textiles. For example, supply chains of organic cotton can be detected using blockchain with transaction laws and smart contracts. Advantages of using blockchain-based trackers include flexibility, transparency,

TABLE 5.1
NFT Digital Strategies (Alexander and Bellandi 2022)

Stages	Phase 1	Phase 2	Phase 3
Value creation firm	Collaboration, marketing, and promotion	Financial exchange value evolving	Several equitable revenue stream
Digital asset	Collectibles	Virtual good collections	Resale and sale visual goods
Adoption	Lower (adopters), uniqueness, security	Increasing (early majority and adopters), reducing barriers	Higher, no adoption hurdle
Digital strategy	Education, marketing, experimentation	Scalable revenue stream	Metaverse captivation

distinctive opportunity, and giving partners involved powers to be able to find their supply and maintain supply chain (Agrawal et al. 2021).

Advances in technology, changes in retailing (Grewal et al., 2017), and the global pandemic have led customers to prefer online purchasing over traditional in-store shopping, which has also contributed to the emergence of non-fungible tokens (NFTs) (Alexander and Bellandi 2022). Pandemic and advanced technology increases online retailing by more than 14% which is equivalent to US$ 1.819 billion (Global Data 2020). Buying luxury goods through digital channels, instead of visiting shops, has increased 27% in 2021, meaning US$ 70 billion when estimated (WGSN 2022). NFTs is a technology used to find the perceptions of customers when buying luxury goods, and it is divided into NFT digital strategies and NFT value creation. It is driven by millennial and generation Z people between 18 and 44 ages. The development of NFT digital strategy is shown in three phases in Table 5.1. NFTs can be utilized across both the fashion sector and auction markets. Gucci, one of the luxury brands, increases the number of young shoppers by inviting them to play games online while familiarizing them with online shopping.

5.5 OPPORTUNITIES AND CHALLENGES

As mentioned earlier, production of some of the sustainable materials negatively impacts the environment. For example, NaOH, $NaC7H_5O_2$, and $NaC7H_5O_2$ are used in the chemical retting process. They are successfully used to separate fibers and improve their properties, but affect the environment negatively as they end up as a source of hazardous leachates (Tshifularo et al. 2020a; Sahu and Gupta 2020). Utilization of toxic chemicals causes ground, air, and water pollutions and deforestation. Sandin and Peters reported that majority of toxic chemicals are mostly from treatment processes, such as dyeing, printing, and finishing (Sandin and Peters 2018).

Raw materials are pivotal in the global economy as they support the development of latest technologies, such as green economy. It becomes a challenge in either developing or developed countries to obtain authentic raw materials. Due to this, several critical raw materials (CRMs) have been created to support the manufacturing of products with lower carbon content. Lee and colleagues reported that lower carbon economy is possible when decision makers from industries, politics, and government agencies promote modern technology that is environmentally friendly. Creation of several CRMs was introduced by countries due to exorbitant market prices and supply shortages (Diemer et al. 2018; Lee et al. 2018).

5.6 FUTURE OUTLOOK

Manufacturers produce genuine products and sustainable supply chains using blockchain-based tracking systems and NFTs. Pandemic and advanced technology have changed the retail markets; therefore, small brands and retailers must utilize technology to attract customers.

5.7 CONCLUSION

Excessive utilization of fossil fuels, which results in global warming, pollution, and threats to human life and health, can be minimized by using environmentally friendly materials that are biodegradable and renewable. In addition, these materials must not be hazardous to the environment during their life span from raw materials, manufacturing processes, discarding to utilization. Instead of using fossil fuels, carbon capture and decarbonization can be used to lower the emission of GHGs; however, it is not sustainable. Natural fibers (cotton, flax, and hemp) can be used to minimize the emission of GHGs. Nonetheless, some of the natural fibers require chemical fertilizers during cultivation, for example, cotton. In addition, chemicals are also successfully used in separation and finishing processes of fibers to improve the properties of natural fibers.

There is a higher possibility that chemicals used during cultivation and for enhancing fiber properties may pollute ground water and rivers. This results in water pollution that negatively impacts human and animal health. Continuous research is needed to address the impact of chemicals utilized during cultivation on the environment. Instead of chemicals, biotechnology can be used due to its ability to produce crops that are resistant to pests and diseases. On the other hand, biotechnology supports sustainable agriculture in nutrient-deficient soils by enabling the development of crops that are more resilient and vigorous.

Post-consumer waste from cotton and polymers (polyester, polypropylene, and polyethylene) is recycled and re-purposed to manufacture new products. Majority of scholars agreed that lower mechanical properties are obtained in the recycled products in comparison to virgin materials. However, few researchers reported that the properties of recycled materials are either the same or higher than that of original products, and therefore, detailed research coupled with laboratory tests is required to substantiate the two different statements. Green economy is supported by the availability of genuine raw materials globally. Acquiring such raw materials in both developing and developed countries is challenging. Due to the challenges encountered, various CRMs have been created to enhance the manufacturing of products without releasing carbon into the atmosphere. Creation of CRMs will help to lower GHG emissions, market prices, and supply shortages; however, the participation of government agencies and decision makers is a must. On the other hand, the use of models to analyze utilization of fibers or materials and their impact on the environment is encouraging. However, the majority of them did not address how the required materials will be supplied in relation to population growth. For example, higher population will use land more to live and grow food than for cultivating natural fibers.

REFERENCES

Addo, P. W., Brousseau, V. D., Morello, V., MacPherson, S., Paris, M. and Lefsrud, M. 2021. Cannabis chemistry, post-harvest processing methods and secondary metabolite profiling: A review. *Industrial Crops and Products* 170: 113743.

Agrawal, T. K., Kumar, V., Pal, R., Wang, L. and Chen, Y. 2021. Blockchain-based framework for supply chain traceability: A case example of textile and clothing industry. *Computers and Industrial Engineering* 154: 107130.

Ahmed, A. F., Islam, M. Z., Mahmud, M. S., Sarker, M. E. and Islam, M. R. 2022. Hemp as a potential raw material toward a sustainable world: A review. *Heliyon* 8(1): e08753.

Ahmed, M. A., Roy, P., Shah, M. H., Argha, D. P., Datta, D. and Riyad, R. H. 2021. Recycling cotton dust for organic farming is a pivotal replacement of chemical fertilizers by composting and its quality analysis. *Environmental Research and Technology* 4(2): 108–116.

Alexander, B. and Bellandi, N. 2022. Limited or limitless? Exploring the potential of NFTs on value creation in luxury fashion. *Fashion Practice* 14(3): 376–400.

Alfarisi, S. and Sutopo, W. 2019. Simulation of green operations supply chain in waste of plastic. *IOP Conference Series: Materials Science and Engineering* 495(1): 012080.

Alkaya, E. and Demirer, G. N. 2014. Sustainable textile production: A case study from a woven fabric manufacturing mill in Turkey. *Journal of Cleaner Production* 65: 595–603.

Aryal, J. P., Sapkota, T. B., Khurana, R., Khatri-Chhetri, A., Rahut, D. B. and Jat, M. L. 2020. Climate change and agriculture in South Asia: Adaptation options in smallholder production systems. *Environment, Development and Sustainability* 22(6): 5045–5075.

Atilhan, S., Park, S., El-Halwagi, M. M., Atilhan, M., Moore, M. and Nielsen, R. B. 2021. Green hydrogen as an alternative fuel for the shipping industry. *Current Opinion in Chemical Engineering* 31: 100668.

Azam, F., Ahmad, F., Ahmad, S. and Haji, A. D. 2023. Sustainable raw materials. In *Circularity in Textiles* (Eds. Batool, S. R., Ahmad, S., Nawab, Y. and Hussain, M.), Springer, Cham, 59–128.

Barbosa, L. G., Piaia, M. and Ceni, G. H. 2017. Analysis of impact and tensile properties of recycled polypropylene. *International Journal of Materials Engineering* 7(6): 117–120.

Bonou-Zin, R. D., Allali, K. and Fadlaoui, A. 2019. Environmental efficiency of organic and conventional cotton in Benin. *Sustainability* 11(11): 3044.

Boschmeier, E., Archodoulaki, V. M., Schwaighofer, A., Lendl, B., Ipsmiller, W. and Bartl, A. 2023. New separation process for elastane from polyester/elastane and polyamide/elastane textile waste. *Resources, Conservation and Recycling* 198: 107215.

Burns, E. 2020. *Creating a Weed Management Plan for Hemp*; Michigan State University Extension, Department of Plant, Soil and Microbial Sciences: East Lansing, MI. https://www.canr.msu.edu/news/creating-a-weed-management-plan-for-hemp [Accessed 15 January 2026].

Cao, G., Rong, M. Z. and Zhang, M. Q. 2020. Continuous high-content keratin fibers with balanced properties derived from wool waste. *ACS Sustainable Chemistry and Engineering* 8(49): 18148–18156.

Chen, H. J., Bai, Q. Y., Liu, M. C., Wu, G. and Wang, Y. Z., 2021a. Ultrafast, cost-effective and scaled-up recycling of aramid products into aramid nanofibers: mechanism, upcycling, closed-loop recycling. *Green Chemistry* 23(19): 7646–7658.

Chen, X., Memon, H. A., Wang, Y., Marriam, I. and Tebyetekerwa, M. 2021b. Circular economy and sustainability of the clothing and textile industry. *Materials Circular Economy* 3(1): 12.

Choudhary, K., Sangwan, K. S. and Goyal, D. 2019. Environment and economic impacts assessment of PET waste recycling with conventional and renewable sources of energy. *Procedia CIRP* 80: 422–427.

Chowdhury, M. F. M., Dip, T. M., Al Mamun, M. A., Uddin, M. B., Rahman, M. T., Adee, T., Chanda, A., Das, A., Hoque, M. E. and Asif, M. M. 2025. Bio-enzymatic treatment of flax fiber waste: Fostering sustainable reusability by transforming waste to resource. *Environmental Technology and Innovation* 40: 104538.

Das, S., Ray, M. K., Panday, D. and Mishra, P. K. 2023. Role of biotechnology in creating sustainable agriculture. *PLOS Sustainability and Transformation* 2(7): e0000069.

De Jong, E. and Jungmeier, G. 2015. Biorefinery concepts in comparison to petrochemical refineries. In *Industrial Biorefineries and White Biotechnology* (Eds. Pandey, A., Höfer, R., Taherzadeh, M., Nampoothiri, K. M. and Larroche, C.), Elsevier, London, 3–33.

Debnath, S. 2021. Flax fiber extraction to textiles and sustainability: A holistic approach. In *Sustainable Fashion and Textiles in Latin America* (Eds. Gardetti, M. A. and Larios-Francia, R. P.), Springer, Singapore, 73–85.

Delate, K., Heller, B. and Shade, J. 2021. Organic cotton production may alleviate the environmental impacts of intensive conventional cotton production. *Renewable Agriculture and Food Systems* 36(4): 405–412.

Diemer, A., Nedelciu, E., Schellens, M. and Gisladottir, J. 2018. Challenges for sustainability in critical raw material assessments. *International Journal of Management and Sustainability* 7(3): 156–179.

Dong, M., Xue, Z., Liu, J., Yan, M., Xia, Y. and Wang, B. 2018. Preparation of carrageenan fibers with extraction of Chondrus via wet spinning process. *Carbohydrate Polymers* 194: 217–224.

El Seoud, O. A., Kostag, M., Jedvert, K. and Malek, N. I. 2020. Cellulose regeneration and chemical recycling: closing the "cellulose gap" using environmentally benign solvents. *Macromolecular Materials and Engineering* 305(4): 1900832.

H&M. 2021. H&M group leads new investment round in Infinited Fiber Company. https://hmgroup.com/news/hm-group-leads-new-investment-round-in-infinited-fiber-company/ [Accessed 20 January 2026].

Eyupoglu, S. 2020. Sustainable plant-based natural fibers. In *Sustainability in the Textile and Apparel Industries: Sourcing Natural Raw Materials* (Eds. Muthu, S. S. and Gardetti, M. A.), Springer Nature, Cham, 27–48.

Filimonova, L., Matys, E., Skvortsova, N. and Valiullina, E. 2022. Finding ways to solve problems of waste recycling: Biodegradable hemp materials. *IOP Conference Series: Earth and Environmental Science* 988(3): 032040.

Gedik, G. and Avinc, O. 2020. Hemp fiber as a sustainable raw material source for textile industry: Can we use its potential for more eco-friendly production? In *Sustainability in the Textile and Apparel Industries: Sourcing Natural Raw Materials* (Eds. Muthu, S. S. and Gardetti, M. A.), Springer Nature, Cham, 87–109.

Global Data. 2020. Global online retailing 2019–2023 with COVID-19 impact. https://www.globaldata.com/media/retail/multichannel-retail-covid-19-report-consumer-retailer-behavior-pandemic-demonstrate-online-physical-sales-channels-form-highly-competitive-single-retail-market-according-lates/ [Accessed 25 August 2025].

Godde, C. M., Mason-D'Croz, D., Mayberry, D. E., Thornton, P. K. and Herrero, M. 2021. Impacts of climate change on the livestock food supply chain; A review of the evidence. *Global Food Security* 28: 100488.

Grewal, D., Roggeveen, A. L. and Nordfält, J. 2017. The future of retailing. *Journal of Retailing* 93(1): 1–6.

Griggs, A. T., Roulier, A., Granda, N. A. and Burek, J. 2025. Life cycle assessments of residual hemp biomass from cannabidiol production focusing on disposal pathways and biomethane recovery via anaerobic digestion. *Resources, Conservation and Recycling* 223: 108525.

Harshitha, A. G., Kumar, S. and Jain, A. 2021. December. A review on organic cotton: Various challenges, issues and application for smart agriculture. *10th International Conference on System Modeling and Advancement in Research Trends (SMART),* IEEE, Moradabad, India, 2021, 143–149.

Hustvedt, G. and Dickson, M. A. 2009. Consumer likelihood of purchasing organic cotton apparel: Influence of attitudes and self-identity. *Journal of Fashion Marketing and Management: An International Journal* 13(1): 49–65.

Islam, S. 2020. Sustainable raw materials: 50 shades of sustainability. In *Sustainable Technologies for Fashion and Textiles* (Eds. Nayak, R.), Woodhead Publishing, Cambridge, 343–357.

Jamaludin, S. I. S., Zaini, M. A. A., Sadikin, A. N. and Jani, W. N. F. A. 2024. Textile waste valorization as potential activated carbon precursor for the removal of water contaminants: Commentary. *Materials Today: Proceedings* 96: 110–115.

Jem, K. J. and Tan, B. 2020. The development and challenges of poly (lactic acid) and poly (glycolic acid). *Advanced Industrial and Engineering Polymer Research* 3(2): 60–70.

Jia, F., Lu, Z., Zhao, H., Huang, T., Yang, P., Li, C., Zhang, J., Liu, J., Pang, R. and Hua, L. 2026. Valorization of waste aramid fibers: A comparative review of deprotonation, depolymerization, and beyond. *Polymer Reviews* 10:1–25.

Kumar, V., Hallqvist, C. and Ekwall, D. 2017. Developing a framework for traceability implementation in the textile supply chain. *Systems* 5(2): 33.

Lee, R. P., Reinhardt, R., Keller, F., Gurtner, S. and Schiffer, L. 2018. A raw materials transition for a low-carbon economy: Challenges and opportunities for management in addressing the trilemma of competitiveness, supply security and sustainability. In *Managing Natural Resources* (Eds. George, G. and Schillebeeckx, S. J. D.), Edward Elgar Publishing, Cheltenham, 61–87.

Li, J., Carvajal, R., Bruner, L. and Kaminski, N. E. 2021. The current understanding of the benefits, safety, and regulation of cannabidiol in consumer products. *Food and Chemical Toxicology* 157: 112600.

Li, Y., Fang, X., Shen, C., Jiang, W., Huang, S. and Ma, G. 2024. Review of bio-enzyme for soil improvement. *Biogeotechnics* 4(1): 100143.

Liu, W., Liu, S., Liu, T., Liu, T., Zhang, J. and Liu, H. 2019. Eco-friendly post-consumer cotton waste recycling for regenerated cellulose fibers. *Carbohydrate Polymers* 206: 141–148.

Lozano-González, M. J., Rodriguez-Hernandez, M. T., Gonzalez-De Los Santos, E. A. and Villalpando-Olmos, J. 2000. Physical–mechanical properties and morphological study on nylon-6 recycling by injection molding. *Journal of Applied Polymer Science* 76(6): 851–858.

Mannheim, V. and Simenfalvi, Z. 2020. Total life cycle of polypropylene products: Reducing environmental impacts in the manufacturing phase. *Polymers* 12(9): 1901.

Medronho, B. and Lindman, B. 2015. Brief overview on cellulose dissolution/regeneration interactions and mechanisms. *Advances in Colloid and Interface Science* 222: 502–508.

Meran, C., Ozturk, O. and Yuksel, M. 2008. Examination of the possibility of recycling and utilizing recycled polyethylene and polypropylene. *Materials and Design* 29(3): 701–705.

Mu, B., Yu, X., Shao, Y. and Yang, Y. 2024. High-quality acrylic fibers from waste textiles. *Science of the Total Environment* 931: 172752.

NCL Innovations. 2018. CSIR-National Chemical Laboratory (NCL), Pune, India. www.petrecycling.in [Accessed 11 September 2025].

O'Connor, S., Ehimen, E., Pillai, S. C., Black, A., Tormey, D. and Bartlett, J. 2021. Biogas production from small-scale anaerobic digestion plants on European farms. *Renewable and Sustainable Energy Reviews* 139: 110580.

Okafor, C. C., Madu, C. N., Ajaero, C. C., Ibekwe, J. C., Nzekwe, C. A., Okafor, C., Madu, C., Ajaero, C., Ibekwe, J. and Nzekwe, C. 2021. Sustainable management of textile and clothing. *Clean Technologies and Recycling* 1(1): 70–87.

Phan, K., Ügdüler, S., Harinck, L., Denolf, R., Roosen, M., O'Rourke, G., De Vos, D., Van Speybroeck, V., De Clerck, K. and De Meester, S. 2023. Analyzing the potential of the selective dissolution of elastane from mixed fiber textile waste. *Resources, Conservation and Recycling* 191: 106903.

Qin, Y., Qu, M., Kaschta, J., Allen, V. and Schubert, D. W. 2019. Studies on recycled polyester. In *Recycled Polyester: Manufacturing, Properties, Test Methods, and Identification* (Eds. Muthu, S. S.), Springer, Singapore, 29–67.

Radhakrishnan, S. 2017. Sustainable cotton production. In *Sustainable Fibers and Textiles* (Ed. Muthu, S. S.), Woodhead Publishing, Cambridge, 21–67.

Ragaert, K., Delva, L. and Van Geem, K. 2017. Mechanical and chemical recycling of solid plastic waste. *Waste Management* 69: 24–58.

Ramos, M., Dias, A. P. S., Puna, J. F., Gomes, J. and Bordado, J. C. 2019. Biodiesel production processes and sustainable raw materials. *Energies* 12(23): 4408.

Reganold, J. P. and Wachter, J. M. 2016. Organic agriculture in the twenty-first century. *Nature Plants* 2(2): 1–8.

Rissanen, M., Schlapp-Hackl, I., Sawada, D., Raiskio, S., Ojha, K., Smith, E. and Sixta, H. 2023. Chemical recycling of hemp waste textiles via the ionic liquid based dry-jet-wet spinning technology. *Textile Research Journal* 93(11–12): 2545–2557.

Russell, S., Swan, P., Trebowicz, M. and Ireland, A. 2016. Review of wool recycling and reuse. In *Natural Fibers: Advances in Science and Technology Towards Industrial Applications: From Science to Market* (Eds. Fangueiro, R. and Rana, S.), Springer, Dordrecht, 415–428.

Sahu, P. and Gupta, M. K. 2020. A review on the properties of natural fibers and its bio-composites: Effect of alkali treatment. *Proceedings of the Institution of Mechanical Engineers, Part L: Journal of Materials: Design and Applications* 234(1): 198–217.

Sakthivel, R., Ramesh, K., Purnachandran, R. and Shameer, P. M. 2018. A review on the properties, performance and emission aspects of the third-generation biodiesels. *Renewable and Sustainable Energy Reviews* 82: 2970–2992.

Sandin, G. and Peters, G. M. 2018. Environmental impact of textile reuse and recycling: A review. *Journal of Cleaner Production* 184: 353–365.

Sauer, M. and Kuhnel, M. J. C. C. 2019. Composites market report 2019. *Carbon Composites* 2: 1–11.

Sayyed, A. J., Deshmukh, N. A. and Pinjari, D. V. 2019. A critical review of manufacturing processes used in regenerated cellulosic fibers: Viscose, cellulose acetate, cuprammonium, LiCl/DMAc, ionic liquids, and NMMO based lyocell. *Cellulose* 26(5): 2913–2940.

Schneiderman, D. K. and Hillmyer, M. A. 2017. 50th anniversary perspective: There is a great future in sustainable polymers. *Macromolecules* 50(10): 3733–3749.

Senthil Kumar, P. and Saravanan, A. 2018. Organic cotton versus recycled cotton versus sustainable cotton. In *Organic Cotton: Is It a Sustainable Solution?* (Eds. Gardetti, M. A. and Muthu, S. S.), Springer, Singapore, 141–155.

Silva, G. S. and Rosenbach, M. 2021. Climate change and dermatology: An introduction to a special topic, for this special issue. *International Journal of Women's Dermatology* 7(1): 3–7.

Stanwix, W. and Sparrow, A. 2014. *The Hempcrete Book: Designing and Building with Hemp-Lime*. Bloomsbury Publishing.

Statista. 2026. Leading cotton-producing countries worldwide in 2024/2025. https://www.statista.com/statistics/263055/cotton-production-worldwide-by-top-countries/?srsltid=AfmBOoqrcJegTaeoZf1lRs9PyRfu_z4iRwacDmbx1hxg1K-yDDyky9 [Accessed 09 February 2026].

Suen, D. W. S., Chan, E. M. H., Lau, Y. Y., Lee, R. H. P., Tsang, P. W. K., Ouyang, S. and Tsang, C. W. 2023. Sustainable textile raw materials: Review on bioprocessing of textile waste via electrospinning. *Sustainability* 15(15): 11638.

Suzuki, S., Togo, A., Gan, H., Kimura, S. and Iwata, T. 2021. Air-jet wet-spinning of curdlan using ionic liquid. *ACS Sustainable Chemistry and Engineering* 9(11): 4247–4255.

Taib, N. A. A. B., Rahman, M. R., Huda, D., Kuok, K. K., Hamdan, S., Bakri, M. K. B., Julaihi, M. R. M. B. and Khan, A. 2023. A review on poly lactic acid (PLA) as a biodegradable polymer. *Polymer Bulletin* 80(2): 179–1213.

Textile Exchange. 2021. Preferred fiber and materials market report 2021. https://textileexchange.org/news/textile-exchange-preferred-fiber-and-materials-market-report-2021 [Accessed 15 January 2026].

Thill, S., Schmidt, T., Wöll, D. and Gebhardt, R. 2021. A regenerated fiber from rennet-treated casein micelles. *Colloid and Polymer Science* 299(5): 909–914.

Tonsi, G., Maesani, C., Alini, S., Ortenzi, M. A. and Pirola, C. 2023. Nylon recycling processes: A brief overview. *Chemical Engineering Transactions* 100: 727–732.

Tshifularo, C., Anandjiwala, R. and Patnaik, A. 2020a. Effect of process parameters on properties of polypropylene and kenaf fibers needle-punched nonwoven geotextiles. *Indian Journal of Fiber and Textile Research* 45(3): 274–285.

Tshifularo, C. A. and Patnaik, A. 2020b. Recycling of plastics into textile raw materials and products. In *Sustainable Technologies for Fashion and Textiles* (Ed. Nayak, R.), Woodhead Publishing, Cambridge, 311–326.

Wang, L., Huang, S. and Wang, Y. 2022. Recycling of waste cotton textile containing elastane fibers through dissolution and regeneration. *Membranes* 12(4): 355.

WGSN. 2022. Luxury brand strategies 2022. WGSN Insights Team, 24 February.

Xu, C. K., Cheng, H., Liao, Z. and Hu, H. 2019. An account of the textile waste policy in China (1991–2017). *Journal of Cleaner Production* 234: 1459–1470.

Yan, L., Chouw, N. and Jayaraman, K. 2014. Flax fiber and its composites: A review. *Composites Part B: Engineering* 56: 296–317.

Yang, B., Li, W., Zhang, M., Wang, L. and Ding, X. 2021. Recycling of high-value-added aramid nanofibers from waste aramid resources via a feasible and cost-effective approach. *ACS Nano* 15(4): 7195–7207.

Yang, L. and Dong, S. 2017. Sustainable product strategy in apparel industry with consumer behavior consideration. *Sustainability* 9(6): 920.

Yarger, J. L., Cherry, B. R. and Van der Vaart, A. 2018. Uncovering the structure–function relationship in spider silk. *Nature Reviews Materials* 3(3): 1–11.

Zhang, S., Chen, C., Duan, C., Hu, H., Li, H., Li, J., Liu, Y., Ma, X., Stavik, J. and Ni, Y., 2018. Regenerated cellulose by the lyocell process, a brief review of the process and properties. *BioResources* 13(2): 4577–4592.

Zimniewska, M. 2022. Hemp fiber properties and processing target textile: A review. *Materials* 15(5): 1901.

6 Sustainable Yarn Formation Processes

Ashvani Goyal

6.1 INTRODUCTION

The global textile industry is at a very important point in its history. The need for environmental sustainability is forcing the textile industry to change the traditional measures of success in manufacturing, such as speed, strength, and cost-effectiveness. Yarn formation is the middle step that turns raw fibers into yarn structures. It is also the step in the textile supply chain where significant energy is used, waste is created, and value is added. Industrial sector is the main driver of industrial energy demand and a key factor in the lifecycle durability and microfiber shedding characteristics of the final garment. Agricultural phases often have the biggest effect on land and water use, and wet processing (dyeing and finishing) has the biggest effect on chemical toxicity.

Life Cycle Assessment (LCA) methodologies offer a quantitative framework for comprehending these impacts. An LCA typically evaluates a product's environmental performance across four stages: raw material acquisition (on-farm or extraction), processing and manufacturing, the use phase, and end-of-life disposal. In the context of wool production, for instance, while the farming phase contributes approximately 88% of climate change impacts due to methane (CH_4) emissions from livestock, the energy utilized in yarn conversion accounts for a notable 11%, with transport and packaging playing negligible roles. Similarly, for cotton, the impact profile shifts depending on cultivation methods (conventional vs. organic) and the energy mix of the spinning location, with factors such as freshwater eutrophication and global warming potential (GWP) varying significantly between fiber types like silk, jute, and polyester.

The push for sustainability in yarn production is not just a way to follow the rules; it is a complete overhaul of the way things are made. It includes using energy-efficient spinning technologies like rotor and air-jet, improving mechanical processing to work with recycled feedstocks, reducing microplastic shedding through structural engineering (compact spinning), and digitizing mill management (Industry 4.0). This chapter gives a complete technical analysis of these sustainable yarn formation processes. It uses information from mechanical engineering, polymer science, and environmental science to show how the spinning industry will change in the future.

DOI: 10.1201/9781003689225-6

6.1.1 LCA Framework in Spinning

In order to use LCA in making yarn, the system boundaries must be very clearly defined. A "cradle-to-gate" analysis, which looks at the whole life cycle from extraction of raw materials up to the factory gate, is the most common way to measure spinning efficiency. However, the fact that there aren't standardized Product Environmental Footprint Category Rules for certain types of clothing makes it hard to compare studies.

Recent evaluations of cotton yarn production underscore the variability influenced by geographic factors. The environmental effects of virgin cotton yarn, for instance, change a lot depending on how the country that manufactures it effectively uses water and electricity. In China and India, where coal is the main source of energy, spinning has a much higher GWP than in areas that use renewable energy. Additionally, comparative LCAs of conventional and organic cotton show trade-offs. For example, organic cotton may still cause freshwater eutrophication because it uses manure-based fertilizers.

When it comes to synthetic fibers, the LCA focus changes to fossil resource depletion and energy intensity. Recycled polyester (rPET) has a much smaller carbon footprint than virgin polyester. This is mostly because it doesn't require the energy-intensive polymerization of crude oil derivatives. But the mechanical recycling process itself makes things more complicated when it comes to fiber quality and the possibility of microplastic release. These factors must be considered in a full cradle-to-cradle assessment.

6.2 OPTIMIZATION OF RAW MATERIALS AND MANAGEMENT OF BALES

Optimizing raw materials is the most important first step in making yarn in a way that is good for the environment. During spinning, especially with natural fibers like cotton, raw materials represent the most significant factor influencing both cost and environmental impact. Growing cotton involves heavy use of land, water, chemical fertilizers, and pesticides. The supply chain for cotton goes thousands of kilometers before it gets to the spinning mill (Akhteruzzaman 2020). The spinner takes over the technical responsibility for the cotton once it gets to the mill. Sustainability now depends on how smartly cotton is handled, not how it has grown.

Cotton is not the same all the time. Even on the same farm, the crop can change because of things like soil type, irrigation patterns, changes in rainfall, pest exposure, and harvest conditions (Sandin and Peters 2018). Natural fibers change on their own, while man-made fibers are made to stay within narrow tolerances. This change in behavior affects how machines work while spinning. Bales that have slightly different staple lengths or micronaires can cause spinning breaks, too much waste to be removed, or defects that only show up in the woven or knitted fabric. When mills can't control the composition of bales very well, they use more energy or raw materials than they need to. This is an inefficient but historically common way to deal with the problem (Behera and Singh 2021).

Therefore, knowledge is the first step to spinning sustainably. The more a mill knows about its raw materials, the less it needs to use "safety mechanisms" that add to the cost of production and harm the environment. This chapter looks at how data, technology, and process re-engineering can change the way raw materials are handled from a reactive to a predictive, scientifically based field.

6.2.1 Optimizing Digital Bale Management and Mixing

For most of the 20th century, visual inspection and broad grade categories were used to manage bales. These categories helped match cotton to the right yarn count, but they weren't very useful for diagnosis. Two bales with the same grade could still act very differently because they are at different stages of growth, have different amounts of neps, or have different amounts of trash that has already been picked. A spinning master would only find out about this difference when the quality of the carding went down or the drafting became unstable. At this point, the choice was either to separate the bales and put them in a lower-grade mix or to increase the twist and lose productivity. Both responses were solutions, but neither would work in the long run (Subramani and Murthy 2020).

Digital bale management establishes a novel framework based on measurement instead of intuition. Using intelligent data analysis and accurate fiber measurement to maximize raw resource utilization, Uster 360Q assists in handling cotton bales (Uster 2025). The Uster HVI 1000 (High-Volume Instrument) is the most crucial component. For all major cotton quality attributes, including fiber length, strength, micronaire, and color, it provides incredibly dependable and repeatable test results, either from representative samples or by looking at all the bales. In order to produce dependable cotton laydown plans that are appropriate for the mill's inventory, quality standards, and end-use requirements, the Uster FiberQ software automatically collects these particular quality measures (Uster 2025). The FiberQ and FiberQ Bale modules reduce waste and quality disparities while increasing productivity. Additionally, they facilitate consistent mixed decisions, facilitate laydowns, and improve warehouse management. The combined HVI + FiberQ method provides mills with data to help them make decisions that improve the consistency, efficiency, and profitability of handling cotton bales.

Then, software tools like FiberQ or Uster Bale Manager look at the database of bales to make a better "laydown." The laydown isn't random; it's a carefully chosen mix in which each bale makes up for the weaknesses of another. The mixing plan makes sure that the blending level is always the same, instead of trying to make each bale perfect. The goal is to keep the machine settings stable during production by reducing the differences between the feedstock inputs (Rieter 2025a). This plan has many benefits for the environment. First, mills can use a wider range of raw cotton, which means they don't have to reject bales often. According to industry surveys, digital bale management can cut down on the number of bales that are rejected by 20%–30%. This directly increases the effective use of materials that were bought. Second, when fiber variability is kept to a minimum, machine settings can be optimized for efficiency instead of safety. Mills can lower the amount of twist in the yarn and lower the amount of energy used per kilogram of yarn (Behera and Singh 2021). This is better than running with too many twists to protect against weak fibers. In the same way, rotor or ring frame delivery speeds can be raised without making the ends-down frequency too high. Third, preventing defects becomes something that happens on its own. Digital management stops defects from happening in the first place, while traditional methods only find them after they happen.

The effects go all the way down the line. Barre effects, which are some of the most common fabric flaws in knitted and dyed goods, usually come from differences in the fibers themselves rather than the machine settings. When bale blending gets rid of this inconsistency, mills don't have to throw away whole lots of dyed fabric. The environmental benefit is huge: every rejected fabric roll means not only wasted fiber, but also wasted dye, chemicals, heat energy, and water (Napper and Thompson 2016).

Digital bale management also makes the supply chain stronger. When cotton from a certain farm or supplier consistently has problems, mills can send this information back up the chain. Farmers and aggregators who know about quality trends can change how they harvest and gin. So, sustainability is no longer a way to improve things in one place; instead, it becomes a way for everyone in the value chain to work together.

In summary, digital bale management means that mills don't have to rely on fixes much and can make better yarn with less energy, less waste of raw materials, and fewer defects downstream. It changes how raw materials are handled from a reactionary process to a scientific way that can predict what will happen.

6.2.2 Effectiveness of Materials in the Blowroom

After the laydown is set up, the fiber goes into the blowroom, which is the first step in spinning that uses machines. The blowroom breaks up compressed bales into tufts, gets rid of trash, and makes sure that fiber feed is the same for carding. It is also where the first mistakes happen, and they are often the ones that last the longest. Once a staple is broken, the fiber stays short through every stage after that. This shows how important it is to set the blowroom so that cleaning and preservation are in balance (Devaraja and Natarajan 2016).

The only thing that older mills wanted their blowrooms to do was to get rid of trash. It was thought that hitting cotton harder would make the machines work better, since a cleaner feed always helps. This wasn't completely wrong, but it didn't consider the cost of collateral damage. A poorly calibrated blowroom might also remove 0.3–0.5 g of good lint for every gram of trash it takes away. Even though this seemed small on its own, the effect over thousands of kilograms was huge. Also, too much beating made neps and short fibers, which would later make carding noil go up and yarn strength go down.

The philosophy of modern blowrooms is more all-encompassing. Lint is not an unlimited resource, so don't throw it away just to keep things clean. New machines like the Rieter VARIOline or Trützschler CL-X have variable-speed drives, closed-loop sensors, and material intelligence that changes the opening intensity based on how the fiber responds. When tests on raw cotton show that there is not much trash in it, the blowroom automatically lowers the beating force. When the amount of trash goes up, the system increases the cleaning action, but only to the point where staple damage is not likely (Rieter 2021). This controlled modulation keeps the quality of the fiber without hurting the performance of the machine.

A gentle opening has more benefits than just saving materials. Mills make fiber streams that draft more efficiently by keeping longer fibers and cutting down on nep formation. Drafting stability cuts down on end breaks and delivers faster with less twist. Every factor that speeds up or slows down the twist uses less energy per unit of yarn, which strengthens the link between material efficiency and carbon efficiency (Behera and Singh 2021). Also, smoother drafting makes yarns that are more even, which cuts down on waste in weaving and knitting.

It is very important to keep the blowroom clean when using recycled cotton. Recycled fibers come into the blowroom with some problems: they have a lot of short fibers, are brittle, and have more neps. These fibers can't handle a rough opening. When exposed to normal blowroom forces, their usable fraction decreases even more, which makes recycling less valuable for the environment. So, a sustainable blowroom serves as both a quality gatekeeper and a way to make things circular. It is possible to add recycled fibers without losing yarn quality or making spinning waste too expensive (Sandin and Peters 2018).

Economic self-evidence may be the strongest reason to use sustainable blowroom practices. Even a 2% increase in yarn realization at a medium-sized mill can pay for system upgrades in a matter of months. Blowroom optimization is one of the easiest ways to make the textile industry more sustainable because it balances economic efficiency with environmental responsibility.

6.3 A COMPARISON OF DIFFERENT SPINNING TECHNOLOGIES

Spinning technology is the most important part of making yarn, and the choice of which technology to use—ring, rotor (open-end, OE), or air-jet—has a big impact on both the environmental and economic aspects of a textile mill. This choice is never just technical; it is a strategic one that affects how much energy is used, how much carbon is released, how raw materials are used, how well products work with recycled feedstocks, and, in the end, how sustainable are the products that reach the market (Subramani and Murthy 2020).

As the global textile industry moves away from traditional, resource-intensive ways of making things and toward cleaner, more circular ones, the spinning department becomes a place where changes need to be made. The spinning stage alone may use up to 40%–45% of all the electricity used by the mill, so it is very important to make it more efficient. So, choosing a spinning system is not just a technical issue; it is also part of being environmentally responsible in line with international standards like the UN Sustainable Development Goals (SDGs), especially SDG 7, SDG 12, and SDG 13 (UNEP 2023).

6.3.1 Ring Spinning

Ring spinning is still the most common way to make high-quality yarns that need to be strong, soft, and even. Its mechanical principle gives yarn a better structure, but it uses a lot of energy because of the rotating mass that goes into twist insertion (Behera and Singh 2021). The bobbin needs more torque as it is filled with yarn, which means it uses more electricity. The yarn balloon that forms while traveling from the drafting zone also increases aerodynamic drag, which leads to more power losses (Rieter 2021).

Depending on the count, twist, and machine design, the specific energy consumption for ring spinning can be anywhere from 3.5 to 11.6 kWh/kg (Devaraja and Natarajan 2016). Finer yarns need a lot more energy because the amount of energy needed doesn't go up in a straight line as the fiber fineness and twist levels go up. This performance is due to the system's basic thermodynamic limits, not just inefficiencies.

Even so, steady improvements in engineering have made it possible to cut energy use by almost 30%–37% over the past 20 years (Rieter 2021). Compact spinning and other technologies that help with sustainability goals by making fibers more aligned, less hairy, and less likely to shed and pill (Pastore 2019). These improvements help reduce microfiber pollution, which is a new environmental problem that comes from washing clothes (Napper and Thompson 2016).

Ring-spun yarns make clothes last longer because they are stronger, more resistant to wear and tear, and feel better in the hand. This means that people don't have to wash them often and there is less waste, which is a benefit for sustainability even when energy demand is higher (Subramani and Murthy 2020; Senthilkannan and Subramani 2020). When looked at from a LCA point of view, a longer service life can help lessen the effects of production (Fletcher and Tham 2019). This interpretation suggests that ring spinning promotes sustainability through durability rather than energy efficiency, thereby complementing circular manufacturing strategies rather than opposing them.

6.3.2 Rotor (Open-End) Spinning

Rotor or OE spinning is a different way of making things than ring spinning. Instead of a rotating bobbin, a high-speed rotor creates a twist. This reduces mechanical resistance and lets the delivery speeds be much higher (Senthilkannan and Subramani 2018). This means that this technology uses the least amount of energy per kilogram of yarn compared to other major commercial spinning technologies. Since the early 2000s, energy efficiency has improved by more than 50% (Rieter 2025b).

Rotor spinning is one of the best technologies for mechanically recycling cotton because recycled fibers are usually shorter (15–22 mm) and they get damaged during reprocessing (Sandin and Peters 2018). Ring spinning can't reliably process these types of fibers because they break too often and don't stay the same, but rotor spinning's centrifugal consolidation and wrapper fiber reinforcement make it possible to make stable yarn (Wakelyn et al. 2007). This feature directly supports models for closed-loop recycling and a circular economy. This is especially true in the denim and toweling industries, where the look of the yarn is less important. So, rotors make it possible for material loops to cut down on the need for new cotton. This saves water, pesticides, farmland, and energy which are all used to grow cotton. This helps lower the carbon footprint of textiles at many points along the supply chain. The yarns made this way are usually thicker and less smooth than ring-spun yarns. However, in industrial and workwear settings, where durability and cost-effectiveness are more important than tactile softness, rotor spinning has both functional and environmental benefits.

In discussions about sustainability, the most important thing about rotor spinning is that it turns waste into feedstock, which lets mills switch from linear resource flows to circulating systems. So, it's not just a way to save energy; it's also a way to help save resources on a larger scale.

6.3.3 Air-Jet and Vortex Spinning

Murata vortex spinning and other types of air-jet spinning use high-speed air streams to add twist instead of mechanical rotation. It takes a lot of energy to make compressed air, but the very high delivery speeds (often 450–550 m/min) make it more energy-efficient on a per-kilogram basis (Muratec 2025). Air-jet systems have a low operational carbon intensity in buildings that use solar or wind power, which improves their environmental profile (Behera and Singh 2021).

The structural properties of air-jet yarns are a major reason why they are good for the environment. The yarns have a core that runs parallel to the fibers that wrap around them in a spiral shape. This makes them very low in hairiness (Pastore 2019). This means that less sizing is needed during weaving, which is important because applying and removing sizing uses a lot of chemicals and creates a lot of wastewater (Kasim et al. 2018). Fabrics made from air-jet yarns are also less likely to pill and wear down, which helps the product last longer.

One of the most important things about vortex-spun yarns is that they release a lot fewer microfibers when they are washed than ring-spun yarns. This helps reduce microfiber pollution at the consumer level (De Falco et al. 2020). Microfiber pollution has now been found in freshwater systems, oceans, and even the human food chain. This makes solutions that work during the manufacturing process even more important (Napper and Thompson 2016). So, air-jet spinning is not only a way to make things more efficiently, but it also helps protect the environment. Air-jet systems can also mix recycled cellulose (like viscose and lyocell) with rPET, but they don't work as well with very short fibers as rotor systems. Their strength in terms of sustainability is in reducing chemicals and pollution, which works well with the rotor's role in making materials more circular.

6.3.4 A Look at Sustainability from a Different Point of View

When looked at from a sustainability point of view, it's clear that each spinning technology has its own effect on the environment. Ring spinning helps materials last longer and be used more effectively, rotor spinning makes energy usage more efficient and waste more valuable, and air-jet spinning makes downstream processing cleaner and cuts down on microfiber emissions. So, the best way to keep spinning is not to rely on just one system, but to use a mix of systems that fit the type of product, the availability of fibers, the energy conditions, and the planned life cycle. Table 6.1 shows how energy use has changed over time for the three main spinning systems (Maung et al. 2025; Textile School 2011).

TABLE 6.1
Comparative Analysis of Spinning Technologies

Technology	Production Speed (m/min)	Relative Energy Efficiency	Fiber Flexibility	Suitability for Recycled Fibers	Use Phase Impact (Shedding)
Ring spinning	15–25	Low (highest kWh/kg)	High (all fibers)	Low (needs long staple)	Highest
Rotor spinning	100–200	High (lowest kWh/kg)	Medium (coarse counts)	High (tolerates short fibers)	Medium
Air-jet/ Vortex	400–550	Medium-high	Low (synthetics/ blends)	Low (needs clean/long fibers)	Lowest

6.4 TECHNICAL PROBLEMS AND SOLUTIONS FOR MECHANICAL RECYCLING OF COTTON

Mechanical recycling of cotton has become a major way to make textile production more circular. This process converts both post-consumer clothing and pre-consumer textile waste into fibers that can be used for spinning. This keeps a lot of material out of landfills and cuts down on the need for new cotton. The circular goal is exciting, but right now, the limits of material use are set by the physics of fibers and the performance of spinning. To fully realize the sustainability potential of recycled cotton in mainstream textile markets, it is necessary to comprehend the nature of this degradation and the associated technological solutions.

6.4.1 The Science behind Fiber Quality Degradation

Fiber damage is an unavoidable part of the mechanical recycling process. Garnett machines with metal or toothed rollers can tear and open fabrics. These rollers tear the cloth apart, pulling the fibers apart without caring about how strong they are. Each point of contact weakens and breaks down the fibers a little bit more, which lowers the average fiber length and the internal cellulose cohesion. Virgin cotton typically has an upper half mean length of about 29–30 mm, but recycled fibers often drop to 25 mm or less, which means that the usable staple length is much shorter (Devaraja and Natarajan 2016). As this happens, the short fiber content goes up. Depending on how bad the shredding is, it could go up to 30%–45% by weight (Sandin and Peters 2018). In the drafting zone, short fibers are harder to control because they slip between drafting rollers. This causes floating fibers, mass variation, thick-and-thin faults, and uneven twist distribution.

Mechanical deterioration doesn't just shorten things; it also weakens them. Virgin cotton with values around 35 g/tex may drop to ranges closer to 27–28 g/tex after being recycled. This makes the fibers very weak during the stresses of ring spinning (Wakelyn et al. 2007). This flaw shows up as frequent end breaks, poor winding performance, and fabric that doesn't last. Furthermore, recycled materials tend to exhibit a higher degree of nep formation. Virgin fiber samples may have about 100 neps/g, but recycled stock often has more than 400–600 neps/g. This is mostly because the fibers get tangled up and the blended materials break up, like polyester or elastane parts found in clothes that have been worn before (Pastore 2019). Neps make yarn look bad, and these flaws can still be seen in finished fabrics, which makes them less appealing and less likely to be accepted by the market.

The fact that textile waste comes in many different forms makes the mechanical problems even worse. Recycled feedstock often has things in it that shouldn't be there, like sewing thread scraps, zipper pieces, elastane fibers, printing residues, and finishing chemicals. These things make processing harder. These contaminants build up in the rotor groove while the rotor is spinning, and cause wear on the ring travelers while the ring is spinning. This leads to more wear on the machine and uneven yarn formation (Behera and Singh 2021). Because of this, recycled cotton has mostly been used in low-value products like industrial wipes, mops, and stuffing materials. It has mostly been used in coarse rotor yarns, usually in the Ne 6–20 range. Downcycling is a way to keep materials from going to waste, but it doesn't keep their economic or performance value (Fletcher and Tham 2019).

6.4.2 Using Combing as a Way to Upcycle

Combing has recently been added to the process of making recycled fiber, which is an important step in turning recycled cotton from low-value output into high-value yarn. In traditional spinning, combing is only done on the best fiber lots to make them easier to spin by removing the shortest and most tangled fibers. When combing is used on recycled blends, it takes on a new role: it becomes a selective refinement tool that separates usable recycled fibers from ones that are too damaged to use.

A typical industrial workflow involves making a blend, like 50% recycled and 50% virgin cotton, and then adding it to the combing process. The comb disc takes away some of the recycled fraction as noil while keeping the fibers that are still mostly intact. The recycled content in the sliver may go down from 50% to about 40%–42%, but the recycled fibers that are still there have better fiber parallelism, more consistent length, and a much lower nep count (Rieter 2021). Mills often report a 10%–18% decrease in unevenness (CVm%) and a noticeable decrease in the number of thin places, thick places, and neps when compared to carded-only recycled blends. This means that the properties of the yarn get better. Tensile testing shows similar improvements in performance, with strength values going up by as much as 15% because structurally weak fibers were removed. These improvements together make it possible to make finer yarns, which lets ring spinning of recycled blends in the Ne 30–40 range, and they also make it possible to use recycled cotton in clothing areas where it wasn't possible before.

The innovation doesn't stop at better quality; it also includes smart ways to use waste. The noil, which is mostly made up of short, recycled fibers, goes to a rotor spinning line, where its shorter staple lengths are useful for further processing. The use of centrifugal force for yarn formation in rotor spinning is naturally better at handling short fibers. It can turn noil into coarse yarns that are good for denim weft, upholstery fabrics, towels, or fleece products (Subramani and Murthy 2020). This system makes it possible to use two streams of recycled fibers: longer fibers go into ring-spun yarns for clothing, and shorter fibers go into rotor yarns for utility markets.

Instead of being thrown away, each fiber fraction is matched to the use that best fits its technical properties.

The outcome is a vertically integrated model that keeps value and makes the most of fiber. Recycled cotton is not just a way to get rid of waste; it is also a way to make money. The effectiveness of this method shows that circularity in textiles isn't just about recycling; it's also about quality stratification, process design, and carefully matching fiber properties with the right spinning technologies.

6.4.3 Optimizing Rotor Spinning for High Recycled Content

When combing isn't done, rotor spinning is still the most common way to handle a lot of recycled fibers. There are built-in mechanical benefits to rotor systems. For example, the centrifugal field that forms inside the rotor helps to combine short fibers into the yarn body, making up for the fact that the fibers are shorter and less cohesive. But to use recycled cotton well, the spinning parameters need to be carefully adjusted to avoid problems with processing.

The first thing to think about is how fast the rotor spins. For recycled blends, a rotor speed of between 65,000 and 75,000 rpm usually works well. The centrifugal forces in this range help the fibers stick together in the yarn structure without putting too much stress on them, which could cause more fibers to break. Going above this range could cause more defects and worse tensile performance, while going slower often doesn't lead to good yarn consolidation (Behera and Singh 2021). Every recycled batch needs to go through multiple tests because the fiber performance is affected by the previous fabric life cycle, dye chemistry, and finishing treatments.

The speed of the opening roller is another parameter that needs to be changed. This speed controls how the fibers are separated before they enter the rotor. Recycled fibers are more likely to break again if they are opened too aggressively because their structure is weaker. So, mills that process recycled blends usually have roller speeds of about 8,000–8,500 rpm, which is a good balance between separating fibers and being sensitive to mechanical changes (Sandin and Peters 2018). The chosen roller wire profile also affects fiber damage; finer wire surfaces help prevent fragmentation. Recycled waste can contain things like dust, microplastics, and elastane residues that build up in the rotor groove and cause problems. Specialized rotor designs, like those used in Rieter R37 systems, have airflow channels and dust extraction ports that are designed to keep the rotor clean and lower the number of yarn defects that happen when particles build up (Rieter 2021).

Even after optimization, rotor yarns made from recycled materials will often have lower tensile values than those made from new cotton. These results don't mean failure; instead, they show that product alignment needs to be more focused. Denim weft, towels, knit fleece, socks, and upholstery are some of the things that can use yarns with slightly lower tensile strength. This is especially true when the sustainability features of recycled material add value to the market. Mills that fine-tune machine settings and use quality-based material segregation can make yarns with 60%–90% recycled cotton. This cuts down on the need for new resources and has a smaller effect on the environment (Akhteruzzaman 2020).

6.5 ADVANCED SPINNING ARCHITECTURES

As the world moves toward more environmentally friendly textile production, there is an urgent need for spinning systems that do more than just turn fiber into yarn. Even though traditional spinning infrastructures work well, they are best for high-quality virgin cotton and fiber inputs that are easy to predict. They don't work as well for today's resource problems, like when raw materials change, when recycled feedstocks are added, or when yarns are made with certain performance characteristics for technical or environmental uses. Advanced spinning architectures, especially compact spinning, friction spinning, and core-spun configurations, have not replaced mainstream production but rather enhanced it strategically. These systems give mills structural and mechanical benefits that let them make yarns that shed less, use materials more efficiently, and absorb more waste (Lawrence 2003).

In terms of sustainability, these systems are best understood as engineering solutions that fix environmental problems, not just spinning technologies. Compact spinning cuts down on microfiber pollution, friction spinning turns trash into something useful instead of throwing it away, and core-spun structures fill the gap between the limits of mechanical recycling and the needs of products. These architectures, when combined, add new words to the design language of yarn manufacturing. This lets the industry respond positively to the problems of climate change and circular economy goals.

6.5.1 Compact Spinning

Compact spinning has gone from being a niche innovation to a common feature in modern ring spinning plants. The idea is simple: before adding the twist, condense the fiber strand to make better use of the fiber. But the mechanism is subtle and very much based on how fibers work. When fibers leave the front roller in classical ring spinning, they come out in a loose bundle. Before twists can bind them, they are in a tapered shape called the spinning triangle for a short time. In this triangular area, edge fibers are under uneven tension and tend to slip out of the yarn body's structural control (Pastore 2019). These stray fibers make the fabric hairy, weaken it, and make it more likely to get worn out when used again.

With compact spinning, a focused air stream or mechanical condensing channel gets rid of the triangle by making the fiber path narrower. The condensing field lines up the fibers into a tight, parallel bundle so that the twist comes in evenly. There are benefits for both the structure and the environment. Compact yarns have higher tenacity, longer elongation at break, and less mass variation, which lets mills make finer counts with more consistency. From an environmental point of view, the biggest benefit is that it makes hair much less noticeable. Hairiness is not just a cosmetic problem; it is directly related to microfiber shedding during washing, which is becoming more linked to synthetic microplastic pollution in aquatic ecosystems (Napper and Thompson 2016).

After washing, millions of microfibers are released and mixed with wastewater. Some of these are filtered out at treatment plants, but the majority (90%) end up in rivers and oceans. These particles, which are usually less than 5 mm in size, are now present in marine organisms, sediments, and the human food chain (Cesa et al. 2020). By making sure that fewer fibers sit on the yarn's surface, compact spinning cuts down on shedding at the source. For mills that make polyester, viscose, or blended yarns, compact spinning is a main way to reduce problems, not a way to fix them later, like laundry filters or wastewater screening systems.

The energy profile of compact spinning is another important aspect of sustainability. Stronger compact yarns need less twist to work the same way. Reductions in twist multipliers (TMs) lead directly to lower spindle power because adding twist is one of the most energy-intensive parts of ring spinning (Behera and Singh 2021). The compaction system uses energy to suck things up, but the overall energy balance usually favors compact spinning. This is especially true when mills use less traveler friction and higher production speeds.

Compact spinning makes fiber use more accessible to everyone, in addition to being better for the environment and using less energy. Recycled cotton is harder to work with because it has short fibers and is weaker, but it behaves more predictably in a compact zone. The better alignment cuts down on drafting problems and makes it easier for recycled fibers to mix into yarns that will be used for clothing instead of low-quality uses. In this way, compact spinning makes it possible to upcycle, not just recycle, by putting damaged fiber back into markets where it is worth more (Sandin and Peters 2018).

6.5.2 DREF (Friction) Spinning

Friction spinning, which is often sold under the DREF brand, uses a completely different method to make yarn. DREF doesn't use the twist that comes from the yarn package moving around; instead, it uses the surface friction that comes from rotating perforated drums. These drums have individual fibers put on them, and the twist is added by rubbing instead of rotating (Lawrence 2003). This change from standard spinning geometry gives you a unique advantage: the ability to spin very short or uneven fibers that would break or float away in ring or rotor systems.

The DREF system is great at making core-sheath yarns, which are made up of a strong, continuous filament that runs through the center of the yarn and a sheath of short staple fibers that wrap around it. Low-quality cotton, shredded textile scraps, reclaimed acrylic, elastane-contaminated pieces, or even fibers that have been damaged during mechanical recycling (Devaraja and Natarajan 2016) could all be used to make the sheath.

This quality makes DREF spinning a special part of the textile circular economy. Ring and rotor spinning need fibers that are stable, but DREF systems can turn materials that are close to being thrown away into yarns that can be sold. These yarns are not meant for fine shirts or knitwear; instead, they are used in industrial and technical markets, such as filtration media, brake linings, conveyor belts, acoustic panels, roofing felts, carpet backing, and protective canvases (Subramani and Murthy 2020). In these cases, the sustainability benefit is twofold: first, high-value waste fractions are kept out of disposal streams, and second, industries that use virgin high-performance fibers can use upcycled composites instead.

DREF spinning is also good for the environment because it cuts down on the number of steps in the processing. The drum mechanism is what makes fiber cohesion work, not a lot of drafting or high twist insertion. This means that the process needs fewer steps before it starts, which saves energy and wear on the machine. Also, friction spinning doesn't need the same level of environmental control as ring spinning, so it can be done in places with limited infrastructure and still make yarn that works the same way every time.

To sum up, DREF spinning is a way to save materials. It makes fibers last longer, opens up new markets for waste streams that aren't being used, and helps industries that benefit from switching to more sustainable products.

6.6 MICROFIBER SHEDDING

Microfiber shedding has changed from a small scientific problem to a big problem for sustainability. The textile industry is now known to be one of the biggest causes of microplastic pollution around the world, especially when it comes to washing synthetic or synthetic-blend fabrics (Cesa et al. 2020). People often blame fabrics for shedding, but the real problem is the yarn. The amount of shedding depends on how the fibers are put together, how they are bound, and how much stress they are under in the yarn structure. So, spinning technology can tell us about pollution before the fabric even gets to the customer.

Research definitively establishes that fiber liberation initiates at the spinning phase. Loose surface fibers, not twisted enough during spinning, bad fiber alignment, and structural breaks all cause shedding when the fabric is worn or washed (Napper and Thompson 2016). When these fibers are released, they go into wastewater systems, where removal technologies don't work very well because the fibers are so small and their density changes. Their eventual buildup in marine environments poses ecological risks, including ingestion by aquatic organisms and the transfer of chemical additives into biological systems. Therefore, the primary question is how we can prevent fibers from shedding in the first place.

6.6.1 Getting Rid of Hierarchy by Using Spinning Technology

Different spinning methods have very different shedding performance, which makes it easy to see which ones have the most impact on the environment.

Air-jet and vortex spinning always have the lowest shedding rates. Their yarns have a fasciated structure, which means that wrapper fibers hold together a tightly bound central core. This structure is resistant to wear and tear, and the reduced surface hairiness keeps fibers from coming loose even after being washed many times (Muratec 2025). More and more sportswear and athleisure brands are using these kinds of yarns because they meet the needs of both consumers and regulators.

Compact ring spinning is in the second tier. Compact yarns aren't as structurally unified as vortex yarns, but their condensing mechanism makes the fibers on the surface much less visible. Longitudinal studies indicate that compact-spun cotton maintains surface integrity more effectively than traditional ring yarns, thereby reducing the shedding rate throughout the garment's lifespan (Pastore 2019).

When you use a rotor spinning machine, the yarns tend to shed a little bit. Rotor yarns have wrapper fibers around a loosely twisted core. This makes them less stable than compact or vortex yarns, but more stable than standard ring yarns with the same linear density (Devaraja and Natarajan 2016). The shedding rate is affected by the speed of the rotor, how aggressive the opening roller is, and the quality of the raw materials.

Unless they are heavily optimized, regular ring spinning shows the highest shedding rates. The spinning triangle here leaves a lot of fiber ends loose. Even when the twist is increased to make up for it, reducing hairiness costs energy efficiency and tactile comfort, which makes the trade-off not worth it for large-scale commercial production (Behera and Singh 2021).

This hierarchy emphasizes a core tenet: shedding is subject to architectural negotiation. Microfiber pollution doesn't have to be a problem for mills; they just need the right spinning infrastructure and processing philosophy.

6.6.2 Factors in Structure and Process That Affect Shedding

The base structure of spinning architecture is what makes shedding work, while processing parameters make it work even better. The TM is one of the most important factors. A higher twist compresses the yarn body and locks fibers into the structure, making it less likely to wear down. But twist makes the spindle work harder and can make the fabric less soft, which is a problem for clothes that are meant to be comfortable (Rieter 2021). In the same way, faster production speeds may increase throughput, but they can also damage fibers and put them under thermal stress in the drafting zone, which can lead to more fiber fragmentation.

The weather can also affect shedding. Low humidity makes static charge stronger, which makes it harder to control fibers and makes hairier. On the other hand, high humidity can make it harder for travelers to stay lubricated and raise the thermal load. Mills that carefully control environmental factors often see better shedding performance without making any changes to the structure.

6.7 Changing the Surface and Making Chemicals Last Longer

Fiber mechanics and machine engineering are only two parts of the performance equation when manufacturing yarn. The chemistry at the fiber–machine interface often decides if a fiber moves smoothly through the spinning line or breaks down into faults and breaks. Spin finishes, surface treatments, and other extras make high-speed processing possible by reducing friction, static buildup, and fiber damage. But even though these chemicals are used to make processing more efficient, a lot of people are worried about their effects on the environment, especially when they get into wastewater streams or stay in ecosystems.

In the age of sustainability, we need to replace traditional petroleum-based chemicals with formulations and surface modification methods that work well, don't pollute too much, and fit with circular textile flows. This part looks at these kinds of new ideas, like biodegradable spin finishes, plasma treatments that don't use water, and enzymatic bio-polishing. By doing this, it shows how chemical engineering can help with sustainability goals in spinning, dyeing, finishing, and fabric performance.

6.7.1 Biodegradable Spin Finishes

Spin finishes are used on synthetic fibers like polyester and nylon, and sometimes on recycled fibers, to cut down on friction and static electricity during spinning and other processes that come after it. Mineral oil emulsions and surfactants like alkylphenol ethoxylates are common in spin finishes. They improve lubrication but stay in water and soil, harming aquatic life and adding to the number of micro-pollutants in wastewater (Kiron 2022).

The shift toward biodegradable and bio-based spin finishes is a direct response to these environmental issues. Companies like Pulcra Chemicals have made finishes using renewable feedstocks like vegetable oils, sugar derivatives, and natural esters. These formulas are made to meet modern standards like the Global Organic Textile Standard (GOTS) and the Oeko-Tex Standard 100. These certifications require strict limits on harmful or nonbiodegradable materials. Biodegradable spin finishes are now made to break down quickly in normal wastewater treatment conditions, which lowers the risk to the environment without losing lubricity or antistatic performance.

Biodegradable finishes also help save energy during wet processing. To remove many traditional finishes before dyeing and finishing, fabrics need to be washed at high temperatures and for a long time. Bio-based alternatives are made to wash off well at lower temperatures. In industrial practice, this cuts down on the amount of thermal energy used in scouring and dyeing by 10%–25% (Pulcra Chemicals 2025). This decreases greenhouse gas emissions further down the line and helps supply chains use less energy.

In practice, the use of finishes that break down easily in water is part of a bigger trend in the textile industry toward designing for the end of life. Modern mills don't see chemistry as an invisible processing aid anymore. Instead, they see it as a material that needs to be thought about from the start of product design.

6.7.2 Plasma Treatment

Plasma treatment is one of the most interesting new ways to change the surface of textiles. It uses ionized gas to change fiber surfaces without using water or traditional wet chemistry. Plasma is made up of charged particles (ions and electrons) and reactive radicals that interact with the fiber surface, etching and changing its properties (Serrano-Martínez et al. 2024). Plasma exposure adds polar functional groups, like hydroxyl and carboxyl units, which make surfaces more wettable and stickier.

Plasma treatments are especially helpful for synthetic and recycled fibers that have lost their natural crimp or surface chemistry after being mechanically recycled. Plasma increases the surface energy of fibers, which makes them stick together better and increases friction between fibers. This makes drafting and twist insertion easier during spinning. This effect is especially strong in air-jet and vortex spinning, where interactions between the fibers and the surface control how they line up and fit into wrapper structures.

Plasma processing is an important dry method. It doesn't need water or rinsing steps, and it completely stops the production of effluent. Because of this, it fits well with zero liquid discharge (ZLD) goals and environmental rules in areas where water is scarce. In dyeing and finishing, the hydrophilic surface created by plasma also allows for better dye uptake at lower liquor ratios. This means that less water and chemicals are needed, which is a big plus for mills that want to cut costs and their impact on the environment.

New studies have also used plasma to change the surfaces of rPET, making fiber interaction better without lowering molecular weight. This method overcomes the limits of mechanical recycling and meets the needs for secondary processing.

6.7.3 Bio-Polishing with Enzymes

Enzymatic bio-polishing is another chemical way to help the environment. Bio-polishing uses cellulase enzymes to selectively hydrolyze surface microfibrils on cotton and other cellulosic fibers instead of acids or oxidizing agents that are usually used to remove fiber hairs and pills. This biochemical reaction gently gets rid of fibrils that stick out and cause pilling and mechanical fiber release when clothes are washed.

Bio-polishing is gentler on fibers because enzymes work best at moderate temperatures and neutral pH levels. This means that bio-polishing doesn't have the harsh effluent profiles that chemical polishing does. The yarns and fabrics that come out of this process not only have fewer surface hairs, but they also stay stronger, which means that the finishes are both cleaner and last longer. Bio-polished yarns are less likely to release fibers when washed by consumers. This helps reduce microfibers at the source, which is a principle that is now backed up by new research on how to optimize spinning parameters and fiber hairiness as indicators of shedding behavior (Chen et al. 2025).

6.8 NEW IDEAS FOR SYNTHETIC AND REGENERATED FIBERS

Even though a lot of the talk about sustainability is about natural and recycled cotton, synthetic fibers, especially polyester, are still the most common type of fiber made in the world. To deal with their environmental effects, we don't have to stop using synthetics; we just have to change how we use them. rPET and regenerated cellulosic fibers show how innovation can turn petrochemical and cellulosic materials into more circular, sustainable alternatives.

6.8.1 rPET and Chemical Recycling

Polyester is used in more than half of all textiles around the world because it is strong, flexible, and cheap. But it has a big effect on the environment, from getting fossil fuels to polluting the ocean with microplastics to making things hard to throw away (Textile Exchange 2022).

Mechanical recycling of used PET bottles into fibers, commonly known as "bottle-to-yarn," has become an important component of circular economy initiatives. Polyester that has been mechanically recycled saves energy and reduces the need for virgin feedstock, but it does break down over time. Each time a polymer chain is heated or mechanically damaged, its Intrinsic Viscosity goes down, which makes it less effective in demanding textile applications.

To get around these problems, chemical recycling has become a game-changing answer. Chemical recycling, which includes hydrolysis, glycolysis, and methanolysis, breaks polyester down into its monomeric parts (terephthalic acid and ethylene glycol) with very little loss of quality. The monomers can be re-polymerized into "virgin-equivalent" polyester, which completes the cycle for synthetic polymers. This method is better than mechanical recycling because it gets rid of colorants, contaminants, and degradation products, leaving behind raw materials that are good for high-performance uses (Mohtaram and Fojan 2025). Commercial depolymerization facilities that combine these processes with existing textile supply chains are now showing how chemical recycling can be scaled up.

6.8.2 Recycled Cellulosic Fibers and Circularity between Textiles

Historically, wood pulp has been used to make man-made cellulosic fibers (MMCFs) like viscose and lyocell. They avoid a lot of problems that come with petrochemicals, but their production raises concerns about deforestation, chemical waste, and changes in land use. The innovation of textile-based cellulosic regeneration, as shown by products like Circulose®, changes this story by getting dissolving pulp from old cotton textiles instead of wood feedstock (Circulose 2025).

During this process, cotton clothes that people have already worn and are likely to throw away or recycle are shredded, cleaned, and turned back into high-quality cellulose pulp. This pulp then goes through normal MMCF spinning routes, like viscose or lyocell. This makes a real textile-to-textile closed loop that keeps value and cuts down on the need for new agricultural or forestry inputs. Circulose pulp has properties that are similar to dissolving pulp made from wood, such as high purity and uniformity. This means that it can be processed on existing industrial MMCF lines without needing to make major changes to the equipment.

Lyocell is considered one of the most environmentally friendly MMCFs. It uses a closed-loop solvent system (N-methylmorpholine N-oxide) to dissolve cellulose and make fibers (Figure 6.1), with solvent recovery rates close to 99% and very little chemical discharge (Sawiak et al. 2025). The fibers that come out of this process are biodegradable, absorbent, and strong enough to be mixed with cotton, wool, and even rPET without hurting the spinning performance later on. MMCFs made from recycled materials are a new idea that combines caring for the environment with the current industrial infrastructure.

6.9 DIGITALIZATION AND ENERGY MONITORING IN INDUSTRY 4.0

The shift to sustainable yarn production doesn't just depend on materials; it also depends on the information. Modern spinning mills are not just machines anymore; they are ecosystems that run on data. The industry 4.0 paradigm—characterized by sensors, automation, and cloud-based analytics—facilitates the exact monitoring and enhancement of resource utilization at a level previously inconceivable two decades

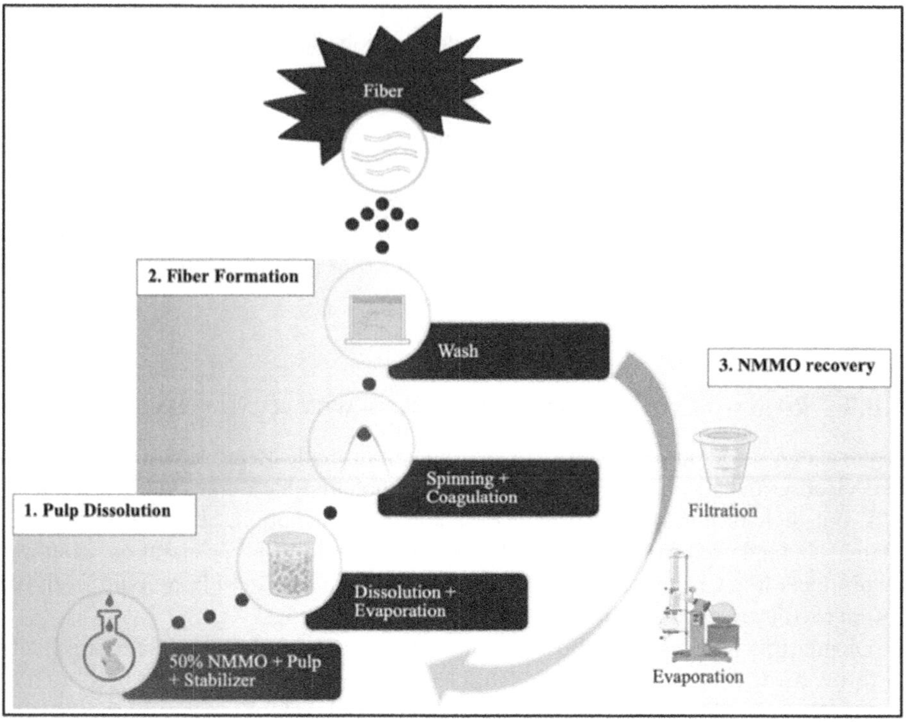

FIGURE 6.1 Lyocell fiber production from cellulose pulp using N-methylmorpholine N-oxide solvent.

Source: **Sawiak et al. 2025.**

ago. In sustainable spinning, this change has one goal: to cut down on invisible waste, which is the loss of resources that happens when machines break down, settings aren't right, or energy drifts over time (Behera and Singh 2021).

Industry 4.0 does not do away with human knowledge. Instead, it shifts the responsibility of monitoring from intuition to smart systems. Supervisors and technicians no longer have to do manual diagnostics, which lets them make higher-level decisions based on real-time performance data. Mills that use modern digital infrastructure say that their productivity, product consistency, and energy intensity all go up at the same time, which is a balance that was once thought to be impossible (Rieter 2025a).

6.9.1 Systems for Monitoring Energy in Real Time

In the past, spinning mills only measured their energy use at the billing meter. This gave them a total amount of energy used, but it didn't show them where or why energy was being lost. Today, systems like Rieter's ESSENTIAL platform and Truetzschler's MyMill give you a view of the whole process by putting sensors on spindles, blowroom motors, compressors, and winding heads. These sensors measure how much power each machine uses per kilogram of yarn, so mills can figure out how much energy each machine uses in real time (Rieter 2025a; Truetzschler 2025).

The value of these systems is that they can find problems:

- A single-ring spindle that draws more energy could mean that the traveler is rubbing against something or that the lubrication has failed.
- If a rotor unit's consumption goes up, it could mean that the bearings are worn out or that the rotor is out of balance.
- Compressors that use too much power are often a sign of air leaks or wrong drafting suction settings.

In the past, these kinds of problems would go on unnoticed until they caused failure or quality complaints that required intervention. The monitoring system sends out predictive maintenance alerts today, which tell technicians to step in before energy losses get worse. This method is a change from "repair after failure" to "prevent failure from happening" (Basak and Choudhury 2021).

The effects are huge when they happen on a large scale. A mill with 25,000 spindles that loses just 2 W/spindle because of friction that isn't noticed wastes 50 kW all the time. This is the same as paying for more than ₹20 lakhs (Indian Rupee) worth of extra electricity each year, depending on the tariff structure. Industry 4.0 systems get rid of this loss, making sustainability not just legal but also profitable.

6.9.2 Fine Control of Twist Multipliers

The TM is the spinning variable that has the most effect on both yarn strength and energy use. More twist not only adds more fiber binding force and makes tensile properties better, but it also

- slows down the speed of production,
- raises the load on the spindle's power,
- speeds up the wear and tear on travelers, and
- makes it easier for thermal energy to escape through friction.

Now, digital systems let mills model the exact twist needed for a certain fiber blend instead of using standard twist heuristics. Machine-learning-assisted modeling provides a "sweet spot" parameter in which the twist is sufficient to achieve the necessary tenacity while simultaneously reducing energy consumption and mechanical stress.

According to research using Response Surface Methodology, lowering twist by just 2%–4% can cut spindle energy use by 5%–8% without hurting the fabric's performance (Subramani and Murthy 2020). These savings add up to millions of kilograms every year for mills that make Ne yarns in the 20–40s range.

This level of accuracy is especially important when spinning recycled cotton blends because the length of the fibers and the number of short fibers need different structural compensation than virgin fibers. Mills can make long-lasting, low-shedding yarns that are energy-efficient by combining compact spinning with TM optimization (Sandin and Peters 2018).

6.10 MODELS FOR MANAGING WASTE AND CASE STUDIES FROM DIFFERENT REGIONS

Sustainability is never the same everywhere; it depends on where you are. The availability of waste, national infrastructure, regulatory environments, and consumer demand all affect which spinning strategies are possible. The following case studies show how sustainability changes in different parts of the world while still working toward the same global goal, keeping value.

6.10.1 The Panipat Model: Large-Scale Mechanical Recycling

Panipat, in northern India, is known around the world as a center for mechanical textile recycling. It processes more than 1 million ton of postconsumer and postindustrial waste each year (Mohtaram 2021). The area was the first to use shoddy spinning, a method that uses a rotor to turn torn wool and cotton scraps into coarse yarns that are mostly used for blankets, carpets, and industrial cloth.

The result for sustainability is strange. Panipat is a material recovery engine because it takes a lot of trash away from landfills. However, the yarns that come from this process often go to lower-value markets because regular spinning systems can't make clothing-grade quality from shredded materials (Sandin and Peters 2018).

There are still problems with work and social life. Historically, informal sorting units did not have enough protection for workers, and the tearing machines that exposed workers to dust made breathing problems worse (Akhteruzzaman 2020). Recent investment and compliance programs, on the other hand, show that operations are moving toward being more formal and automated. The lesson from Panipat is clear: to be sustainable, we need to be efficient with resources and people.

6.10.2 Usha Yarns

Usha Yarns, which is situated in Chandigarh, India, is an excellent example of how recycling clothes may generate better items than downcycling them. Sorting textile waste streams by the type of fiber they contain is the first step in the company's planned recycling program. This makes it easier to obey the rules and get stuff back fast. After mechanical recycling, compact spinning is employed to make the yarn stronger and less fuzzy. This is vital to keep the fibers safe. Usha Yarns is different because they use textile waste that has already been dyed instead of normal wet dyeing procedures. This enables them to combine colors. Mixing colors in this manner without water keeps dye effluent from leaking out, which is terrible for the environment. Making yarn with solar electricity is much better for the environment. Using these technologies together can turn recycled fibers from low-value utilitarian applications into fashion-grade fabrics that are good for everyday use. The global recycled standard makes sure that the business meets criteria for keeping chemicals safe, tracking them, and the chain of possessions.

The Usha Yarns case indicates that the problems with recycled fiber arise from how it is made, not the fibers themselves. If they are designed effectively and the right technology is applied, recycled fibers can be used in commercial clothes instead of new cotton. According to Usha Yarns (2023), recycling is not only good for the environment, but also a way to manufacture products that are good for the environment.

6.10.3 Zero Liquid Discharge in Tamil Nadu, India

Tirupur, a major textile hub in Tamil Nadu, is a great example of how government action can lead to sustainability. Because of serious pollution of groundwater, state officials required textile dyeing and finishing plants to use ZLD as shown in Figure 6.2. With ZLD, mills must (Liang et al. 2021)

- use ultrafiltration and reverse osmosis (RO) to get back at least 95% of the wastewater,
- evaporate the rest of the sludge to make salt crystallize, and
- use the water that was recovered again in production (Akhteruzzaman 2020).

This order separated production from ecosystems that depend on freshwater. But it costs energy to do this, since evaporation and RO systems use a lot of energy. Mills responded by putting money into renewable energy, recovering waste heat, and processing chemistry that is made to handle low-effluent loads. ZLD is a good example of a bigger lesson in sustainability: engineering solutions need to be systemic, not linear.

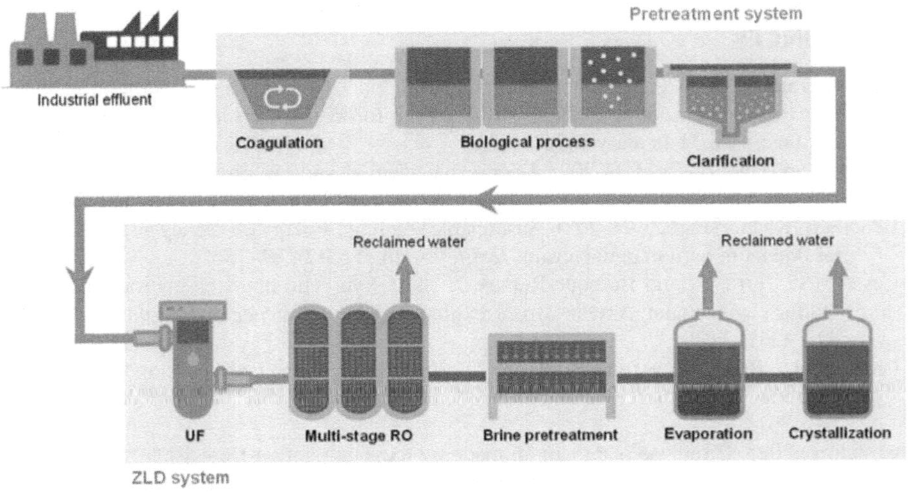

FIGURE 6.2 Typical processes to achieve zero liquid discharge (ZLD) in management of industrial effluent.

Source: Liang et al. 2021.

6.11 CONCLUSION

Sustainable yarn manufacturing can best be understood as a careful balance between three interrelated factors: energy, quality, and ecology. In earlier approaches, spinning primarily emphasized productivity and speed; however, modern sustainability requirements demand a more integrated and balanced perspective. When energy efficiency is achieved at the expense of yarn quality, the resulting fabrics tend to have a shorter service life, leading to frequent replacement and an increased overall environmental burden. Conversely, an uncompromising focus on quality—often relying on inefficient machinery, excessive twist insertion, or energy-intensive processes—results in higher energy consumption and increased emissions. True sustainability lies between these two extremes, where adequate yarn quality is achieved with optimized energy use. Within this framework, two strategic directions are increasingly evident: high-value recycling, which employs advanced and compact spinning techniques to convert recycled fiber blends into yarns with sufficient strength, durability, and controlled hairiness for apparel applications; and high-volume recycling, which utilizes rotor and air-jet spinning systems to process short-fiber feedstocks at relatively low energy costs, thereby diverting large volumes of textile waste from landfills. Looking ahead, the spinning mill is expected to evolve from a conventional production unit into a materials management center, capable of digitally assessing fiber characteristics, predicting yarn behavior, assigning feedstocks based on their suitability, and continuously monitoring both energy consumption and quality parameters. In this future context, waste will no longer be regarded as an indicator of poor production, but rather as a valuable resource that must be intelligently reintegrated into the manufacturing cycle.

REFERENCES

Akhteruzzaman, M. 2020. Sustainable cotton production: Challenges and opportunities. https://icac.org/wp-content/uploads/2025/02/Article-for-World-Cotton-Day-2020-converted.pdf [accessed 31 January 2026].

Basak, S. and Choudhury, A. R. 2021. Energy management and sustainability in textile spinning mills through Industry 4.0 technologies. *Journal of Cleaner Production* 283:124–138.

Behera, B. K. and Singh, S. N. 2021. Sustainable spinning: Energy efficiency and environmental impact in yarn manufacturing. *Textile Progress* 53(2): 89–156.

Cesa, F. S., Turra, A. and Baruque-Ramos, J. 2020. Synthetic fibers as microplastics in the marine environment: A review from textile perspective. *Science of the Total Environment* 598: 1116–1129.

Chen, H. L., Burns, R. S. and Burns, L. D. 2025. Optimization of spinning parameters to minimize microfiber shedding from textile products. *Textile Research Journal* 95(3): 287–301.

Circulose. 2025. Circulose is a material made by recycling cotton from 100% discarded textiles. https://www.circulose.com/about-circulose/ [accessed 31 January 2026].

De Falco, F., Di Pace, E., Cocca, M. and Avella, M. 2020. The contribution of washing processes of synthetic clothes to microplastic pollution. *Scientific Reports* 9: 6633.

Devaraja, R. and Natarajan, K. V. 2016. Material efficiency and quality optimization in blowroom and carding processes. *Indian Journal of Fiber & Textile Research* 41(4): 412–419.

Fletcher, K. and Tham, M. 2019. *Earth Logic: Fashion Action Research Plan*. The Design Research Institute, London.

Kasim, R., Ismail, M. R. and Ibrahim, N. A. 2018. Environmental impact of sizing agents in textile weaving: A review. *Journal of Environmental Chemical Engineering* 6(4): 4555–4565.

Kiron, M. I. 2022. Spin finish in textile: Properties, components and problems. https://textilelearner.net/spin-finish-in-textile/ [accessed 24 November 2025].

Lawrence, C. A. 2003. *Fundamentals of Spun Yarn Technology*. CRC Press, Boca Raton, FL.

Liang, Y., Lin, X., Kong, X., Duan, Q., Wang, P., Mei, X. and Ma, J. 2021. Making waves: Zero liquid discharge for sustainable industrial effluent management. *Water* 13: 2852.

Maung, K., Ali, M. and Ansary, A. I. 2025. Power consumption of ring spinning frame using plastic t-in cylinder. *International Journal of Textile Engineering and Management* 1:1.

Mohtaram, F. and Fojan, P. 2025. From waste to value: Advances in recycling textile-based PET fabrics. *Textiles* 5(3): 24.

Mohtaram, S. 2021. Textile waste management in Panipat: A case study of mechanical recycling infrastructure. *Waste Management and Research* 39(6): 823–835.

Muratec. 2025. Vortex spinning machine. https://www.muratec-vortex.com/history/ [accessed 26 January 2026].

Napper, I. E. and Thompson, R. C. 2016. Release of synthetic microplastic plastic fibers from domestic washing machines: Effects of fabric type and washing conditions. *Marine Pollution Bulletin* 112(1–2): 39–45.

Pastore, C. M. 2019. Structural mechanics of yarns and fabrics: Performance and sustainability implications. *Textile Research Journal* 89(15): 3127–3145.

Pulcra Chemicals. 2025. Process chemicals for the production of man-made fibers. https://www.pulcra-chemicals.com/kundenloesungen/faserhilfsmittel/ [accessed 20 January 2026].

Rieter. 2021. Digital bale management and fiber optimization systems. Technical documentation. Rieter Machine Works Ltd, Winterthur.

Rieter. 2025a. Digital spinning mill management system essential. https://www.rieter.com/products/smart-spinning-mill/spinning-mill-digitization/essential [accessed 24 November 2025].

Rieter. 2025b. Energy efficiency. https://www.rieter.com/sustainability/planet/climate/energy-efficiency [accessed 30 January 2026].

Sandin, G. and Peters, G. M. 2018. Environmental impact of textile reuse and recycling: A review. *Journal of Cleaner Production* 184: 353–365.

Sawiak, M., Souto, B. A., Lawson, L., Lo, J. and Dolez, P. I. 2025. Recovery of N-methylmorpholine N-oxide (NMMO) in lyocell fiber manufacturing process. *Fibers* 13(1): 3.

Senthilkannan, K. and Subramani, T. N. 2018. Comparative analysis of ring and rotor spinning technologies: Energy consumption and yarn properties. *Autex Research Journal* 18(3): 220–229.

Senthilkannan, K. and Subramani, T. N. 2020. Optimization of spinning processes for recycled cotton feedstocks. *Journal of Natural Fibers* 17(8): 1156–1168.

Serrano-Martínez, V. M., Ruzafa-Silvestre, C., Hernández-Fernández, C., Bañón-Gil, E., Arán-Ais, F. and Orgilés-Calpena, E. 2024. Improving cotton fabric dyeability by oxy gen plasma surface activation. *Surfaces* 7: 1079–1095.

Subramani, T. N. and Murthy, P. V. S. 2020. Response surface methodology for optimization of twist multiplier in ring spinning: Energy and quality implications. *Textile Research Journal* 90(11–12): 1389–1402.

Textile Exchange. 2022. Preferred fiber and materials market report 2021. https://textileexchange.org/app/uploads/2021/08/Textile-Exchange_Preferred-Fiber-and-Materials-Market-Report_2021.pdf [accessed 24 November 2025].

Textile School. 2011. Power consumption in spinning mills. https://www.textileschool.com/161/power-consumption-in-spinning-mills/ [accessed 24 November 2025].

Truetzschler. 2025. MyMill, cloud based Mill monitoring system. https://www.truetzschler.com/fileadmin/user_upload/truetzschler_spinning/downloads/broschuere/fact_sheets/Digital_FactSheet_My_Mill_EN.pdf [accessed 24 November 2025].

UNEP. 2023. Sustainability and circularity in the textile value chain. https://wedocs.unep.org/rest/api/core/bitstreams/07737510-2d2b-44c2-a458-5ddd4e5fdf0d/content [accessed 24 November 2025].

Usha Yarns. 2023. *Recycled fiber spinning process*. https://ushayarns.com/recycled-fiber-spinning-mills-process/ [accessed 24 November 2025].

Uster. 2025. Think quality at the next level. https://www.uster.com/fileadmin/user_upload/Downloads/0_360Q/0.2_Raw_Material_Mgmt/Fiber2Yarn/Uster_360Q_digital_brochure_web_en_25.pdf [accessed 24 November 2025].

Wakelyn, P. J., Bertoniere, N. R., French, A. D., Thibodeaux, D. P., Triplett, B. A., Rousselle, M. A., Goynes Jr., W. R., Edwards, A. D., Hunter, L., McAlister, D. D. and Gamble, G. R. 2007. *Cotton Fiber Chemistry and Technology*. CRC Press/Taylor & Francis Group, Boca Raton, FL.

7 Sustainability in Fabric Manufacturing Processes

Vinay Kumar Midha and Monica Sikka

7.1 INTRODUCTION

The global textile industry is undergoing a transformative shift toward sustainability in response to growing environmental concerns, resource limitations, and changing consumer expectations. Although the textile industry has experienced rapid growth over the past few years, it still faces significant environmental challenges and ranks among the world's most polluting industries. The environmental impact includes water pollution, carbon dioxide (CO_2) emissions, the release of harmful chemicals and dyes into the environment, and increased garment waste due to shorter product life cycle (Naqvi et al. 2024).

Textile and clothing production plays a significant role in global climate change, largely because of its dependence on fossil fuels, particularly for electricity and thermal energy. Fabric manufacturing consumes substantial energy, water, chemicals, and fibers sourced from both natural and petroleum-based materials. Weaving is especially energy-intensive, requiring 2–5 MJ/kg of cloth, compared to 1–4 MJ/kg for knitting (Sustainability Directory 2025a, 2025b).

Although the textile industry has experienced rapid growth, it faces significant environmental challenges, ranking among the most polluting industries and contributing to water pollution, CO_2 emissions, the release of harmful chemicals, and increased garment waste due to shorter product life cycles. Sustainability in fabric manufacturing encompasses more than just eco-friendly production; it integrates technical innovation, regulatory support, and social responsibility across the textile value chain. Advances in weaving technologies, energy-efficient machinery, digital monitoring, and waste-minimizing production strategies have enabled reductions in environmental footprints while maintaining the high performance and aesthetic quality of fabrics. Moreover, global and regional policy frameworks, eco-labels, and standards, such as the Global Organic Textile Standard (GOTS), are encouraging manufacturers to adopt circular economy principles and implement transparent supply chains. At the same time, rising consumer demand for sustainable textiles and the rapid growth of the sustainable fabrics market are reshaping business models, motivating companies to align ecological integrity with economic viability.

This chapter discusses sustainability in fabric manufacturing, focusing on environmental impacts, sustainable fabric manufacturing technologies, and the influence of policies and consumer preferences. It also highlights challenges and future directions for enhancing sustainability in the industry.

DOI: 10.1201/9781003689225-7

TABLE 7.1
Textile Life Cycle and Energy Consumption

Stages	Main Activities	Key Energy-Intensive Points
Raw material	Farming natural fibers and extracting raw materials for synthetic fibers	Land preparation, planting, nurturing growth, harvesting, and transportation of raw materials
Fiber production	Ginning and spinning for cotton; polymerization and extrusion for synthetics	Electrical and thermal energy for machinery operation and material transformation
Fabric production	Weaving/knitting; dyeing, printing, finishing	Heat energy for dyeing and drying, as well as electrical energy, powers the machinery
Product manufacturing	Cutting, sewing, assembly, pressing	Energy for power sewing machines, cutting tools, lighting in factories, and for pressing or ironing finished goods
Distribution and retail	Transport to warehouses/retail, storage, and display	Energy consumption, in the form of fuel for transportation (trucks, ships, and airplanes), and also for lighting, heating, and cooling
Consumer use	Washing, drying, and ironing	Washing machines, dryers, and iron are significant energy consumers in households
End-of-life management	Collection, transport, landfilling, incineration, recycling	Transport to facilities, shredding/sorting, thermal treatment, recycling operations

7.2 OVERVIEW OF FABRIC MANUFACTURING

Textile fabrics play a crucial role in the fashion industry. Textiles are unique materials because they are often robust, moldable, permeable, lightweight, and flexible (easily bent, sheared, and twisted). Fabric manufacturing converts fibers into functional textiles via interrelated processes such as yarn preparation, weaving or knitting, and finishing (Pattanayak 2020). Table 7.1 outlines the key stages of the textile life cycle, highlighting main activities and dominant energy-intensive processes.

Weaving and knitting are two textile production methods. Weaving is more energy-intensive, using 2–5 MJ/kg, while knitting uses 1–4 MJ/kg. Nonwoven fabrics may reduce energy consumption but can still require significant energy for bonding. The choice of method affects overall energy use (Sustainability Directory 2025b).

7.3 WEAVING TECHNOLOGY

Woven fabrics are created by interlacing warp and weft threads. The process involves preparing yarn, consolidating it into beams, applying protective coatings, and then drawing the warp through the loom components (Goswami et al. 2004). The flowchart of the preparatory processes in weaving is shown in Figure 7.1.

The fabrics are produced on a loom after the preparatory procedures. Earlier, handlooms and shuttle looms were employed. To address the drawbacks of these looms, shuttle-less looms have been developed. Shuttle-less looms are favored in the industry due to their superior fabric production rate and quality. The shuttle looms are the most adaptable devices on the market. These looms experience multiple issues,

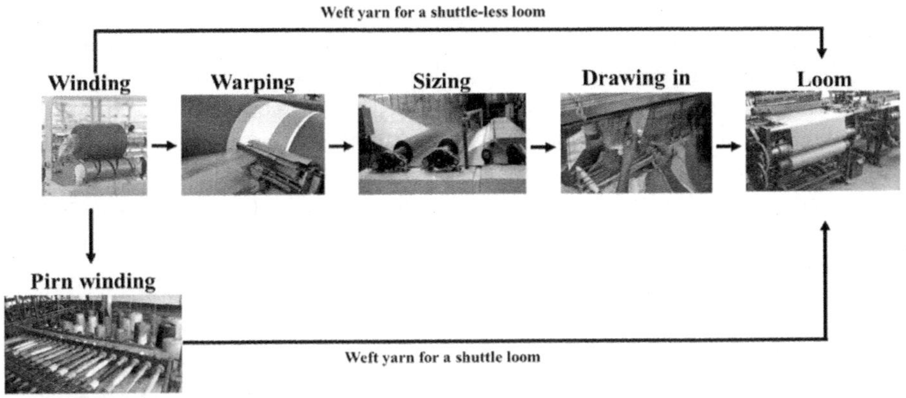

FIGURE 7.1 Preparatory processes of weaving.

Source: **Goswami et al. 2004.**

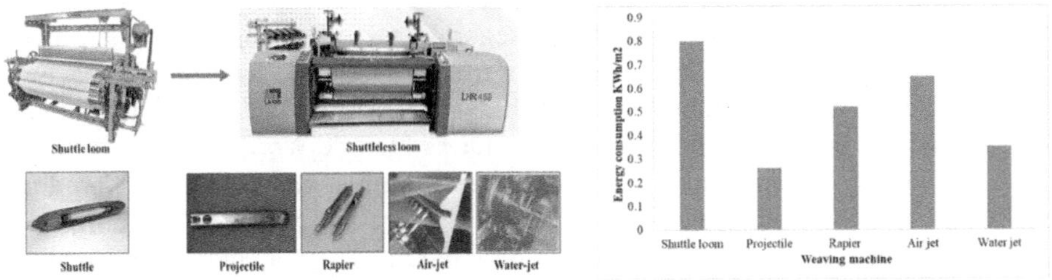

FIGURE 7.2 Energy consumption of various weaving machines.

Source: **Pattanayak 2020.**

including the requirement for a substantial medium (shuttle) to transport the weft, the absence of guided weft carriage, increased power consumption per square meter of fabric production, and elevated noise levels. Shuttle-less looms, including rapier, projectile, air-jet, and water-jet looms, are used to produce fabrics with a higher production rate and better quality. Rapier looms produce a variety of fabric types with less noise due to reduced vibration. Projectile looms are ideal for weaving wide fabrics because of their high productivity and reliability. Air-jet looms have the highest weft insertion rates (WIRs) and are suitable for light to medium weight cotton. Water-jet looms are ideal for light- to medium-weight fabrics made from hydrophobic fibers, such as nylon and polyester, as they combine rapid WIRs with low energy usage (Castelli et al. 2010). Figure 7.2 shows the comparative energy consumption of various weaving machines (Pattanayak 2020).

7.4 KNITTING TECHNOLOGY

Knitting technology is the second most widely used method for fabric manufacture, after weaving. Knitted fabrics can be manufactured by two techniques: the weft-knitting method and the warp-knitting method (Figure 7.3). Weft-knitted fabrics are made from horizontal knitted loops using flat-bed and circular knitting machines. Flat machines produce fashionable and seamless textiles, while circular machines create garment-length innerwear and cut-and-sew fabrics. Warp-knitted fabrics consist of loops created by warp threads that extend along the fabric's length. This method uses multiple yarns with tricot or Raschel machines, resulting in robust structures. Although warp knitting has limited adaptability due to its beam supply, it offers a solid fabric structure and high productivity, making it suitable for domestic applications, technical textiles, and composite reinforcement.

Warping, sizing, drawing in, and desizing are eliminated in knitted fabric manufacture, making it simpler than weaving. Consequently, knitting technology is more environmentally sustainable as it requires less energy and water. Furthermore, knitting produces less wastewater and requires less effluent treatment. Consequently, the production of knitted cloth requires less electrical energy. The electrical energy consumption in knitting is influenced by several aspects, including yarn linear density, yarn tension, machine speed, gauge, garment style and size, and the garment production method. Compared to a weaving mill, a warp-knitting mill uses substantially less electricity overall for warp preparation, fabric manufacturing, and ambient parameter tuning. In terms of product categories, warp knitting offers the greatest benefit over weaving in the manufacture of trouser fabrics. Warp-knitting machines use around one-tenth of the energy that weaving machines use. The potential electricity savings are 0.57 kWh/m^2 in absolute terms (Textile World 2022).

The industry is experiencing a trend toward ultrafine gauge knitwear, predominantly in sportswear and undergarments. Fine-gauge knitwear is less environmentally sustainable than coarser options, as it requires higher energy and material consumption, resulting in approximately a 14% decrease in productivity. Circular machines currently attain gauges of 60 needles/inch, whereas flat-bed technology can accomplish up to 18 needles/inch. Seamless garment manufacturing can reduce environmental impact by creating 3D clothes without conventional fabric production, cutting, and sewing, resulting in 30%–40% savings in time and energy. This technology

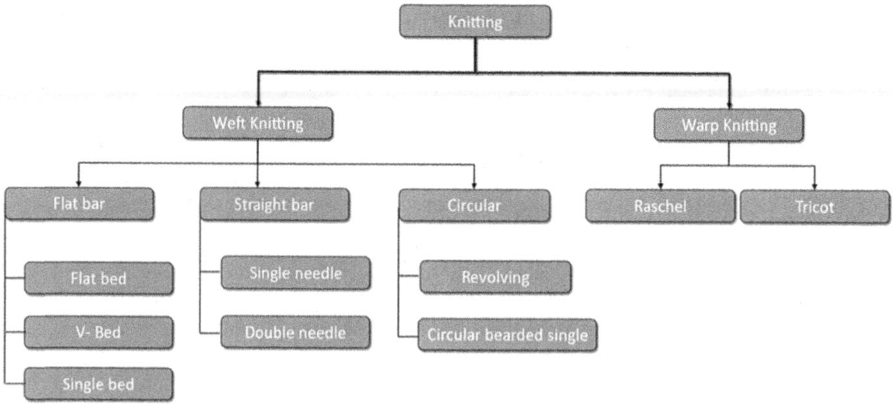

FIGURE 7.3 Knitting machine classification.

is applicable solely to specific knitted garments and may be prohibitively expensive for smaller manufacturing runs, thereby constraining its widespread adoption (Lau and Yu 2016; Nayak and Padhye 2015).

7.5 NONWOVEN TECHNOLOGY

Nonwoven fabrics are textile structures produced directly from fibers or polymers without spinning, weaving, or knitting. Their manufacturing involves two main steps: web formation and web bonding, allowing high production speed, design flexibility, and efficient material use. Web formation is carried out using dry-laid, wet-laid, or polymer-laid techniques. Dry-laid processes use carding or air-laying of staple fibers, wet-laid processes form webs from fiber suspensions in water, and polymer-laid processes such as spunbond and meltblown extrude filaments directly from molten polymers. The formed web is then bonded using mechanical, thermal, or chemical methods to provide strength and durability, with the bonding technique determining the final properties of the nonwoven fabric.

The manufacturing process, whether it is a spunbond, meltblown, or needle-punching line, typically consists of several fundamental stages: polymer melting and extrusion, filament drawing and web construction, bonding, and, lastly, winding and finishing. Each of these processes contributes to the total energy bill, but the proportions are far from equal. Extrusion and heating systems can account for 60%–70% of the overall power consumption per ton of nonwoven fabric, according to industry estimates. One of the main factors is the type of raw material, as energy consumption is strongly influenced by the different melting points and processing properties of polypropylene and polyethylene terephthalate (Al Nonwoven, 2025). Figure 7.4 shows the comparative energy consumption of various nonwoven processes (Textile Exchange 2023).

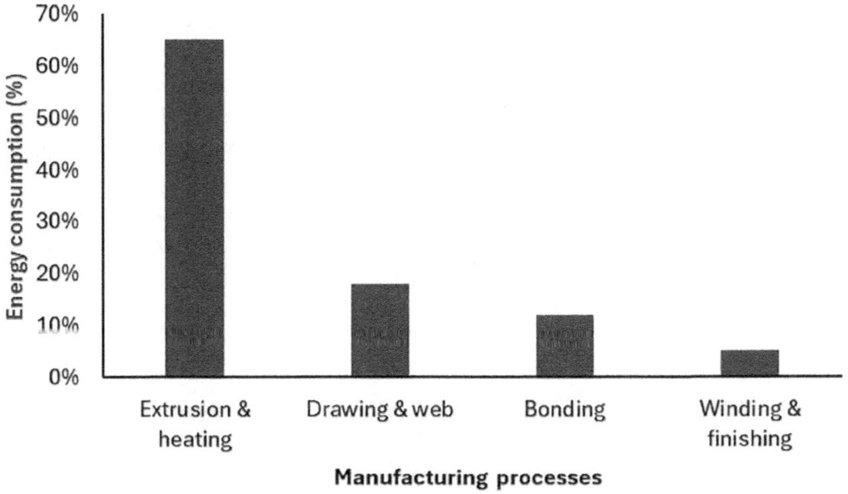

FIGURE 7.4 Energy consumption by nonwoven processes.

Source: **Textile Exchange 2023.**

7.6 BRAIDING TECHNOLOGY

Braiding is one of the most ancient crafting methods known to humanity. It has been utilized to create various forms of braids, including round types such as cords or ropes and flat types such as strands. Originating from manual crafting techniques, maypole braiding (or circular braiding) has evolved into an automated processing technology. Braided fabrics are typically manufactured using braiding machines such as horn-gear, maypole, and track-and-column braiders. In these machines, yarn carriers move along predetermined paths, crossing over and under one another to generate the braided structure (Mishra 2023).

7.7 ENVIRONMENTAL IMPACT OF FABRIC MANUFACTURING TECHNIQUES

Fabric production methods such as weaving, knitting, and nonwoven fabrication differ in their environmental impacts because of their distinct requirements for energy, water, emissions, and waste. High-speed weaving and knitting require a significant amount of electricity to operate the machines and process the yarn. The preliminary and finishing phases, which require a substantial amount of water, contribute significantly to the environmental impact of each method. Using efficient technologies and renewable energy sources lessens these effects. The following sections examine energy and resource consumption, emissions and waste generation, and the influence of manufacturing technology on the environmental footprint of fabric production.

7.8 ENERGY AND RESOURCE CONSUMPTION IN FABRIC MANUFACTURING

Fabric production methods vary considerably in their energy and resource demands. High-speed weaving and fine-gauge knitting processes require substantial electrical energy due to the continuous operation of machines and the increased demands for yarn processing. Electricity serves as the primary energy source for textile production. The energy is utilized for mechanical processes, lighting, and for regulating relative humidity and temperature.

Energy use is unevenly distributed across textile processing stages. The textile sector's total energy consumption is 35%, with yarn manufacturing accounting for 22%, wet processing (including pretreatments, dyeing, printing, and garment wet processes) accounting for 37%, and 6% for finishing treatments (Palamutcu 2010). Beyond energy use, textile manufacturing also generates dust, fiber fly, airborne particulates, and foreign matter. Advances in spinning, weaving, and knitting machinery have helped mitigate these issues by improving cleanliness and process control.

Conventional textile and garment manufacturing machinery predominantly utilizes fossil fuels, including petroleum and natural gas. These energy sources cannot be sustained continuously due to their scarcity and the waste they produce. Moreover, the combustion of these fuels generates various air pollutants that exacerbate environmental degradation (Vachon and Klassen 2008). This led to significant improvements in engine design, fuel enhancements, and the use of renewable fuels such as natural gas and liquefied petroleum gas (LPG), which reduce primary and secondary air pollutants. Furthermore, solar thermal energy can effectively replace fossil fuels (Hosseini et al. 2023).

Fabric manufacturing is an energy-intensive operation. The total electrical energy utilized per linear meter of fabric (encompassing preparation and production) is 0.45–0.55 kWh, whereas the thermal energy expended per linear meter of fabric (encompassing preparation and manufacturing) is 18.8–23 MJ. Additionally, the overall energy usage for manufacturing 1 kg of cloth is predicted to range from 1.7 to 4.2 kWh (Fletcher 2012).

Additionally, different loom motions use different amounts of energy. The three primary motions are shedding, picking, and beating; the secondary motions are let-off and take-up. Additionally, auxiliary motions, including stop motions, i.e., warp stop motion, weft stop motion, and warp protector motion, regulate fabric quality. Figure 7.5 shows the comparative energy consumption of various weaving processes (Koç and Çinçik 2010). Researchers found that primary motions consume most of the energy, while other actions utilize the remaining energy (Koç and Çinçik 2010). Among shuttle-less weaving machines, air-jet, rapier, and water-jet looms are the most commonly used for fabric production. Air-jet looms are considered more sustainable than other types due to their high WIR and low production costs. However, the warp for air-jet looms needs to be carefully prepared due to the maximum speed. Moreover, the use of compressed air in air-jet looms is energy-intensive.

The sustainability potential of alternative fabric-forming technologies is highlighted by industrial case studies. An Indian enterprise operating 400 air-jet weaving machines, which produces around 134,000 m^2 of textiles daily, can reduce its CO_2 emissions by approximately 55 tons/day by transitioning to warp knitting, based on India's 2019 CO_2 emissions rate of 725 g/kWh. Furthermore, reduced environmental

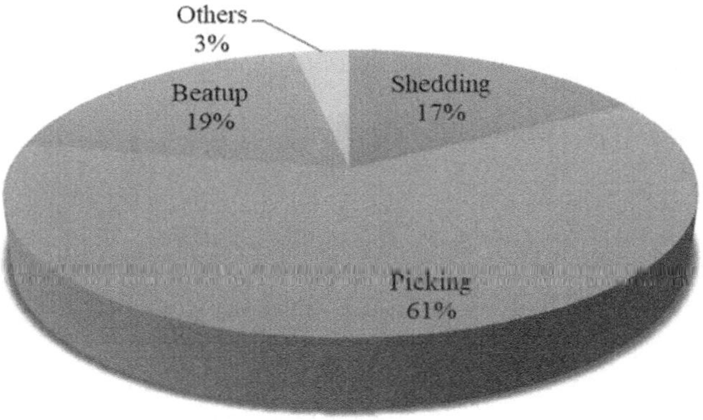

FIGURE 7.5 Distribution of energy consumption among loom motions.

***Source*: Koç and Çinçik 2010.**

impact and lower electricity expenses are achieved. The potential savings are around \$8,000/day at a cost of \$0.104/kWh by the end of 2021 (Textile World 2022).

Knitting processes further demonstrate significant variability in energy consumption across machine types and fabric constructions. According to Allwood et al. (2006), the energy needed for production alone is 24 MJ, while the total energy consumption for a T-shirt manufactured with circular knitting is 109 MJ. The energy required to manufacture with a flat-bed knitting machine ranges from 3 to 10 MJ. The efficiency and productivity of knitting machines can be enhanced by utilizing energy-efficient drive systems, maintaining consistency, and integrating electronic devices in place of mechanical alternatives, thereby promoting sustainability.

7.9 EMISSIONS, WASTE GENERATION, AND ENVIRONMENT POLLUTION

A variety of environmental pollutants, including solid waste, wastewater effluents, and airborne emissions, are generated throughout the fabric manufacturing process. Particularly in traditional weaving and finishing processes, fiber fly, lubricants, sizing agents, and chemical residues all contribute to air and water pollution. Landfill burdens are increased by rejected materials, waste fabric, and edge trimmings. To address these concerns, sustainable manufacturing practices prioritize cleaner production methods, waste minimization, textile waste recycling, and enhanced effluent treatment systems.

Energy consumption further intensifies the environmental footprint of fabric manufacturing, particularly during weaving preparatory processes such as sizing. Steam is typically utilized to warm the size liquor in the size box and heat several drying rollers. This thermal energy is generated at the boilers for steam production and conveyed to the sizing department via insulated pipes. Generally, fossil fuels such as petrol, diesel, and gas are employed to run the boiler. In addition to energy usage in various manufacturing processes, energy is also consumed for material handling, administrative tasks, and temperature regulation.

Alongside energy use, weaving preparatory processes also generate toxic compounds and waste from the finishing process, particularly during the formulation of conventional sizing. These formulations often require the use of harmful chemicals to prepare the effluents. Therefore, the adoption of an eco-friendly sizing agent, such as polyvinyl alcohol (PVA), has gained attention (Moll et al. 2009). The environmental impact becomes even more pronounced during chemical processing stages, where nearly 3,000 different chemicals and auxiliary agents are used for operations such as cleaning, bleaching, mercerization, dyeing, printing, and finishing. About 30% of the world's production of dyes, pigments, and auxiliary chemical compounds also comes from these procedures. Many of these substances contribute to the release of toxic effluents, including heavy metals and persistent organic compounds. Due to their harmful effects, several countries have imposed strict regulations or outright bans on specific chemicals. Programs and technologies, such as enzymes, can help reduce the use of harmful chemicals and resource use.

7.10 INFLUENCE OF MANUFACTURING TECHNOLOGY ON ENVIRONMENTAL FOOTPRINT

The manufacturing technology used to produce the cloth significantly impacts its overall environmental footprint. Compared to traditional technologies, shuttle-less looms, seamless knitting, and advanced nonwoven techniques frequently use less material and are more productive. Automating and monitoring processes digitally enhances control, reducing faults and resource losses. Utilizing environmentally friendly technologies helps to reduce carbon emissions and enhance the sustainability of fabric production. Traditional clothing manufacturing techniques, such as weaving, knitting, spreading, cutting, and stitching, can be replaced by 3D seamless garment production. This innovation leads to a 30% decrease in manufacturing duration, along with reductions in lead time, labor costs, and energy consumption. Additionally, waste from cutting out pattern pieces is minimized (Moll et al. 2009).

7.11 ADVANCEMENTS IN SUSTAINABLE FABRIC MANUFACTURING

Recent advancements in fabric manufacturing focus on reducing environmental impacts while maintaining high productivity and material performance. Innovations in machinery, process optimization, and digital technologies have enabled more efficient use of energy, water, and raw materials, supporting the transition toward sustainable and responsible textile production.

7.12 SUSTAINABLE WEAVING TECHNOLOGIES

The latest advancements in sustainable weaving focus on reducing energy and water consumption and lowering emissions through shuttle-less looms, process optimization, and Industry 4.0 integration. These technologies increase productivity while reducing resource use per meter of fabric and enhancing overall environmental performance.

7.12.1 Shift from Shuttle to Shuttle-Less Looms

Modern sustainable weaving has transitioned from shuttle looms to shuttle-less looms, such as rapier, air-jet, and water-jet looms, which offer higher productivity and lower specific energy consumption. Modern weaving technologies utilize less energy and offer greater efficiency than traditional weaving machinery, while producing less waste. Therefore, adopting advanced weaving technology is a crucial step in the pursuit of sustainability (Pattanayak 2020).

Rapier looms, driven by electric motors and mechanical components, are reported to be more energy-efficient than air-jet looms, which rely on compressed air, making them attractive for mills targeting reduced electricity consumption per unit of output. Air-jet looms eliminate water use during weft insertion and operate in a completely dry environment, thus conserving freshwater and avoiding wastewater generation from the weaving stage. Design improvements, such as variable-frequency drive compressors, optimized air-pressure settings, and smart controls, have significantly reduced air consumption and idle energy losses in air-jet systems. Water-jet looms, primarily used for hydrophobic synthetic yarns, can consume up to 50% less energy than air-jet looms for equivalent production, as pump-driven water jets are more efficient than compressed air systems. However, this advantage must be balanced against water management requirements. Newer water-jet machines incorporate closed-loop circulation and filtration to recirculate noncontact water, reducing water withdrawal and aligning with sustainability criteria when adequate water infrastructure is available (Pattanayak 2020).

7.12.2 Energy Efficiency and Emission Reduction

Weaving machines consume 50%–60% of the total energy consumption in a weaving plant, whereas the remaining energy is used for humidification, compressors and lighting. The extent of this energy use varies depending on the type of loom and the weft insertion system.

Energy consumption in weaving can be minimized by enhancing process efficiency through routine maintenance, cleaning, and the implementation of efficient drive systems. Minimizing the waste yarn produced during the weaving process can enhance sustainability. The weaving process can be rendered more sustainable by optimizing drive systems and enhancing productivity. Measures implemented during the weaving process to diminish energy usage can enhance process efficiency and productivity, thereby contributing to sustainability (Pattanayak 2020). To improve the efficiency of weaving processes, a few measures have been proposed (Hasanbeigi and Price 2012):

- The loom utilization rate should exceed 90%. A 10% decrease in loom machine utilization will result in a 3%–4% rise in specific energy usage.
- Replace the loom's electric motor with an energy-efficient one.
- The energy consumption per unit of product can be significantly affected by the type of weaving machine used. Therefore, energy efficiency should be considered when purchasing new looms. However, not all looms can produce all fabrics, and some are limited to specific types. Thus, we cannot recommend a loom without analyzing each condition.
- Warp and weft yarn quality impacts weaving productivity and efficiency. Thus, employing higher-quality yarns, which may be more expensive, may reduce yarn breakage and stoppage during weaving and ultimately prove cheaper in the long run.

7.12.3 Digitalization, Industry 4.0, and Smart Weaving

The design, production, and use of textiles are being revolutionized by technologies such as artificial intelligence (AI) and the Internet of Things (IoT). The integration of automation and robotics represents a significant technological advancement in the textile industry. Traditional manual labor is being replaced by automated machinery and robotics, improving production efficiency and reducing errors. Computerized weaving machines can now create complicated designs with minimum human involvement. Moreover, AI-driven quality control systems can identify fabric problems, thereby guaranteeing elevated production standards (Middle East Africa

Textile News 2025). Researchers have demonstrated that integrating IoT into textile manufacturing can reduce downtime by 15% and increase energy efficiency by 20%. IoT-enabled monitoring solutions enhance uptime and reduce waste by detecting machine issues early (Sabareesh et al. 2024).

7.13 SUSTAINABLE KNITTING TECHNOLOGY

Unlike conventional weaving, which requires multiple processes and machines, a single knitting machine can produce an entire fabric, making it more resource-efficient and a sustainable alternative. Studies have shown that weaving looms consumes more energy and produces more CO_2. The innovation of knitting technologies is discussed below (Maiti et al. 2022).

7.13.1 Seamless and Three-Dimensional Knitting Technologies

Three-dimensional knitting technologies are also transforming the textile production industry. Furthermore, integrating digital design and knitting systems enables precise patterning and customization, reducing the need for excessive dyeing and chemical treatments. Similarly, 3D knitting machines create seamless, zero-waste garments, boosting efficiency and sustainability (Middle East Africa Textile News 2025). The development of 3D and 4D knitting technologies is essential to creating affordable, sustainable clothing by reducing waste, transportation emissions, and post-knitting processes. These methods eliminate the cut-and-sew stages, using only the necessary amount of yarn and avoiding leftover fabric. By enabling on-demand production and reducing overproduction and excess inventory, they enhance sustainability. Additionally, they streamline manufacturing by eliminating multiple steps and minimizing reliance on manual labor, leading to greater efficiency and less waste from unsold items. As a result, these technologies are more environmentally friendly than traditional manufacturing, lowering energy consumption and carbon emissions while promoting sustainable fashion without sacrificing functionality.

7.13.2 Sustainable Advances in Circular Knitting

Recent advancements include circular weft-knitting machines specifically engineered to produce fabric for cut-and-sew shoe production. Contemporary circular knitting machines are equipped with computerized control systems that enable rapid style alterations and stitch transfer. These technological advancements help minimize material waste while also improving efficiency, versatility, and accuracy. The Italian company Santoni remains at the forefront of seamless knitting technology, particularly in the manufacturing of intimate apparel, beachwear, and fine-gauge sportswear, all of which require fewer seams (Knitting Industry 2024).

7.13.3 Digitalized and Computerized Flat Knitting

Digital knitting has transformed production by enhancing accuracy and sustainability. Utilizing computer-aided design (CAD) and sophisticated knitting software, intricate designs can be created efficiently while minimizing waste. Digital knitting enables seamless garment manufacturing, reducing material waste by up to 30% compared to conventional techniques. This supports sustainable fashion objectives and helps reduce the environmental footprint of clothing production (Grumpy Ginger Yarn Co 2024).

Computerized flat knitting machines use servomotors, electromagnets, and needle-selection mechanisms to perform knitting tasks. These advanced machines control individual needles to create complex textures and designs with minimal human input, enhancing customization through digital programming. Design data is prepared on a computer, converted to specific formats, and transferred to the machine via USB drives for processing and control of knitting operations. This digital control enables precise manipulation of yarns, allowing the production of tailored clothing and fabrics with varied stitch patterns and colors, while reducing waste more effectively than circular knitting. Modern flat knitting machines feature multi-gauge systems, split cam-carriage systems, and electro-optical tension control, enabling the creation of versatile patterns with greater accuracy (Shima Seiki 2026).

7.13.4 Specialized Warp Knitting

Techniques in specialized warp knitting, including crochet knitting, circular warp knitting, and multiaxial knitting, expand the range of warp-knitted fabric structures and improve functionality. For instance, crochet machines can create lace, ribbons, and technical textiles using electronically controlled guide bars. Circular warp knitting enables the production of seamless, tubular fabrics, making it particularly suitable for hosiery and medical textiles. Multiaxial warp knitting incorporates reinforcement yarns in multiple directions, boosting the strength and stability of fabrics for technical applications such as composites and protective textiles. With the adoption of computerized controls and automation, specialized warp knitting is advancing continuously, offering creative solutions for both the fashion and industrial sectors (Woolmark 2026; Hahn et al. 2023).

7.14 SUSTAINABLE NONWOVEN

The nonwoven production industry is increasingly embracing eco-friendly practices to meet environmental regulations and consumer expectations. Developments in nonwoven materials include the emergence of biodegradable alternatives and the incorporation of recycled fibers to reduce environmental impacts.

7.14.1 Use of Recycled Fibers

Manufacturers frequently use recycled materials to produce nonwoven fabrics, thereby considerably reducing the environmental impact of sourcing virgin fiber. By using recycled fibers in their products, producers reduce their environmental impact and encourage resource conservation. Nonwoven manufacturers improve resource efficiency and environmental sustainability by emphasizing recycled fibers (Ningbo 2026). Although the use of recycled polyester (r-PET) is a major advancement in sustainability, further processing issues may arise. Moisture and contaminants could affect filament quality; therefore, more rigorous drying and filtering may be required before extrusion. The energy equation gains an additional layer from this preprocessing. According to machinery vendors, contemporary r-PET lines are specifically designed to effectively address these issues by using cutting-edge drying and filtration systems optimized for minimal energy impact (Ningbo 2026).

7.14.2 Energy-Efficient Technologies

In nonwoven manufacturing, cutting-edge technologies are being used to reduce energy use in production. Energy-efficient manufacturing techniques that drastically reduce carbon emissions and resource consumption are the goal of innovations like the Energy First project. These solutions support global sustainability objectives while improving the efficiency of nonwoven production (Table 7.2). Adopting energy-efficient technologies increases the economic feasibility of production processes while also lessening the environmental impact of nonwoven manufacturing (Al Nonwoven 2025).

TABLE 7.2
Energy-Efficient Technologies in Nonwoven Manufacturing

System Area	Key Issue/Energy Challenge	Improvement Measures	Expected Benefits
Extrusion system	Largest energy consumer in nonwoven manufacturing; heat loss from poorly insulated extruder barrels; worn screw and barrel components increase friction and motor load; use of outdated DC or fixed-speed motors	Install high-quality removable insulation jackets; replace worn screws and barrels with modern designs; upgrade to high-efficiency AC motors with variable-frequency drives	Up to 50% reduction in heater energy use; lower motor power demand; 20%–40% savings in motor energy; improved process stability; payback often within 1 year
Thermal bonding systems	High heat loss from hot oil circulation systems due to poor insulation; inefficient temperature control	Improve insulation and maintenance; adopt zones or induction heating where feasible; better temperature regulation	Reduced thermal energy losses; improved operational efficiency and energy control
Drying systems (recycled PET processing)	Drying stage is highly energy-intensive; older desiccant dryers consume excessive energy; risk of over-drying	Replace conventional dryers with infrared or vacuum drying systems; integrate real-time moisture sensors for closed-loop control	Lower energy consumption; precise moisture control; avoidance of unnecessary over-drying
Compressed air systems	High energy uses due to excessive pressure and inefficient operation	Reduce system pressure; use variable-speed compressors to match demand	Significant electricity savings; improved system efficiency

(Continued)

TABLE 7.2 (*Continued*)
Energy-Efficient Technologies in Nonwoven Manufacturing

System Area	Key Issue/Energy Challenge	Improvement Measures	Expected Benefits
Cooling/chiller systems	Major electrical load; inefficiency from dirty condensers and low chilled water temperatures	Regular condenser cleaning; maximize chilled water temperature where possible	Reduced power consumption and improved cooling efficiency
Pneumatic conveying systems	Energy losses due to long pipelines and sharp bends	Optimize layout by minimizing pipe length and bends	Lower conveying energy requirements and better overall sustainability

7.15 SUSTAINABLE BRAIDING

Braiding is naturally more efficient in terms of material use than many other textile methods because of its near-net-shape manufacturing capabilities, which help reduce waste. Additional sustainability enhancements can be achieved through the design of energy-efficient machinery and the optimization of processes. Contemporary braiding machines that feature servo-driven systems, enhanced carrier paths, and intelligent tension management use less energy while maintaining consistent fabric quality. The adoption of digital control systems and automation facilitates accurate regulation of yarn tension, carrier speed, and take-up rates, minimizing yarn breakage and the need for reprocessing. Furthermore, optimizing braid angle and structural design during simulation helps eliminate trial-and-error manufacturing, thereby reducing material and energy use (Emonts et al. 2021; Mishra 2023).

7.16 OTHER SUSTAINABLE APPROACHES

Alternative sustainable methods in fabric manufacturing emphasize eco-friendly materials and cleaner processing technologies, aiming to minimize environmental impact. These innovations significantly boost sustainability within the industry. The adoption of sustainable fibers, such as organic cotton, hemp, bamboo, and recycled fibers, helps lower resource consumption and reduce environmental pollution. Organic cotton reduces the use of harmful agrochemicals, while hemp and bamboo offer fast-growing, renewable alternatives with strong mechanical and antimicrobial properties. Recycled fibers further contribute by minimizing waste and reducing dependence on virgin raw materials (Guerena and Sullivan 2003). In parallel, innovations in dyeing and finishing processes play a vital role in sustainability by reducing water, energy, and chemical usage. Technologies such as low and ultralow liquor ratio dyeing, continuous and combined processing, waterless and low-water dyeing methods, and enzyme-based treatments significantly decrease environmental burden. Additionally, advancements like digital inkjet printing, CO_2 dyeing, and energy-efficient heating systems enhance resource efficiency and reduce emissions, making textile processing more sustainable overall (Rana et al. 2015).

7.17 EMERGING SUSTAINABLE TECHNOLOGIES

Recent advancements in textile machinery reflect a strong shift toward sustainable manufacturing, where energy efficiency, waste reduction, and high productivity are achieved simultaneously. Modern weaving, knitting, and nonwoven technologies increasingly integrate resource-efficient designs and digital optimization to minimize environmental impact without compromising fabric quality or production performance. The following examples illustrate key sustainable machines and processes shaping next-generation textile manufacturing.

- The Phoenix projectile weaving machine embodies the principle of sustainability without any sacrifices. According to the company, this innovative machine has set a new performance record among projectile weaving machines, boasting a WIR of 1,600 m/min. Additionally, with a consumption of 6.5 kW/h, it is considered the most energy-efficient model in the entire spectrum of weaving systems (IQ-SPS GmbH 2023).
- The R9500EVO rapier weaving machine is an ideal machine for producing a complete variety of technical textiles, including those made from the finest monofilament and high-tenacity multifilament yarns, as well as fabrics with multiple pick insertions. The R9500EVO can be equipped with the Itema iSAVER®tech, which removes waste selvedge on the left side of the fabric, significantly reducing raw material waste and resource consumption, resulting in cost savings and enhancing sustainable weaving practices. The iSAVER®tech is compatible with a wide selection of technical yarns and can be retrofitted as an upgrade kit on all models in the R9500 series (Itema 2022).
- Wellknit Textile Technology Co., Ltd. has established itself as a front-runner by creating circular knitting machines that merge precision engineering with innovative energy-efficient functionalities. Wellknit's offerings allow manufacturers to optimize energy usage, reduce waste, and improve overall production efficiency, supporting the textile sector in its transition to a more sustainable future while maintaining high performance (Wellknit 2026).
- The digital stitch control system utilized in Shima Seiki machines accurately calculates the yarn needed for each stitch, effectively reducing yarn waste. Additional advantages of seamless knitting include the ability for mass customization, production on demand, enhanced trim-ability for finished edges, and greater consistency in product quality. In contrast to traditional V-bed flat knitting machines, Santoni® circular knitting machines consume less energy and produce garments with fewer seams. These machines also have faster turnaround times, which boosts their sustainability and efficiency. As a result, due to their benefits for sustainability and production speed, the market for seamless clothing, particularly undergarments like bras and briefs, is expected to grow significantly (Ahmed et al. 2025).

- The innovative Lenzing Web Technology is a process for forming nonwoven webs that begin with wood pulp and generates a fabric composed entirely of continuous Lyocell filaments. This web can be combined with conventional nonthermal nonwoven bonding methods, including hydroentangling and needle punching. Furthermore, the patented technology introduces a distinctive self-bonding process that allows filaments to adhere into a fabric during the laying process. This self-bonding feature facilitates a product lineup that offers a significantly broader range of surface textures, drapeability, and dimensional stability compared to other nonwoven technologies that utilize 100% cellulosic fibers (Textile Network 2018).
- A.Celli's E-Wind Winders and Slitter Rewinders offer precise tension control, reducing material waste by 15%. Their cloud-based Industry 4.0 solutions enable real-time monitoring, improving energy efficiency by 20%. The Smartlogistix brand distinguishes itself in the market with modular automated packaging systems and automated guided vehicles designed for efficient material handling. Andritz's neXline spunlace systems deliver high-speed production (up to 25,000 tons annually) while employing energy-efficient hydroentanglement that cuts down water usage by 30%. The Wetlace™ carded-pulp (CP) process combines wet-laid and spunlace technologies to create flushable wipes, offering a distinctive solution for biodegradable nonwovens (Textile School 2025; ANDRITZ 2022).

7.18 POLICY, REGULATIONS, AND INDUSTRY STANDARDS FOR SUSTAINABLE TEXTILES

Sustainability in fabric manufacturing is regulated by evolving policies, regulations, and standards that prioritize circularity, chemical safety, and supply chain transparency. These policies ensure environmental compliance while promoting the adoption of cleaner and more efficient fabric manufacturing technologies.

7.18.1 European Union Regulations

The EU's circular-economy plan aims to improve recycling and sustainable design in textiles, introducing REACH legislation to limit hazardous chemicals and an eco-label for products. QR codes on care labels can help consumers access repair and care instructions, highlighting the environmental impact of washing and drying textiles (European Commission 2020).

7.18.2 United States Regulations

The U.S. government is encouraging textile producers to take responsibility for end-of-life products through initiatives like California's SB 707. This requires producers to support recycling and reuse efforts, shift waste management from local governments to industry, and promote sustainability (California Legislative Information 2024).

7.18.3 Indian Regulations

The Environmental Protection Act 1986 of India sets standards for emissions, effluents, and waste management in textile processing, supporting cleaner dyeing, finishing, and solid waste practices. The zero liquid discharge law requires the textile sector to process and recycle wastewater at a zero-discharge level. They promote the utilization of organic cotton and natural dyes for sustainable textile and fashion items. The Bureau of Indian Standards and eco-labeling initiatives, aligned with ISO 14024, promote reduced water and chemical use, as well as increased use of organic or low-impact fibers (Central Pollution Control Board 2024).

7.18.4 Bangladesh and China Regulations

Bangladesh has the most certified green textile factories, while China promotes green manufacturing and recycling in its textile sector.

7.19 GLOBAL INDUSTRY TRENDS

The textile industry supports the use of organic natural fibers and recycled synthetic fibers to achieve net-zero emissions by 2050, in addition to national environmental protection laws. The majority of fashion firms are adopting reusable materials and ecological techniques for their textile and clothing products (Hui 2025). Several internationally recognized benchmarks have been established to raise awareness about the three pillars of sustainability.

- GOTS ensures sustainable practices in processing organic fibers, mandating organic materials, limiting harmful chemicals, and enforcing fair labor. It offers two certification levels: Organic (95% fibers) and Made with Organic (70%).
- OEKO-TEX® Standard 100 certifies textiles based on the presence of harmful substances, evaluating everything from raw materials to finished products. All components, including fabrics and accessories, must be tested for safety with no exception.
- Eco-labels verify the environmental performance of products, encouraging sustainable choices and reducing harmful practices. They comply with ISO standards and help consumers support environmentally friendly businesses (Rahaman et al. 2024).

7.20 CONSUMER AWARENESS AND MARKET DEMAND FOR SUSTAINABLE FABRICS

The textile sector significantly contributes to environmental problems, such as water pollution, greenhouse gas emissions, and textile waste. A significant number of customers remain oblivious to these repercussions, and this ignorance can lead to unsustainable purchasing behaviors. By increasing awareness, we can empower individuals to understand the true cost of fast fashion and encourage a shift toward more mindful consumption. When consumers are cognizant of the difficulties, they can advocate for improved practices from brands. This demand can catalyze transformation within the sector, compelling companies to embrace more sustainable materials, production techniques, and labor standards. Furthermore, consumer awareness can enhance the concept of circularity in fashion. When customers understand the importance of recycling, reusing, and repairing garments, they can help reduce textile waste. This entails exercising greater mindfulness in purchasing decisions and selecting durable, high-quality things (Sustainability Directory 2025a). Mature consumers can be educated to purchase only what is necessary, with an emphasis on recycling and the sustainability of the items they desire. The second issue involves improving transparency and environmental labeling for all consumers. It is advisable to provide consumers with information on the environmental impact of their textile and fashion items, including carbon dioxide emissions and water use, on product labels (Hui 2025).

According to a 2023 Nielsen survey, 41% of Indian customers are willing to spend more for eco-friendly textile products, and 73% of them are aware of environmental problems (Celestin et al. 2024). Sustainable shopping in India is largely about affordability, given rising prices and limited selection, especially at major retailers. When it comes to buying organic products, customer purchasing behavior falls under the category of "green consumer attitude." In major urban centers of the country, including Mumbai, Delhi, and Bangalore, a survey revealed that merely 26% of individuals are inclined to pay a premium for sustainable items in 2023 (Minhas 2025). The market for sustainable textiles was estimated at $37.26 billion in 2025 and is projected to reach $115.72 billion by 2035. Consumer awareness of the ethical aspects of textile production is driving a significant shift in the sustainable fabrics industry. Millennials and Generation Z have begun to exhibit shifts, demonstrating alterations in purchasing behaviors, favoring products that incorporate sustainability (Research Nester 2025).

7.21 CHALLENGES AND FUTURE SCOPE

Sustainable fabric manufacturing faces several challenges, including high production costs associated with the use of eco-friendly raw materials and cleaner processing technologies, limited availability of organic or recycled fibers, and complex global supply chains that hinder transparency and traceability. Technological barriers, such as the need for energy-efficient machinery and waterless dyeing systems, along with inconsistent regulations and certification standards, further hinder the adoption of sustainable textiles, while consumer awareness and willingness to pay for them remain variable.

Despite these challenges, the future of sustainable fabric manufacturing is promising, driven by advances in bio-based and recycled materials, circular-economy practices that promote recycling and upcycling, and the integration of Industry 4.0 technologies, such as AI and IoT, for energy and resource optimization. Innovations in water and energy management, combined with strengthened global policies, eco-labeling, and collaborative research, can significantly reduce environmental impact, enhance transparency, and align consumer demand with environmentally responsible production, ultimately shaping a more sustainable and resource-efficient textile industry.

7.22 CONCLUSION

Fabric manufacturing transforms fibers into functional textiles through an energy- and resource-intensive process. Weaving generally consumes more energy than knitting. These techniques, including weaving, knitting, and nonwovens, impose significant environmental burdens due to high energy consumption (up to 4.2 kWh/kg), water usage, chemical effluents, and emissions. Recent technological advancements—such as shuttle-less looms, fine-gauge and seamless knitting, 3D knitting, and Industry 4.0-enabled smart monitoring—are enhancing efficiency, reducing waste, and lowering environmental impact, often achieving up to 50% energy savings compared to conventional methods.

Sustainability in fabric manufacturing is further reinforced by regulatory frameworks, including the EU REACH, India's Zero Liquid Discharge policies, and certifications such as GOTS and OEKO-TEX, alongside a growing consumer willingness to pay premiums for eco-friendly products. The adoption of sustainable fibers, such as r-PET, organic cotton, hemp, and bamboo, is accelerating market growth, with projections estimating the sustainable fabrics sector to reach $115.72 billion by 2035. These combined technological, regulatory, and market-driven trends signal a decisive shift toward more environmentally responsible and resource-efficient textile production. Ultimately, producing textile fabrics from eco-friendly raw materials, renewable energy, and biodegradable chemicals ensures their role as sustainable materials in advancing the journey toward sustainable fashion.

REFERENCES

Ahmed, D. M., Mobin, A., Pritha, N. M., Ador, A. I., Ahmed, S. and Ahmed, Md. R. 2025. Advanced knitting technology for sustainable fashion. In *SDG 12 and Global Fashion Textiles Production* (Eds. Sayem, A. S. M. and Goswami, P.), Singapore: Springer Nature, 123–139.

Al Nonwoven. 2025. A proven 5-step guide to reducing power consumption per ton nonwoven fabric in 2025. https://www.alnonwoven.com/a-proven-5-step-guide-to-reducing-power-consumption-per-ton-nonwoven-fabric-in-2025/ [Accessed 26 December 2025].

Allwood, J., Laursen, S., Rodríguez, C. and Bocken, N. 2006. *Well dressed? The Present and Future Sustainability of Clothing and Textiles in the United Kingdom.* Cambridge: University of Cambridge, Institute of Manufacturing.

ANDRITZ. 2022. ANDRITZ neXline spunlace eXcelle. https://www.andritz.com/products-en/nonwoven-textile/spunlace/excelle-range/spunlace-excelle-range [Accessed 10 February 2026].

California Legislative Information. 2024. SB 707 Responsible Textile Recovery Act of 2024 https://leginfo.legislature.ca.gov/faces/billNavClient.xhtml?bill_id=202320240SB707 [Accessed 10 February 2026].

Castelli, G. S., Maietta, G., Sigrisi and Slaviero, I. M. 2010. *Reference Books of Textile Technology: Weaving.* Vol. 1. Milano: Fondazione ACIMIT Italian Association of Textile Machinery Producers Moral Body via Tevere.

Celestin, M., Vasuki, M., Sujatha, S. and Kumar, A. D. 2024. Implementing green technologies to reduce environmental impact: Economic and competitive advantages of ecofriendly practices. *International Journal of Scientific Research and Modern Education* 2(9): 33–39.

Central Pollution Control Board. 2024. The Environment (Protection) Act, 1986, amended 1991. https://www.understandupsc.com/environment-protection-act-1986/ [Accessed 10 February 2026].

Emonts, C., Grigat, N., Merkord, F., Vollbrecht, B., Idrissi, A., Sackmann, J. and Gries, T. 2021. Innovation in 3D braiding technology and its applications. *Textiles* 1: 185–205.

European Commission. 2020. Circular economy, The EU aims to transition to a circular economy for a cleaner and more competitive Europe. https://environment.ec.europa.eu/strategy/circular-economy_en [Accessed 10 August 2025].

Fletcher, K. 2012. *Sustainable Fashion and Textiles: Design Journeys*. London: Routledge.

Goswami, B. C., Anandjiwala, R. D. and Hall, D. 2004. *Textile Sizing*. Boca Raton, FL: CRC Press.

Grumpy Ginger Yarn Co. 2024. The future of knitting: Technology and innovation. https://www.grumpyginger.com.au/blogs/news/the-future-of-knitting-technology-and-innovation?srsltid=AfmBOopy_MYQ53gE3fhpahyk5bJWuJE_ps9p3RVEu0KuSn1zQhoQQBG6 [Accessed 26 December 2025].

Guerena, M. and Sullivan, P. 2003. Organic cotton production. *Appropriate Technology Transfer for Rural Areas (ATTRA)*, National Center for Appropriate Technology (NCAT).

Hahn, L., Zierold, K., Golla, A., Friese, D. and Rittner, S. 2023. 3D textiles based on warp knitted fabrics: A review. *Materials* 16: 3680.

Hasanbeigi, A. and Price, L. 2012. A review of energy use and energy efficiency technologies for the textile industry. *Renewable and Sustainable Energy Reviews* 16(6): 3648–3665.

Hosseini, D., Shahabaddin, S., Dehshiri, S. J. H. and Firoozabadi, B. 2023. Evaluation of using solar energy in Iran's textile industry towards cleaner production: Sustainable planning and feasibility analysis. *Journal of Cleaner Production* 421: 138447.

Hui, C.-l. Ed. 2025. Introductory chapter: Textile Industry and the Environment: Challenges, recent development and future. In *Textile Industry and the Environment: Challenges, Recent Development and Future*, London: IntechOpen.

IQ-SPS Gmb H. 2023. New weaving machines. https://www.iq-sps.com/ [Accessed 04 March 2026].

Itema. 2022. The Itema iSAVER® Range: The future of sustainable weaving. https://www.itemagroup.com/isaver-range/ [Accessed 04 March 2026].

Knitting Industry. 2024. Hosiery/seamless: The continuing evolution of Santoni seamless technology. https://www.knittingindustry.com/the-continuing-evolution-of-santoni-seamless-technology/ [Accessed 20 December 2025].

Koç, E. and Çinçik, E. 2010. Analysis of energy consumption in woven fabric productions. *Fibers and Textiles in Eastern Europe* 18(2): 14–20.

Lau, F. and Yu, W. 2016. Seamless knitting of intimate apparel. In *Advances in Women's Intimate Apparel Technology* (Ed. Yu, W.), Cambridge: Woodhead Publishing, 55–68.

Maiti, S., Maity, S., Pandit, P., Maulik, S. R. and Singha, K. 2022. Sustainability analysis for knitting process and products. In *Advanced Knitting Technology* (Eds. Maiti, S., Rana, S., Pandit, P. and Singha, K.), Cambridge: Woodhead Publishing, 657–671.

Middle East Africa Textile News. 2025. Latest technologies in the textile industry 2025. *Kohan Textile Journal*. 2025. https://kohantextilejournal.com/latest-technologies-in-the-textile-industry/ [Accessed 26 December 2025].

Minhas, A. 2025. India: Readiness to pay for sustainable products by age. *Statista*. https://www.statista.com/statistics/1320631/india-consumer-readiness-to-pay-for-sustainable-products-by-age/?srsltid=AfmBOoq536S3LBJUx5caqr_34F1cjxeeyrdQgQCQslXpTD8ZmLOFwytR. [Accessed 20 December 2025].

Mishra, R. 2023. *Advances in Textile Structural Composites*. Basel: MDPI.

Moll, P., Schütte, U., Zöll, K., Molfino, R., Carca, E., Zoppi, M., Bonsignorio, F., Callegari, M., Gabrielli, A., Principi, M., Montorsi, R., Eberhardt, C. P. and Engel, F. 2009. Automated garment assembly and manufacturing simulation. In *Transforming Clothing Production into a Demand-Driven, Knowledge-Based, High-Tech Industry: The Leapfrog Paradigm* (Eds. Walter, L., Kartsounis, G.-A. and Carosio, S.), London: Springer, 9–59.

Naqvi, S. L. H., Nadeem, M., Ayub, F., Yasar, A., Naqvi, S. H. Z. and Tanveer, R. 2024. Social and environmental impacts in textile production. In *Dye Pollution from Textile Industry: Challenges and Opportunities for Sustainable Development* (Ed. Singh, P.), Singapore: Springer Nature, 423–453.

Nayak, R. and Padhye, R., Eds. 2015. Introduction: The apparel industry. In *Garment Manufacturing Technology*, Cambridge: Woodhead Publishing, 1–17.

Ningbo, M. H. 2026. Top nonwoven manufacturing processes explained. https://www.mh-chine.com/blog/products/nonwoven-manufacturing [Accessed 22 January 2026].

Palamutcu, S. 2010. Electric energy consumption in the cotton textile processing stages. *Energy* 35(7): 2945–2952.

Pattanayak, A. K. 2020. Sustainability in fabric manufacturing. In *Sustainable Technologies for Fashion and Textiles* (Ed. Nayak, R.), Cambridge: Woodhead Publishing, 57–72.

Rahaman, Md. T., Pranta, A. D., Repon, Md. R., Ahmed, Md. S. and Islam, T. 2024. Green production and consumption of textiles and apparel: Importance, fabrication, challenges and future prospects. *Journal of Open Innovation: Technology, Market, and Complexity* 10(2): 100280.

Rana, S., Pichandi, S., Karunamoorthy, S., Bhattacharyya, A., Parveen, S. and Fangueiro, R. 2015. Carbon footprint of textile and clothing products. In *Handbook of Sustainable Apparel Production* (Ed. Muthu, S. S.), Boca Raton, FL: CRC Press, 141–166.

Research Nester. 2025. Sustainable fabrics market trends, size and forecast to 2035. https://www.researchnester.com/reports/sustainable-fabrics-market/6818 [Accessed 23 October 2025].

Sabareesh, N., Karthik, S. and Nagarajan, L. 2024. Impact of Industry 4.0 on textile production and supply chain. *International Journal of Engineering Research and Technology* 13(11): 1–7.

Shima Seiki. 2026. Flat knitting machine. https://www.shimaseiki.com/product/knit/machine/ [Accessed 06 March 2025].

Sustainability Directory. 2025a. Why is consumer awareness important for sustainable textile adoption? https://fashion.sustainability-directory.com/question/why-is-consumer-awareness-important-for-sustainable-textile-adoption/ [Accessed 26 December 2025].

Sustainability Directory. 2025b. Textile energy consumption. https://energy.sustainability-directory.com/term/textile-energy-consumption/ [Accessed 25 November 2025].

Textile Exchange. 2023. Material market report. https://textileexchange.org/app/uploads/2023/11/Materials-Market-Report-2023 [Accessed 20 December 2025].

Textile Network. 2018. Sustainability in the nonwoven industry. https://textile-network.com/en/Technical-Textiles/Textile-Flaechen/Sustainability-in-the-Nonwoven-Industry [Accessed 22 October 2025].

Textile School. 2025. Sustainable textile innovation: Top nonwoven machinery manufacturers in the industry. https://www.textileschool.com/30772/sustainable-textile-innovation-top-nonwoven-machinery-manufacturers-in-the-industry/ [Accessed 23 December 2025].

Textile World. 2022. Independent study by Gherzi shows: Warp knitting consumes significantly less energy than weaving. https://www.textileworld.com/textile-world/knitting-apparel/2022/10/independent-study-by-gherzi-shows-warp-knitting-consumes-significantly-less-energy-than-weaving/ [Accessed 26 December 2025].

Vachon, S. and Klassen, R. D. 2008. Environmental management and manufacturing performance: The role of collaboration in the supply chain. *International Journal of Production Economics* 111(2): 299–315.

Wellknit. 2026. Explore the advantages of Wellknit circular knitting machines for modern textile production. https://knittingmachine.com.tw/Explore-The-Advantages-of-Wellknit-Circular-Knitting-Machines-for-Modern-Textile-Production-id44411916.html [Accessed 25 January 2026].

Woolmark. 2026. Warp Knitting Technology. https://www.woolmark.com/industry/product-development/processing-innovations/warp-knitting-technology2/ [Accessed 05 March 2026].

8 Sustainable Garment Manufacturing Processes

Manoj Kumar Paras, Antaleena Banerjee, and Trisha Kumari

8.1 INTRODUCTION

The clothing and textile industry has been one of the largest contributors to the global economy, as it provides jobs and facilitates international trade. With the increasing demand for clothing, the industry has evolved from traditional handloom weaving to factory production. However, this rapid growth has resulted in major environmental and social issues. The conventional manufacturing methods are water- and energy-intensive, with high chemical usage and significant waste generation, resulting in resource depletion and harming the environment and workers (Gereffi and Frederick 2010; Dissanayake 2026).

The increasing awareness of climate change, environmental issues, labor concerns, and resource depletion has put the clothing industry in the spotlight. Consequently, there is a growing need for transparency, accountability, and responsible manufacturing. The clothing industry is increasingly adopting sustainable and lean manufacturing, which focuses on minimizing waste, maximizing resource utilization, and improving social responsibility while ensuring productivity and profitability (Allwood et al. 2006).

This chapter explores the major innovations in clothing manufacturing that can meet these challenges. The chapter starts with an introduction to conventional clothing manufacturing and its sustainability issues, followed by a description of the major clothing manufacturing centers. Then, the chapter continues with the exploration of the new innovations in clothing manufacturing, such as sustainable production, green processing, and sustainable materials. Finally, it concludes with the opportunities, challenges, and future outlook of sustainable clothing manufacturing.

8.2 GARMENT MANUFACTURING PROCESS

The garment and apparel industry is a significant contributor to employment, export revenue, and technological innovation in the global economy. With several different aspects of the industry within the total supply chain, the industry's value chain extends all the way from fibers through production, design, sourcing, logistics, and retailing. The value of the global textile and apparel market at this time will most likely exceed several billion dollars, while the textile and apparel sectors continue to be at a strong and robust stage within global trade. China, Bangladesh, Vietnam,

DOI: 10.1201/9781003689225-8

and India have been known to dominate garment exports because they have the economics of scale, labor capacity, and are cost-competitive (Gereffi and Frederick 2010; Allwood et al. 2006).

While the garment and apparel sector has historically been a primary driver of industrial growth, it has become a focal point for the sustainability movement. Traditionally manufactured goods require the consumption of tremendous amounts of natural resources (in particular, the consumption of water and energy), while relying heavily on fossil fuel inputs, and producing large amounts of waste throughout their entire production process – from cotton cultivation through dyeing and finishing. The need for a sustainable transformation has necessitated that stakeholders rethink traditional processes to optimize current methodologies, and integrate modern, lean, and eco-efficient practices into all levels of the industry (Niinimäki et al. 2020; Chen et al. 2024).

8.2.1 Traditional Garment Manufacturing Process

Garment manufacturing consists of a series of steps that convert fabrics into everyday fashion. In the past, most garment manufacturers arranged their operations around high production volume, maximizing profitability and production cycle time at the expense of considering environmental/societal impacts. The historical framework of the traditional garment manufacturing industry assists in creating a clear understanding for the introduction of sustainable and "lean" manufacturing approaches (Shah and Ward 2007; Dissanayake and Sinha 2012).

8.2.1.1 Design and Develop Products

Design is the first phase in the manufacturing model. The designer will develop new ideas for products based on the current fashion trend, market demand, and customer requirements, for example, through sketches, fabric selections, color selection, pattern creation, sample development, etc. However, traditional approaches do not adequately consider sustainability and its components, and designers may not integrate eco-design principles or focus on producing attractive, sustainable products; therefore, each year of creating a new product, designers create excess fabric, which gets discarded and has no opportunity to be recycled (Fletcher 2014; Rissanen and McQuillan 2016).

8.2.1.2 Sourcing of Materials

Getting fabric and materials needed to create garments like buttons, zippers, thread, and labels is called material sourcing. In traditional garment manufacturing, production relies heavily on virgin fiber production, which requires substantial amounts of energy, water, and chemicals. If a manufacturer does not make quality sourcing decisions, this may create problems such as excess inventory, excessive waste, and fabric waste (Blackburn 2009; Choudhury 2017).

8.2.1.3 Spreading and Cutting Fabrics

Fabric spreading is defined as the laying out of fabrics on the floor of the cutting table by cutting through multiple layers to cut out different parts of the garment.

In addition to fabric spreading, there are many different fabric cutting techniques that can be used to cut through the layers of a fabric. However, traditional fabric cutting methods often generate significant waste due to poorly optimized fabric markers, fabric defects, and inefficient cutting layouts, making this stage a major contributor to textile industry waste (Rissanen and McQuillan 2016; Sandin and Peters 2018).

8.2.1.4 Sewing and Assembling of Garments

The major function of sewing is to join different pieces of fabric cut out from patterns cut on various kinds of machines to create ready-to-wear garments. While manufacturers of ready-to-wear garments typically have very labor-intensive and energy-intensive sewing processes, the traditional methods used by garment manufacturers resulted in imbalanced lines, excessive machine downtime, low production rates, and high amounts of work-in-progress, as well as fatigue among workers (Shah and Ward 2007; Bruce et al. 2004).

8.2.1.5 Finishing Processes

Finishing activities include the final steps in the finishing process. Examples include removing excess threads, washing the garment, heat-treating the garment before it ships, and quality-inspecting the garment before it is shipped. Conventional finishing methods use a lot of resources (water, chemicals, and energy) when washing and dyeing garments due to their high resource use. If the finishing process is not coordinated and inspected, it will cause defects that may either require additional resources to repair or reject; both will incur additional resource usage (Kant 2012; Hasanbeigi and Price 2015).

8.2.1.6 Packaging and Shipping

After garments go through finishing processes, they are placed into packaging that is used for transport, such as poly bags, cardboard, tags, and immediate product packaging. Poly bags (plastic) and cardboard packaging constitute most of the conventional packaging; poly bags cannot be recycled, which generate hazardous greenhouse gases. Along with the fuel used to ship products, CO_2 is also generated when transporting and distributing products (Allwood et al. 2006; European Commission 2022).

8.2.2 Resource Consumption and Waste Generation

Traditional garment manufacturing has been known to consume more resources such as water, energy, and materials. At the same time, it produces a significant amount of waste by virtue of the leftover fabrics, damaged goods, or defective items created during the production process and by what are referred to as hazardous materials, including chemicals used in the manufacture of clothing. Therefore, all these factors clearly demonstrate that traditional garment manufacturing is not a sustainable way to produce clothing that has every possibility of causing damage to the environment (Allwood et al. 2006; Kozlowski et al. 2012).

In addition to the reasons previously outlined for adopting sustainable and green manufacturing methods in traditional garment production, the sector also faces significant challenges. such as excessive consumption of resources (including water) and waste production, extended timeframes associated with garment production, and negative impacts resulting from the garment manufacturing sector on both the environment and society in general. Sustainable and lean manufacturing can be incorporated by garment manufacturers as an approach to enhance efficiency, reduce resource consumption, and minimize waste through the use of more effective production methods (Shah and Ward 2007).

8.2.3 Sustainability Challenges in Traditional Practices

Inefficient use of materials in textile manufacturing causes substantial waste of fabric, which in turn results in increased energy consumption and environmental pollution. The main cause of this effect is the excessive use of water in dyeing and finishing operations, leading to the generation of substantial amounts of untreated wastewater that are released into rivers and water catchments. The above-mentioned problems clearly indicate the need for efficient use of resources and sustainable processing practices in the textile chain (Kant 2012; Nayak and Patnaik 2021).

The production of handloom textiles has always had immense social and cultural significance in various societies, especially in India, where it still sustains the traditional skills and practices. However, despite its cultural importance, the handloom industry has limitations in terms of price competitiveness and production capacity compared to the mass production of textiles in factories. This disparity again emphasizes the importance of sustainable interventions that maintain a balance between efficiency and the preservation of traditional practices (Ministry of Textiles 2022).

8.3 MODERN GARMENT MANUFACTURING PRACTICES

To address growing concerns surrounding sustainability, manufacturers are increasingly utilizing new technologies to develop sustainable products and services. The three primary technologies being utilized by manufacturers today to manufacture their products and services are lean production methodologies, digitalization, and green technologies (Ahmadi-Gh and Bello-Pintado 2024).

The lean production methodology looks at creating value by eliminating all forms of waste (people, material, process, and energy) within a business' entire value chain; manufacturers are able to utilize existing lean tools such as value stream maps, continuous improvement processes, just-in-time production, and others to eliminate as much as 50% of the water and energy footprints of their operations, as well as to improve the quality and timeliness of the products delivered to their customers, all by utilizing lean processes (Thanki and Thakkar 2018).

Modern apparel manufacturers have taken advantage of the latest technologies and are utilizing many of the latest developments in these technologies, including the automation of many manufacturing processes such as automated cutting machines, intelligent robotic sewing machines, AI-based balancing systems for production lines, and real-time quality control. Numerous studies have shown that the use of these automated processes has resulted in an increased acceptance rate of products, decreased cycle time, decreased labor cost, decreased waste of materials, and as a result, an increase in scale values due to improved sustainable practices (Ahmadi-Gh and Bello-Pintado 2024; Muthu 2014; Niinimäki et al. 2020).

The ability for manufacturers to reduce the number of product prototypes and shorten the lead time between design and production of products due to the sophistication of digital technologies (computer aided design [CAD], virtual prototyping, and data analysis) has increased the efficiency of creating products by allowing for the design of the product on CAD first, as opposed to having numerous prototypes built and tested during the manufacturing process (Fletcher 2014).

Textile manufacturers are developing and utilizing new dye technologies (i.e., wet and dry dyeing methods), as well as better techniques for treating wastewater and recycling yarns. Using these new materials and processes, manufacturers worldwide are producing textiles with fewer resources than before and are adopting the principles of a circular economy (Muthu 2014).

Many fashion brands are now using innovative ways to reduce their water usage and CO_2 emissions related to the dyeing of textiles. One company has developed a new dyeing technology that allows them to reduce their dyeing water usage by 97% compared to conventional cotton dyeing (Niinimäki et al. 2020). Table 8.1 provides a comparative insight into the traditional and modern systems (Kant 2012; Nayak and Patnaik 2021).

In contrast to the historical approach of garment manufacturing, which has existed within cultures for centuries, sustained manufacturing practices appear less frequently but offer higher results than their traditional counterparts when looked at through the lens of industrial performance metrics (export competitiveness). In addition, contemporary manufacturing practices offer increased efficiency of resources, increased controls of quality, speed in responses to market needs, and increased compliance for environmental impact, when compared to traditional practices; therefore, there is no single prescribed way to manufacture clothing using a sustainable method.

The clothing sector must utilize a system that is sustainable, scalable, inclusive, and of high quality, to manufacture and sell its products. In the quest of businesses to find ways forward, frequently referred to as "sustainable", an important question emerges regarding the traditional method of garment manufacturing. Will this method be successful in an economy that increasingly expects the production of many articles with minimal impact on the environment? Should we instead consider utilizing modern

TABLE 8.1
Comparative Insights into Traditional and Modern Systems

Aspect	Traditional	Modern/Sustainable
Resource use	High water and energy	Reduced consumption via technology
Waste	Little recovery	Circular systems and reuse
Productivity	Labor-intensive	Automated + lean-driven
Quality and scalability	Varies with skill	Standardized and scalable
Environmental impact	High production	Lower emission and eco-treatments
Social impact	Informal labor risk	Better compliance + training

technology in manufacturing garments? The complexity of this subject does not imply simplicity of the answer. To begin with, while a "traditional" method of manufacturing versus a "modern" method represents two opposing ends of an avoidable continuum, the most successful solution may emerge through the combination of both methods of manufacturing in the best way possible, utilizing combination-type manufacturing methods to create hybrid garments that achieve the desired durability and sustainability character of both types of technology (Fletcher 2014).

8.4 SUSTAINABLE GARMENT MANUFACTURING PROCESSES

Sustainability is very important to the clothing industry now and is a top priority for many companies. As manufacturers become more aware of their responsibility to meet environmental regulations, pledge to produce clothing sustainably, and increase customer awareness of the environmental effects of clothing, they are starting to develop their sustainable business models. Along with this increased focus on sustainability, the garment industry is also beginning to adopt new operational strategies, such as lean manufacturing, to increase the efficiency of production processes and reduce waste, while enhancing the quality of products being produced. Lean manufacturing practices are derived from auto manufacturing and have provided an efficient operational method for the clothing industry through their focus on ongoing improvement/efficiency (Thanki and Thakkar 2018).

8.4.1 Waterless and Low-Water Dyeing Technologies

Dyeing and finishing processes are among the most environmentally harmful processes of clothing manufacturing because of the large amounts of water and chemicals required to complete the process. The conventional dyeing method consumes very large volumes of water and creates wastewater that includes residual dyes, salt, and toxic chemicals. To deal with these environmental issues, developments in the field of dyeing have led to waterless and low-water dyeing technologies that are more sustainable than their predecessors (Kant 2012).

The supercritical carbon dioxide (CO_2) dyeing technology is one such example of a recent innovation in dyeing technology developed to provide a method of dyeing fabrics without the need for water; this dyeing process uses high-pressure CO_2 gas instead of water as the dyeing medium. The closed-loop nature of the supercritical CO_2 dyeing process allows for the complete recovery and reuse of carbon dioxide, eliminating the need for wastewater to be produced and lowering the energy used to operate the process. Research results show that in addition to being a more environmentally friendly dyeing process, CO_2 dyeing provides increased dye penetration into the fibers of clothing materials and improves the consistency of the dye and color applied to the fibers. Unfortunately, the high cost of purchasing the specialized equipment needed to utilize CO_2 dyeing, combined with the lack of fiber compatibility for many materials, has been a limiting factor for widespread acceptance of CO_2 dyeing as a general dyeing process (Hasanbeigi and Price 2015).

Foam dyeing and air dyeing technologies also promote water conservation through the reduction of the liquor ratios used during the dyeing and finishing processes.

TABLE 8.2
Comparison of Conventional and Sustainable Dyeing Technologies

Dyeing Technology	Water Consumption (L/kg Fabric)	Energy Consumption	Chemical Usage	Wastewater Generation	Environmental Impact
Conventional reactive dyeing	100–150	High	High	Very high	Severe
Supercritical CO_2 dyeing	0	Low–Moderate	Low	None	Minimal
Foam dyeing	10–15	Moderate	Low	Low	Low
Air dyeing	0–5	Low	Very low	Negligible	Minimal
Digital printing	5–10	Low	Controlled	Very low	Low

Both technologies use less water and consume less energy because there is virtually no rinsing and drying required after the dyeing process has been completed. Digital textile printing gives manufacturers a sustainable alternative by allowing them to accurately apply dyes directly to the surface of fabric without the use of liquid, thereby minimizing chemical waste and shortening the overall time to create finished products. Notwithstanding the benefits of these technologies to the environment, the scalability of these technologies differs depending on the manufacturing setting. High capital costs, the requirement for expert personnel, and resistance to process change continue to be major challenges, especially in developing nations and small-scale manufacturing facilities (Hasanbeigi and Price 2015). Comparison of conventional and sustainable dyeing processes is shown in Table 8.2 (Kant 2012; Hasanbeigi and Price 2015).

8.4.2 Sustainable Fibers and Raw Materials

There is an increasing trend globally to use organic (naturally grown), recycled, and plant-based (bio-based) textiles as sustainable textile options, which utilize fewer natural resources and produce less impact on nature than their traditional counterparts (conventional cotton, polyester). Due to the elimination of synthetic chemicals (i.e., pesticides, fertilizers), organic cotton reduces the effects of soil erosion and water pollution, while also providing for increased biodiversity. Research has shown that organic cotton farming can reduce water use by approximately 80% when compared to conventional cotton farming, especially in rainfed agricultural areas (Niinimäki et al. 2020).

With an increasing number of recycled plastics, such as recycled polyester, being produced from postconsumer PET (polyethylene terephthalate) bottles, making recycled polyester is a popular choice for manufacturers looking to reduce the amount of plastic waste in landfills and oceans. Life cycle assessment (LCA) studies have shown that producing recycled polyester requires significantly less energy and generates fewer greenhouse gas emissions than producing virgin polyester. In addition, regenerated cellulosic fibers such as lyocell and modal, which are made

TABLE 8.3
Comparison of Water, Energy, and Sustainability Impacts of Different Fibers

Fiber Type	Raw Material Source	Water Requirement	Energy Requirement	Biodegradability	Sustainability Advantage
Conventional cotton	Agricultural crops	Very high	Moderate	Biodegradable	Low
Organic cotton	Certified organic farms	Low	Moderate	Biodegradable	High
Recycled polyester	Postconsumer PET waste	Very low	Low	Nonbiodegradable	High
Lyocell (Tencel)	Wood pulp	Low	Low	Biodegradable	Very high
Hemp	Natural plant fiber	Very Low	Low	Biodegradable	Very high

using closed-loop solvent processes to recover and reuse processing chemicals, have a lower environmental impact than their virgin equivalents. Advancements in bio-based fibers derived from algae, agricultural waste, and food waste are expected to continue driving the industry toward more inclusive material recycling and a circular economy (Shen and Patel 2010).

Although these fibers have environmental advantages, there are trade-offs associated with each choice that need to be taken into account in material selection. Organic cotton decreases the use of chemicals and water but can lead to lower production and higher costs. Recycled polyester decreases plastic pollution and greenhouse gas emissions but still leads to microplastic shedding. Regenerated cellulosic fibers lead to cleaner production methods but require sustainable forestry and chemical practices. By understanding trade-offs, manufacturers, students, and professionals can make informed choices in selecting sustainable fiber alternatives (Niinimäki et al. 2020; Shen and Patel 2010). Table 8.3 represents sustainability impacts of different fibers (Niinimäki et al. 2020; Shen and Patel 2010).

8.4.3 Energy-Efficient Machinery and Green Processing

The textile industry consumes a lot of energy in manufacturing clothing. This has led to the increased use of manufacturing processes that are environmentally friendly and consume less energy. Studies have shown that the use of technologies such as energy-efficient sewing machines, automated cutting machines, and high-efficiency heat exchangers is increasing. These technologies use less electricity and are environmentally friendly. The use of renewable energy sources such as solar panels and biomass energy systems can further reduce the emission of greenhouse gases and the cost of energy in textile manufacturing plants (Hasanbeigi and Price 2015; Nayak and Patnaik 2021).

Green processing technologies such as enzymatic desizing, bio-polishing, and ozone bleaching are being used in place of traditional chemical processes. These processes are environmentally friendly and cost-effective. Enzymatic processes use lower temperatures, resulting in lower energy consumption, lower chemical usage, and enhanced worker safety. Digital fabric printing is another technology that has been shown to be environmentally friendly compared to screen printing. Studies have shown that this technology uses up to 90% less water, less chemicals, and has a shorter production time, which can mean faster time-to-market and cost-effectiveness (Hasanbeigi and Price 2015; Muthu 2014).

8.4.4 Zero-Waste Cutting and Technology Application

Cutting processes create significant amounts of fabric waste in garment manufacturing and present critical sustainability issues. Zero-waste cutting methods are designed to minimize fabric waste with better pattern design and layout planning. Developing better computerized pattern generation software will allow for more exact placement of patterns on the fabric so it uses as much of the fabric as possible. In addition, studies show zero-waste patterns reduce fabric waste to under 5% (compared to 15% or more produced by traditional cutting techniques). Automating manufacturing, generating digital patterns, and using data to create production plans are additional examples of how to achieve lean and sustainable practices by improving precision while lowering the risk of human error during the manufacturing process (Rissanen and McQuillan 2016).

8.5 OPPORTUNITIES AND CHALLENGES OF SUSTAINABILITY IN THE GARMENT INDUSTRY

The sustainable manufacturing industry has grown considerably since the publication of the Sustainable Manufacturing Resource Guide in 2006, with many of the world's leading corporations embracing more sustainable manufacturing practices. While the challenges of implementation are still present, the increased commitment of large corporations indicates a profound shift toward innovation in cleaner and more efficient manufacturing systems (Allwood et al. 2006; Niinimäki et al. 2020).

Sustainable manufacturing involves a broad set of practices, such as minimizing energy and water use, using renewable energy, using recyclable and reusable materials, and enhancing social responsibility in supply chains. However, sustainability should not only be considered from an environmental perspective. Sustainability also needs to be assessed from economic and social perspectives. For clothing manufacturers, it is important to define how these efforts can affect costs, brand, risk, and competitiveness (Blackburn 2009).

However, the challenges and opportunities that come with sustainable manufacturing depend on the size of the firm. Large companies tend to concentrate on issues such as supply chain openness, the adoption of renewable energy, and large-scale technology investments, while small- and medium-scale enterprises are faced with more pressing challenges such as limited financial resources, a lack of technical knowledge, and short-term returns on investment. By prioritizing cost-effective opportunities such as energy efficiency, waste management, and process improvements, small companies can start their journey toward sustainable manufacturing, while large companies can drive the change in the industry through investments and innovation (Blackburn 2009).

8.5.1 Opportunities Presented by Sustainability

Sustainable development provides a wide range of business opportunities beyond the environmental benefits associated with sustainability efforts; these opportunities may include financial performance, compliance with environmental regulation requirements, technological advances, and increased stakeholder engagement. Resource efficiency is a prime opportunity for increasing profit margins through sustainable practices. It enhances savings through reduced resource consumption (water, energy, and raw materials). The implementation of lean and green manufacturing practices has been demonstrated to achieve superior environmental performance than traditional manufacturing. In addition, companies that implement lean and green practices will improve their internal utilization of resources and environmental performance, thereby improving their profitability and sustainability performance (Carvalho et al. 2011).

One other significant opportunity is the ability to comply with global sustainability standards. The implementation of standardized frameworks for companies to report on their environmental, social, and governance (ESG) performance provides a structured process for reporting, benchmarking, and evaluating ESG performance against established standards such as the EU's corporate sustainability reporting directive and the global reporting initiative which establish a framework for measurement of corporate sustainability, and the Higg Index for apparel which provides a measurement framework for sustainability in the apparel industry. The implementation of global sustainability standards creates a more transparent corporate sustainability reporting process which helps create confidence among investors and aligns corporate strategy with global commitments to reduce greenhouse gas emissions, protect labor rights, and preserve the use of our natural resources. Furthermore, transparency regarding ESG reporting allows companies better visibility into their supply chain, enabling companies to make better-informed decisions and strategies (Global Reporting Initiative 2021).

The garment industry also uses sustainability as a motivation for creating new products. The use of circular economy principles, for example, using recycled materials to create new products, designing products for long-term use, and creating regenerative production practices, encourages manufacturers to improve existing manufacturing practices into better, more sustainable, and environmentally friendly systems. Examples of this would include using bio-based raw materials to create products and creating closed-loop systems for recycling water and chemicals. The examples from above demonstrate that by embracing sustainability, manufacturers will likely be able to create new, innovative solutions for their products and create new types of business models. Additionally, incorporating sustainability practices can enhance brand reputation and build customer loyalty; as more consumers choose to purchase sustainably and ethically produced products, therefore companies that take advantage of this increasing consumer demand are more likely to have greater overall market share, increased potential for premium pricing, enhanced brand engagement with stakeholders, and the development of long-term brand resilience from positive corporate reputations due to the inclusion of employee welfare, fair trade, and community involvement within corporate sustainability programs (Ellen MacArthur Foundation 2017; Niinimäki et al. 2020).

8.5.2 Challenges Hindering Sustainability Adoption

The textile industry is confronted with many challenges in the adoption of sustainable production despite the numerous benefits. The first and most pressing challenge is the high capital required for the adoption of clean technologies, water recycling, renewable energy, and sustainable waste management. This is particularly daunting for small- and medium-scale textile manufacturers in developing countries, where access to capital, subsidies, and sustainability assistance programs is limited. The long payback period for sustainable production investments is also a major deterrent. In the short term, the most pressing challenges include financial limitations, lack of access to modern technology, and the absence of qualified personnel to manage sustainable production systems (Allwood et al. 2006; Niinimäki et al. 2020).

One of the long-term challenges in sustainable textile production is the complexity of the textile supply chain, which is tiered and spread across several countries. This makes it difficult to track performance in terms of environmental and social sustainability. The lack of data cohesion, transparency, and communication systems is a major hindrance to progress in the textile supply chain (Ellen MacArthur Foundation 2017). This is particularly important in developing countries, where textile manufacturers have limited digital capacity to report their activities.

Government policies and regulatory structures also have an impact on the rate of adoption of sustainability. In areas where there is a lack of enforcement, there may be no motivation for firms to progress beyond the minimum level of compliance, while in other areas, more stringent policies may result in higher costs of doing business. In the long term, the absence of professionals in the area of environmental management, sustainable supply chains, and LCA continues to be a major challenge. Without education and capacity-building initiatives, manufacturers, particularly in developing countries, will continue to struggle with incorporating sustainability into their business practices (OECD 2018). Table 8.4 represents opportunities and challenges of garment industry (Allwood et al. 2006; Ellen MacArthur Foundation 2017; OECD 2018; Niinimäki et al. 2020).

The relationship between challenges and opportunities for garment manufacturers demonstrates the duality of their existence as a source of both opportunity and challenge. While sustainability provides the opportunity for cost savings, global compliance with their earned ethical credentials to be upgraded to a higher level based on their earned ethical credentials, and the opportunity for improved reputation, realizing these opportunities requires overcoming the systemic and institutionalized barriers historically underpinning the garment manufacturing industry. For example, the industry is now able to independently benchmark its environmental and social performance against the rest of the market (an industry best practice in

TABLE 8.4
Opportunities and Challenges Metrics in Garment Industry Sustainability

Sustainability Dimension	Typical Opportunity	Reported Benefit	Noted Challenge	Reported Barrier
Resource efficiency	Energy and water reduction	Cost savings, reduced emissions	High capital outlay for tech upgrades	Slow return on investment in small and medium enterprises (SMEs)
Standards and compliance	Adoption of ESG reporting frameworks	Improved transparency and investor trust	Complex compliance requirements	Inconsistent legal enforcement
Supply chain management	Circular supply chain design	Reduced waste, better traceability	Fragmentation across tiers	Information asymmetry
Workforce development	Sustainability upskilling	Enhanced innovation capacity	Skill shortages	Lack of targeted training programs

the future), by adopting more sophisticated sustainability methodologies (e.g., Higg Index). However, such benchmarking methodologies require significant investment from the garment manufacturers on both data systems and capability development (Sustainable Apparel Coalition 2023).

Additionally, research conducted globally indicates that collaborative strategies involving partnerships with government, industry associations, and civil society organizations could alleviate several of the issues related to the lack of infrastructure and workforce training in many developing countries. It has been established through research that policy incentives (such as tax credits for investing in clean technologies and grants for skills development) can support the economic feasibility of investments in sustainability in developing nations (OECD 2018). Finally, companies that incorporate sustainability into their core business strategies, transitioning from a compliance-based approach to one focused on innovation, are well positioned to leverage the benefits that accompany their transition to a circular economy. Examples of innovative business models incorporating sustainability include product-as-a-service models, take-back schemes, and material reuse and recycling networks, demonstrating that sustainable practices provide businesses with economic opportunities, while reducing their environmental footprints (Bocken et al. 2016).

8.6 FUTURE OUTLOOK ON SUSTAINABLE PRACTICES IN GARMENT MANUFACTURING

The objectives of sustainability at the global level, especially the Sustainable Development Goals (SDGs) of the United Nations, are increasingly influencing the development of the fashion industry because of its large environmental impact. The conventional linear "take, make, dispose" approach to production has resulted in resource depletion, environmental degradation, and waste generation. Thus, the future of the fashion industry will be based on the application of the principles of the circular economy, the use of innovative manufacturing technologies, the increased use of renewable energy, the use of digital technology to enhance supply chain transparency, and global partnerships. These will collectively bring about a change in the way clothes are designed, made, distributed, and recycled (United Nations 2015).

Studies have shown that the adoption of circular economy approaches, like recycling, remanufacturing, and product life extension, can help mitigate environmental challenges while ensuring that industrial activities are consistent with the achievement of SDG goals. On the other hand, the development of new technologies like artificial intelligence (AI), the Internet of Things (IoT), and blockchain can provide new avenues for enhancing data-informed decision-making, resource optimization, and real-time supply chain monitoring. Nevertheless, some research gaps need to be addressed. These include the development of scalable circular economy models for small and medium enterprises, the long-term economic effects of digital transparency platforms, consumer behavior in circular fashion, and the development of policy structures that facilitate industry collaboration (Bocken et al. 2016; Ahmadi-Gh and Bello-Pintado 2024).

8.6.1 Circular Economy and Sustainable Models

The purpose of the circular economy is to separate resources and economic growth by taking all waste out of the life cycle of a product and allowing for the continued use of materials. In the textiles industry, for example, circular economy strategies include the development of durable products, remanufacturing, recycling and remodeling of garments leading to increased resource efficiency and waste reduction. The literature review indicates that the implementation of circular practices in the textile industry would have the capacity to be transformative, allowing for reduction of environmental impact and improvement in sustainability performance. However, there needs to be a high level of supply chain collaboration, development of advanced reuse and recycling infrastructure, and supportive regulatory environments to implement these circular models (Ellen MacArthur Foundation 2017; Bocken et al. 2016; Hayat et al. 2020).

A demonstration of the circular economy is through the reuse of remnant by initiatives to produce new products made from such remnant fabrics with a high profit margin, demonstrating that it is possible to achieve both socioeconomic and environmental benefits through circularity in local garment production. These initiatives achieved high levels of remnant fabric recycling rates (up to 95%) and occurred locally as part of creating employment. Hence, it demonstrates two benefits: the economic benefits of circular economy models and the reduction of environmental impact (Ellen MacArthur Foundation 2017; Bocken et al. 2016).

8.6.2 Blockchain and Supply Chain Transparency

With the advancement of blockchain technology, it is increasingly being recognized as a significant enabler of increased transparency and accountability in garment supply chains. By creating a decentralized and immutable digital ledger that tracks product information from raw materials to finished products, blockchain technology allows stakeholders to develop auditable trails that can prove sustainable behavior throughout the entire value chain. This helps to minimize fraud, enhance compliance with environmental and social criteria, and improve traceability – key considerations for unlocking consumer confidence in sustainable garment products (OECD 2018).

The combination of blockchain technology and smart contracts further amplifies its potential by allowing for automated compliance with preset sustainability criteria and promoting a culture of continuous improvement among supply chain partners. Literature also emphasizes the rising interest in blockchain-based frameworks that can facilitate the SDGs and enhance accountability in the garment industry (United Nations 2015). However, there are still some challenges to its adoption, such as the cost of adoption, data governance and data sharing, and compatibility between different digital platforms used by supply chain partners. Overcoming these will be critical to unlocking the full potential of blockchain technology for sustainable garment manufacturing.

8.6.3 Renewable Energy Adoption and Decarbonization

The future of sustainability in the clothing industry will be shaped by the transition to renewable energy. The increasingly stringent regulations on carbon emissions are pushing the clothing industry to adopt renewable energy technologies such as solar, wind, and bioenergy in the factories, thereby cutting down on the use of fossil fuels. This trend has the potential to greatly reduce CO_2 emissions, improve energy security, and make industry more resilient to energy price volatility. Besides the environmental advantages, the adoption of renewable energy can also improve cost-competitiveness by mitigating long-term energy costs, making the industry less vulnerable to volatile fuel prices, and reducing the risk of energy-related regulatory penalties and carbon pricing. Although there is a growing interest in the topic, there is a lack of empirical research that can quantify the long-term financial risk-reduction value of renewable energy adoption in the clothing industry (Hasanbeigi and Price 2015; Niinimäki et al. 2020).

8.6.4 Global Collaboration and Policy Frameworks

International cooperation and enabling policies are essential in ensuring that garment manufacturers adopt sustainable practices. At the international level, international policies, public–private partnerships, and global sustainability guidelines promote technology transfer and the adoption of clean production processes by developing countries. At the national level, governments establish policies and regulatory frameworks, such as extended producer responsibility, environmental reporting guidelines, and economic incentives, to promote the adoption and innovation of circular economy business models. Increasing consumer demand for sustainably produced and responsibly made products further supports these policy initiatives by encouraging brands to pursue sustainability as a means of differentiation and competitive advantage (OECD 2018).

At the organizational level, sustainable garment manufacturing requires the alignment of technological innovation, circular economy thinking, renewable energy integration, and shared governance in day-to-day operations. The intersection of AI, IoT, and blockchain technology is driving Industry 4.0 and 5.0 strategies that facilitate data-driven, resource-smart manufacturing and supply chain transparency. Yet, issues such as system compatibility, data protection, and employee skills development persist (Ahmadi-Gh and Bello-Pintado 2024). A complete transition to a circular economy will thus demand a collective effort by stakeholders at the global, national, and organizational levels.

8.7 CONCLUSION

From the analysis in this chapter, one thing is clear: traditional garment manufacturing, as important as it was in the past, does not align with the realities of the modern world Water and energy consumption, along with waste generation, have reached levels that have pushed the industry to a point where sustainability is no longer optional but a necessity. Sustainable garment manufacturing provides a clear way forward. Lean manufacturing, technology, and green technology demonstrate that it is possible to reduce the impact on the environment without compromising performance. Sustainable materials, energy-efficient machinery, low-water dyeing, and a circular economy demonstrate that productivity, quality, and competitiveness can be enhanced while being sustainable.

At the organizational level, sustainable fashion production requires the integration of technological advancements, circular economy thinking, renewable energy use, and governance. The intersection of AI, IoT, and blockchain technologies is facilitating the adoption of Industry 4.0 and Industry 5.0, which promote data-informed and resource-smart manufacturing practices. Nevertheless, there are still issues to be addressed, such as system compatibility and data protection, as well as the need for upskill employees. A complete transformation toward a circular economy will thus necessitate the collective efforts of various stakeholders at the global, state, and organizational levels.

REFERENCES

Ahmadi-Gh, Z. and Bello-Pintado, A. 2024. Towards sustainable manufacturing: How does digitalization and development affect sustainability barriers? *Journal of Cleaner Production* 476: 143792.

Allwood, J. M., Laursen, S. E., Malvido de Rodríguez, C. and Bocken, N. M. P. 2006. *Well Dressed? The Present and Future Sustainability of Clothing and Textiles in the United Kingdom*. Cambridge: University of Cambridge, Institute for Manufacturing.

Blackburn, R. S. 2009. *Sustainable Textiles: Life Cycle and Environmental Impact*. Cambridge: Woodhead Publishing.

Bocken, N. M. P., de Pauw, I., Bakker, C. and van der Grinten, B. 2016. Product design and business model strategies for a circular economy. *Journal of Industrial and Production Engineering* 33(5): 308–320.

Bruce, M., Daly, L. and Towers, N. 2004. Lean or agile: A solution for supply chain management in the textiles and clothing industry. *International Journal of Operations and Production Management* 24(2): 151–170.

Carvalho, H., Duarte, S. and Machado, V. C. 2011. Lean, agile, resilient and green: Divergencies and synergies. *International Journal of Lean Six Sigma* 2(2): 151–179.

Chen, H., Long, R., Niu, W., Feng, Q. and Yang, R. 2024. How does individual low-carbon consumption behavior occur? An analysis based on attitude process. *Applied Energy* 116(5): 376–386.
Choudhury, A. K. R. 2017. Environmental impacts of the textile industry and its assessment through life cycle assessment. In *Roadmap to Sustainable Textiles and Clothing* (Ed. Muthu, S. S.), Singapore: Springer, 1–39.
Dissanayake, K. 2026. Fashion re-commerce: A state-of-the-art review and future research agenda. *International Journal of Retail and Distribution Management*, 1–21. https://doi.org/10.1108/IJRDM-03-2025-0217
Dissanayake, K. and Sinha, P. 2012. Sustainable waste management strategies in the fashion industry sector. *International Journal of Environmental Sustainability* 8(1): 77–90.
Ellen MacArthur Foundation. 2017. A new textiles economy: Redesigning fashion's future. Cowes, UK.
European Commission. 2022. *EU Strategy for Sustainable and Circular Textiles*. Brussels: European Union.
Fletcher, K. 2014. *Sustainable Fashion and Textiles: Design Journeys*. 2nd edition. London: Routledge.
Gereffi, G. and Frederick, S. 2010. The global apparel value chain, trade and the crisis. World Bank Policy Research Working Paper No. 5281.
Global Reporting Initiative. 2021. GRI sustainability reporting standards. Amsterdam.
Hasanbeigi, A. and Price, L. 2015. A technical review of emerging technologies for energy and water efficiency in the textile industry. *Journal of Cleaner Production* 95: 30–44.
Hayat, N., Hussain, A. and Lohano, H. D. 2020. Eco-labeling and sustainability: A case of textile industry in Pakistan. *Journal of Cleaner Production* 252: 119807.
Kant, R. 2012. Textile dyeing industry an environmental hazard. *Natural Science* 4(1): 22–26.
Kozlowski, A., Searcy, C. and Bardecki, M. 2012. Environmental impacts in the fashion industry. Life-cycle and stakeholder framework. *Journal of Corporate Citizenship* 45: 17–36.
Ministry of Textiles. 2022. *Annual report 2021–22*. New Delhi: Government of India.
Muthu, S. S. 2014. *Roadmap to Sustainable Textiles and Clothing*. Singapore: Springer.
Nayak, R. and Patnaik, A. 2021. *Waste Management in the Fashion and Textile Industries*. Kidlington: Woodhead Publishing and Elsevier.
Niinimäki, K., Peters, G., Dahlbo, H., Perry, P., Rissanen, T. and Alison, G. 2020. The environmental price of fast fashion. *Nature Reviews Earth and Environment* 1: 189–200.
OECD. 2018. *Due Diligence Guidance for Responsible Supply Chains in the Garment and Footwear Sector*. Paris: OECD Publishing.
Rissanen, T. and McQuillan, H. 2016. *Zero Waste Fashion Design*. London: Bloomsbury.
Sandin, G. and Peters, G. M. 2018. Environmental impact of textile reuse and recycling. *Journal of Cleaner Production* 184: 353–365.
Shah, R. and Ward, P. T. 2007. Defining and developing measures of lean production. *Journal of Operations Management* 25(4): 785–805.
Shen, L. and Patel, M. K. 2010. Life cycle assessment of man-made cellulose fibers. *Lenzinger Berichte* 88: 1–59.
Sustainable Apparel Coalition. 2023. *Higg Index Methodology*. San Francisco: Sustainable Apparel Coalition.
Thanki, S. and Thakkar, J. 2018. A quantitative framework for lean and green assessment of supply chain performance. *International Journal of Productivity and Performance Management* 67(2): 366–400.
United Nations. 2015. *Transforming Our World: The 2030 Agenda for Sustainable Development*. New York: United Nations.

9 Innovation in Textile and Apparel Production that Reduce Negative Environmental Impacts

Case Studies of Evolution, Barriers, and Future Directions

Mary Ruppert-Stroescu

9.1 INTRODUCTION

Reducing the negative environmental impact of the fashion industry has been a strong subject of discussion, research, and action since the dawn of the 21st century. Areas of attention to be addressed in this chapter include eliminating environmental contaminants, reducing waste, and avoiding products derived from petrochemicals. Through select case studies, the bridge between research and widespread adoption of eco-effective textile and apparel production will be examined, exploring innovations that are established, bourgeoning, and incubating (Braungart et al. 2007). For each case, technology adoption analysis will combine with the propulsion theory of creative contributions (Sternberg 1999; Sternberg et al. 2001) to trace the path taken thus far, examine barriers, and envision the future of these environmentally positive efforts.

9.2 TECHNOLOGY ADOPTION AND DIFFUSION

The primary focus of technology adoption literature for the past 10 years has been on information and digital technology, with considerable attention given to the micro, or individual level of adoption (Stoneman and Battisti 2010), the institutional level of information technology adoption (Dastane 2020; Stornelli et al. 2021; Lakhwani et al. 2020), and how information and digital technology diffusion has influenced entire fields and countries (Hooks et al. 2022; Vergragt et al. 2014). Al-Emran and Griffy-Brown (2023) describes the role of technology adoption in sustainable development, again, focusing on information and digital technology. Given this strong emphasis on information and digital technology, there is a gap in recent literature

DOI: 10.1201/9781003689225-9

regarding adoption of technologies other than information and digital technology in terms of sustainable practices.

To address a more expansive understanding of technology, etiologist Carrroll (2017) gives a comprehensive, three-part definition of technology that looks at purpose, function, and intention:

> (a) something that is always inherently intelligent enough either to function, to be used to function, to be imbued with, or to be interpreted as having, a function that only intelligent beings (human or otherwise) have the ability to comprehend; (b) something devised, designed (i.e., primary intention), or discovered (i.e., secondary intention) that serves a particular purpose from a purely secular standpoint, without requiring that mankind be responsible for it, though he may be… (c) a significant beneficiary of rationally-derived knowledge that is "used for" a purpose, without itself necessarily being translated into something physical or material that "does."

Thus, because of the wide range of actions and artifacts that can be considered technology, this chapter will focus on process technology, defined as being developed with primary intention (Carroll 2017), specifically targeting usefulness to fashion industry players, with the function of providing innovation that has the potential to be economically effective and ecologically positive, and with the purpose of reducing the negative impact the fashion apparel industry is contributing to the environment (Pfaffenberger 1992). Due to the nature of the fashion apparel industry, this chapter focuses on the realization of physical, non-digital technology related to the development of a material object.

9.3 THE PROPULSION THEORY OF CREATIVE CONTRIBUTIONS

Incorporation of innovative process technology that can advance the pursuit of eco-effective textile and apparel and a less environmentally destructive fashion industry requires field-level influence. A field is a social organization of a domain. For this analysis, the fashion industry is the domain, and the field is textile development.

Innovation can be defined simply as creativity that is implemented into something useful (Yusuf 2009). Creativity and innovation are distinct yet related concepts, as not all innovation comes from creativity, and not all creative ideas are implemented into innovation. Creativity by itself is insufficient to produce innovation; it is only when novel ideas developed through creative thinking are implemented, functional, and utilitarian does innovation result (Hughes et al. 2018). Creativity in relation to the fashion industry commonly focuses on the fashion aspect, typically recalling elaborate, unusual clothing and textiles that gain the attention of the media, and is adopted by an elite few. However, creativity in the domain of the fashion industry can take many forms (Ruppert-Stroescu and Hawley 2014). Creative output can manifest in the end product, as in the prior example, however, a significant level of creative output is not as visible because it is generated within processes. This process-based creativity is much less obvious; for example, the creative process required to make thousands of one simple, uninteresting product at the required price and distribution level is as integral to the success of that product as its aesthetic qualities, yet when examining the product, it is impossible to perceive the level of creativity that was engaged during the process of development and manufacture.

Sternberg et al. (2013, 2001, 1999) outline a propulsion model of creative contributions to a field within three categories and eight types. The categories include creative contributions that (1) recognize and reinforce the field as it is: *replication* contributes by either reinforcing or refuting aspects of the field, and *redefinition* by changing the perception of the field, not actually advancing the field but reinforcing its current position, *forward incrementation* is perhaps the most prevalent type of creativity where the contribution builds on existing momentum and is readily accepted, and *advanced forward incrementation* builds on existing momentum but the field is not quite ready to accept the contribution; (2) contributions that refuse the current system: *redirection* recognizes where the field is at the moment and takes it into a new direction, *reconstruction/redirection* reaches into the past of the field and pushes it into a new direction, and *reinitiation* finds a new starting point for the field and moves forward from there; and (3) contributions that combine existing systems: *integration* combines completely different and sometimes unrelated paradigms.

These creative contributions have been demonstrated to range from an individual's personal development forward, to moving entire fields forward (Sternberg et al. 2001). The analysis that follows will focus on the creative contributions in the textile and apparel field, under the domain of the fashion industry, and will focus on creative contributions within the context of innovation. The study will analyze creative contributions and show how turning creativity into innovation can make fashion more environmentally sustainable.

The case studies to follow fall into three of the eight types of creative contributions that propel a field forward (Steinberg et al. 2013): Forward Incrementation, with CiCLO textiles pushing the industry into a direction it is ready to embrace; Redirection with RECLÉM textile upcycling and sustainable garment design, pushing the industry into a direction that recognizes the current paradigm while pushing it into a new direction; and Reinitiation with SMARC (Synthetic biology Manufacturing of Advanced materials Research Center at Washington University in St. Louis, Missouri, USA) engaging bio-based fibers that are rejecting current petrochemical dependence, discovering a new source for fiber polymers and carving an entirely new category of fiber production.

The core aspect of the CiCLO textiles relates to forward incrementation, where a common nonbiodegradable textile manufacturing practice is taken an important step forward and renders that same textile biodegradable, an aspect of the synthetic fiber life cycle that is widely recognized as desirable. Adding CiCLO pellets to the production of polyester and nylon stimulates the resulting fiber, under bacterially rich conditions, to biodegrade in three years instead of resisting biodegradability altogether. Such innovation will allow the fashion industry to maintain production of petroleum-based textiles, or synthetic fibers, a business sector with strong growth projections, without contributing massive quantities of textiles and microfibers to the environment (Grand View Research 2024).

The RECLÉM textile upcycling and sustainable garment design innovation is an example of redirection where the innovation builds on aspects of the current textile and apparel field's paradigm but innovates to move it into the new direction of rescuing waste textiles and creating zero waste (Ruppert-Stroescu 2018). The RECLÉM process upcycles textiles without breaking down the material and maintains the design, patterning, and assembly aspects of traditional garment production but introduces a new process at the textile and cutting level.

The creative contribution of SMARC is reinitiating the area of manufactured fibers by rejecting origins in natural or petrochemical resources and moving into an entirely new category of fibers made from unique protein molecules that are engineered for specific characteristic and regenerated.

9.4 CASE NUMBER 1: CICLO TEXTILES: FORWARD INCREMENTATION

9.4.1 Background

In 2012, Andrea Ferris was working in California, managing the product development program to make uniforms for the McDonalds restaurants in the USA, and at one point she was struck with the reality of the process she governed. The uniforms were made of 100% polyester, new uniforms were issued every 2 years, and up to half of the uniforms they produced were never used and went straight to the landfill. Synthetic fibers can sit for up to 200 years before breaking down, and then only become petroleum-based microparticles because polyester will never fully decompose. While the reality of this environmentally toxic situation was sinking in, Ms. Ferris asked herself, if we can make biodegradable doggie bags, cutlery, and other products, why can't we make a biodegradable fabric? She then established a team to develop a biodegradable polyester substitute and formed CiCLO textiles.

On the other side of the American continent in North Carolina, Parkdale Mills, the world's largest producer of spun yarns and the leading manufacturer of cotton products in the Americas developed a consumer products division that made wipes from polyester and polypropylene. In 2014, Parkdale Mills had a similar reality check to Ms. Ferris, and established a research and development branch that was also seeking to develop a biodegradable polyester substitute for their wipes division.

The stars aligned and in 2016, CiCLO textiles and Parkdale learned that they shared the same goal of replacing traditional polyester with a biodegradable alternative, and were approaching the work from different perspectives. Over the next 2 years, the two innovation centers collaborated to determine the best path forward, and in 2018 CiCLO and Parkdale formed a joint venture, with Parkdale taking the lead on translating CiCLO for the market. The next 7 years focused on demonstrating the full biodegradability of the product, building the infrastructure of the global supply chain, and getting buy-in from brands.

Thus, over the course of approximately 10 years, CiCLO has been confirmed to render polyester and nylon, the most common petrochemical fibers (Chen and Burns 2006), biodegradable within 3–5 years. As of mid-2025, CiCLO has demonstrated significant industry penetration. Over 60 global brand partners have contributed to the sale of more than 120 million pounds of CiCLO-based fibers within the global supply chain (Borneman 2025).

Synthetic fibers are nonbiodegradable; they may break down into microfibers over the course of 20–200 years, but they never dissipate into the environment leaving a nonharmful trace (Egan and Salmon 2022). Even when synthetic fibers break down to the micro-level, they are still harmful, infiltrating the food system and thus the human body. A recent study from Oregon that examined the contents of the stomach of 182 fish found that 99% of the fish contained microfibers, and 80% of those microfibers were textile in origin (Traylor et al. 2024). We are not yet certain of the effects of this level of microfibers on the human body, but there is a potential of harm to the body due to the chemicals, finishes, and dyes in synthetic textiles (Periyasamy 2023). Synthetic textile producers need to recognize their role in the diffusion of microplastics and integrate CiCLO into their products.

9.4.1.1 Impact through Evolution, Not Revolution

Manufacturing synthetic fibers typically starts with pellets made from a chemical synthesis creating polymers from nonrenewable and nonbiodegradable petroleum. While CiCLO only addresses the biodegradability of synthetic fibers and not the ecological damage caused by extracting the petroleum and processing the fiber, using CiCLO to reduce the negative environmental impact of synthetic fibers is still very important due a combination of factors: heightened 21st-century knowledge about the negative effects of synthetics on the environment, and three aspects of CiCLO technology: (1) there is no need to adjust current manufacturing practices; (2) the added cost is minimal, can be as low as 1%; and (3) the process of biodegrading happens in a typical biologically active landfill environment.

Often, requiring a small process adjustment in an evolutionary manner can increase the likelihood of technology adoption. CiCLO technology does just that by harnessing existing systems to insert change at one key level by bringing in much needed textile waste–reducing technology and moving the textile fiber production field forward in a direction that is already going, clearly reflecting the forward incrementation category of creative contributions to the textile field and clearly moving creativity to innovation through tangible, useful, industry applications.

9.4.1.2 Future Directions and Challenges

Textile industry expert Cheryl Smire stated that the CiCLO innovation is the most impactful, transformational technology she has seen come to the textile industry in her 31-year career (Cheryl Smire, interview by Mary Ruppert-Stroescu, November 7, 2025). By simply adding the CiCLO pellets to the beginning stages of polyester development, the fashion industry can reduce the negative impact of the fiber on the landfill while embracing the positive qualities of polyester and nylon. Synthetic textile production is expected to grow from the 2022 market size of US$ 66.11 billion to a 2030 projected market size of US$ 117.76 billion, nearly doubling in 8 years (Grand View Research 2024). If all that production included CiCLO, the synthetic fibers in landfills would biodegrade in 3–5 years and the microfibers that are released during use and care would also biodegrade. A revolutionary outcome achieved through incremental evolution.

As an established technology, why, might you ask, is the use of CiCLO not more widespread? The technology was brought to market 8 years ago, fits into the typical manufactured fiber process seamlessly, and increases the wholesale cost only 1%–5%, – yet is not universally adopted. The CiCLO team is addressing multiple factors to foster widespread market penetration with success; for example, a recent announcement revealed that mega-retailer Walmart in the USA has started selling towels made with the CiCLO technology. However, some cultural aspects of the textile and apparel industry will be slower to shift, particularly terms of profit margins. A major polyester manufacturing CEO, who will remain anonymous, was clear in stating that their company will adopt CiCLO when their customers, textile manufacturers, demand it (Bret Schnitker, interview by Mary Ruppert-Stroescu, February 21, 2024).

Consumers driving change at the fiber manufacturing level is a particularly challenging task, because while many end-use consumers have a heightened desire for eco-positive products, paying a premium for them is not generally accepted. Adding to that challenge, those eco-positive consumer voices are most often heard at the retail level, but the complex textile supply chain has many layers between the polyester fiber producer and the retail consumer, and unfortunately the strength of the eco-positive production message gets lost when passing upstream through retail buyers, apparel wholesale distributers, apparel manufacturers, textile wholesalers, and textile manufacturers to arrive with meaning at the front end of the supply chain, the fiber manufacturers.

9.5 CASE NUMBER 2: RECLÉM

9.5.1 Background

The seed for RECLÉM began in Dr. Ruppert-Stroescu's youth, growing up with parents whose upbringing in the poor rural Midwestern USA during the Great Depression framed the family's perspective around "waste not, want not," particularly manifested through her mother's passion for hand quilting. However, having entered the fashion industry in the 1980s when excessive consumption and production dominated, waste reduction practices were not top of mind. It was not until the early 2010s when Dr. Ruppert-Stroescu was challenged to develop a research agenda that she saw a parallel between sustainable fashion industry pioneers and her waste-reducing origins. Inspired by companies such as Patagonia, Stella McCartney, and Eileen Fischer, as well as thought leaders in the sustainable fashion space such as Kate Fletcher (2008) and Sandy Black (2011), put out a call out to undergraduates at an American Midwestern University to research sustainable textile and apparel production.

9.5.2 Activation

A group of enthusiastic apparel design students gathered every 2 weeks for 3 months to address the challenge of finding new ways to work with textile waste. Being in the youth-oriented university setting, the research team decided to collect unwanted t-shirts by putting collection boxes in the College of Human Sciences. Within a few weeks over 200 t-shirts were donated. The research team first sorted the t-shirts by fiber content, keeping the 100% cotton ones and donating any that had synthetic fibers to the local Salvation Army. After laundering and further sorting the t-shirts by color, they experimented with cutting the shirts in ½–¾″ strips with a Gerber Automated Sample Cutter, scissors, and industrial vertical and rotary cutters. With the resulting cotton jersey strips, the researchers engaged classic textile fabrication techniques of knitting, crochet, weaving, and needle felting. Each of these techniques gave satisfactory results in terms of upcycling the postconsumer t-shirt jersey, and creativity could be integrated through the design of the textile, but none of the techniques were creative and innovative in the sense of being something that was not currently done.

During the next phase of development, Dr. Ruppert-Stroescu, reflecting on the resources at hand, revived a Freedom Longarm quilting machine that was stored in the basement of the College. In the span of a few years, Dr. Ruppert-Stroescu and her team obtained university funding through the entrepreneurship division that enabled the part-time hiring of three dedicated students and the purchase of a computer digitizing and automated stitching system for the Freedom Longarm quilting machine. This enabled the team to explore textiles derived from post-consumer knitted fabrics, including cotton jersey, nylon, and polyester, as well as woven materials such as cotton denim and a range of synthetic fiber weaves.

Having access to the university's textile testing lab, the RECLÉM research team performed tests according to American Association of Textile Chemists and Colorists (AATCC) and American Society for Testing and Materials (ASTM) International standards. Absorbency, abrasion, pilling, wrinkle recovery, and tensile strength tests were conducted using four specific stitching techniques and textile types; results were within satisfactory ranges for each test and sample. In addition, the research team continued to experiment making unique textile surface design created with strips, squares, and sections of textile waste, both as yardage and within the garment pattern shape.

Given the success of that research and development, the university's intellectual property office agreed to sponsor the patent application for this process. The patent for US10588369 Textile repurposing and sustainable garment design was filed on 10/23/2017, then issued on 3/17/2020. While obtaining the patent was an important step forward, further research and development combined with a design and production cultural shift are necessary to translate the RECLÉM process into a scalable commercial application that truly enables an important quantity of textile waste to be rescued from landfills. To gain traction that will attract large industry players and potential investors, the RECLÉM team is developing a brand, clé collections, approaching collection development as a form of further inquiry into best practices and demonstrating consumer acceptance of products made with the RECLÉM process.

9.5.3 Future Steps and Challenges to Redirection

The emerging RECLÉM technology for textile upcycling and sustainable garment design has a unique potential because the process maintains traditional aspects of the apparel manufacturing paradigm while shifting the fabric purchasing and pattern layout phases to rescue textile waste and generate no waste, respectively. In addition, the RECLÉM process is distinct as it maintains the integrity of the textile waste without breaking the textile down to the fiber stage, the surface design of the textile is engineered as the textile waste is laid out, and zero waste is generated in the making of textile products, because the textile waste is laid only within the shape of the pattern pieces. And finally, the process can use either pre-consumer cutting room textile waste or postconsumer textile waste.

Challenges for widespread adoption of the RECLÉM method are centered around the technology barriers of risk aversion. Taking a chance to change the status quo design and production mindset of sourcing virgin fabric to make clothing from long, relatively wide pre-made uniform cloth, to textured garment pieces assembled from textile waste, is a cultural and practical leap.

For example, textiles and apparel made with the RECLÉM method have a distinct, textured aesthetic that only works for some markets. Having approached the Green Eileen division of Eileen Fischer with the hope of collaboration, the research team learned that professionals there were intrigued by the process but distinctly mentioned that the texture created with the RECLÉM method did not match the visual and design vocabulary of Eileen Fischer. For brands where cutting virgin fabric is central to their design DNA, zero-waste products can be produced using pre-cut fabric lengths, garments can be made from virgin or deadstock materials, and complementary products can be developed from cutting-room waste using the RECLÉM process.

Working in defiance of the existing design and production system, the RECLÉM process currently functions at a craft, slow fashion level. To increase the overall impact of the RECLÉM process through cost-effective scaling, RECLÉM needs to develop a specific supply chain for acquiring and processing textile waste, laying the textile waste out in the shape of a garment or pattern piece, and stitching the layers together. Specialized machines are required, both some that exist and some that need to be developed. The research team has a specific vision for these components, however, to realize that vision funding is required.

The trajectory for obtaining funding to scale the RECLÉM process further has been complex. While at a micro-level, enthusiasm and acceptance of garments made with the RECLÉM method are high, often when presented to textile and apparel manufacturing professionals, there is difficulty envisioning the practicality of scaling production at a level that provides adequate profit margins. Although the RECLÉM process maintains traditional apparel manufacturing stages such as design, pattern drafting, and garment assembly, it shifts the material flow from purchasing and cutting virgin fabric to processing textile waste and arranging it within pattern pieces.

Future plans include continued efforts to break the technology adoption barriers discussed here by demonstrating the scalability of the RECLÉM process through the proprietary brand, clé collections, reaching out to apparel manufacturers to collaborate by engaging the RECLÉM process to use their cutting room waste to create products for their brands, and by licensing the patented process to technology-adept companies who are also committed to reducing textile waste.

9.6 CASE NUMBER 3: SMARC BIO-BASED REGENERATED TEXTILES: REINITIATING

9.6.1 BACKGROUND

The research group SMARC is working in an emerging field that redirects man-made textiles of petrochemical origin to textiles made from biomaterials. SMARC has manufactured, from bacterial sources, a high-performance thin filament fiber.

This incubating technology, when developed, will reinitiate the fiber development field by having moved to an entirely new place and advancing the field in an environmentally positive direction.

SMARC was founded by two professors in the Department of Energy, Environmental & Chemical Engineering at Washington University: Marcus Foster specializes in polymer chemistry in material science, and Fuzhong Zhang in developing synthetic biology tools and systems for the sustainable production of structurally defined chemicals and high-performance materials. This case study will explore the evolution and potential of this technology, examine challenges, and analyze industry receptivity for the technology.

9.6.2 Evolution and Potential

Silk, a fine, strong, soft lustrous fiber, can come from a variety of sources. The most common is silk fiber from the *Bombyx mori* "silkworm." Making a substitute for this rare and beautiful fiber found in nature was the motivation for the first manufactured fibers at the beginning of the 20th century, rayon and nylon. Currently, researchers have moved beyond silkworm silk to capture the amazing qualities of spider silk, using methods such as developing transgenic silkworms and bioengineering fibers (Dai et al. 2024). When evaluating an approach to engage polymer chemistry and synthetic biology to innovate in the space of high-performance materials, the research team at SMARC targeted criteria including useful mechanical properties, fitting into existing fiber manufacturing processes, and availability of resources.

Zhang was attracted by the mechanical properties of spider silk due to these three key aspects of this natural material: spider silk has the highest breaking strength of any naturally occurring fiber, it absorbs kinetic energy that enables resilience, and it possesses fundamental structural constituents that provided a rich starting point for building a new fiber (Nova et al. 2010). SMARC researchers considered many materials that can be either strong or tough, but the special properties of silk give it the ability not only to tolerate a strong weight, but also to absorb energy. Biomaterials are often transformed into fibers through electrospinning. The SMARC team implemented wet spinning rather than electrospinning, because it is a technique compatible with the traditional textile industry and will most likely facilitate wide-scale adoption of the new fiber. The final criteria for determining a biomass resource in a quantity for mass production still needs attention and will be addressed in the Challenges section.

Respecting the parameters of these criteria, the SMARC team has determined a biomass from which proteins are produced by engineered microbes grown on renewable/biomass-derived feedstocks, then purified and wet spun into a filament fiber. Small quantities of the fiber have been produced and tested for tensile strength and resilience, confirming the ability to reproduce a new silk that has the qualities of spider silk. Through this innovative process, the SMARC team has completely designed a new protein from a natural source that did not exist anywhere in the world, a process outside of any current fiber category.

9.6.3 Challenges

Because these manufactured fibers cannot be situated within the two manufactured fiber categories of regenerated or synthetic, a dominant challenge at the time of the writing of this chapter is terminology; what is a fiber made from proteins fabricated from a grown substance called? Spiber, a Japanese bio-venture, brands its platform as "Brewed Protein" and explicitly positions it as "protein polymers" made by microbial fermentation (Spiber Inc. 2025). Researchers write about their work using terms such as "recombinant spider silk or recombinant spidroin fibers" or "artificial or biomimetic spider silk" when the sequence is silk-like and the fiber is made by wet spinning. This variety of descriptors for a range of production methods makes clear identification and replication nearly impossible. Fiber naming bodies such as the US Federal Trade Commission and the International Standards Organization must address the changing landscape and assign a unifying name that represents biosynthetic silk from which proteins are produced by engineered microbes grown on renewable/biomass-derived feedstocks. Such action would require a proposal to be submitted to the relevant governing body's technical committee, where industry experts would draft language and revise until international agreement would be reached.

Not only is the SMARC fiber unique in being made from biomass, the capability of SMARC to engineer an environmentally positive, uniquely performant fibers are vast. While proving the concept of creating biomass from which proteins are produced by engineered microbes grown on renewable/biomass-derived feedstocks, the SMARC team has realized that their lab can engineer a range of fiber characteristics that go beyond the existing manufactured fiber modifications. To determine the most useful combination of fiber characteristics that would be most appropriately translated to the textile and apparel market, the SMARC team has formed an industry advisory board to determine potential end uses and the characteristics of the fibers that would capture the widest possible range of applications.

9.6.4 Future Directions

A major advantage of the SMARC biosynthetic silk is the capacity to replace fibers from petrochemicals that deplete natural resources and sit in landfills for dozens of years, seeping microfibers and toxic chemicals into the environment. Future directions include attention to the wet processing solvent, working to identify a new eco-positive solvent that can be recycled at a high efficiency without compromising the properties of the fiber, that assures scalability, and that attains a mass-production price point competitive with petrochemical fibers.

Currently producing biomass fiber quantities in milligrams, the next important phase of SMARC's work will be to magnify production at low cost and high yield, so that the SMARC bio-fibers will compete on a level playing field with petrochemical fibers and be adopted by the textile and apparel industry in a significant volume. To this end, the SMARC team is exploring processes that allow large-scale fermentation and verification of the fiber at a large scale and low cost. In addition to physical experimentation in the SMARC laboratories, the team is engaging AI models and simulations to determine and evaluate scalable processes.

9.6.5 Industry Collaboration for Technology Translation

Inventing a new fiber that integrates a range of fiber characteristics that go beyond existing manufactured fibers is in some ways a double-edged sword; when you can do absolutely anything where does one put development effort? To determine the ideal innovative bio-fiber target characteristics, the SMARC research team is cultivating relationships with textile and apparel companies, medical professionals, and other experts who are advising about industry and market gaps that will inform how to set performance targets.

The scientific team needs these industry collaborators because they don't know what properties would be best and they can't afford to spend time on engineering properties that have no value to the textile industry (Marcus Foster, interview by Mary Ruppert-Stroescu, November 13, 2025). The SMARC team's first approach was to work backward from end use through fabric properties, and from fabric properties, into fiber properties. The SMARC team is passionate about developing materials that would give so many new properties that could truly improve the fashion industry and beyond. Applications for transportation, medicine, and space travel are all feasible fields where applications for this technology can be found.

When one can design a result from a blank slate, questions arise that others don't consider. For example, the SMARC team can decide how they want to implement biodegradability into the new bio-fiber. The SMARC team is exploring aspects of the type and length of biodegradability wanted/needed, as they can implement any parameter. Addressing how quickly the fiber should ideally biodegrade, and in what conditions, will be key. For instance, having the bio-fibers biodegrade in a consumer's closet is not ideal, biodegrading in a humid and no light environment, or in salt water, may be desired.

The SMARC team's longer-term goals include further engaging AI to simulate the design of a new fiber that would determine the range of unique properties to meet a significant end use and at the same time drive down the cost of production. Other goals include to identify a market, such as the military, where money is not a limitation and where the performance of the material is so critical and so high that paying a premium for the fiber would be a priority. Overall, SMARC strives to attain a high and unique performance level with the SMARC bioengineered fiber.

9.7 CONCLUSION

Examining these three cases where different levels of creative contributions will propel the fields of textiles and apparel production forward provides insight into the background, research and development, challenges, and future directions of this select group of innovators. The creative contributions of these cases have the potential to shift the fields of fashion apparel and textile production from one that is currently polluting the environment at frightening levels to an eco positive direction. These creative contributions will promote to a healthier fashion industry ecosystem by eliminating environmental contaminants, reducing waste, and avoiding products derived from petrochemicals.

Focusing on process-level creativity and innovation highlights a perspective that often is hidden when observing end-use products but that is nonetheless essential to propelling the textile and apparel industry forward in an eco-effective way by realizing the importance of economic prosperity while at the same time preserving the planet for future generations.

Through the lens of these three cases, we have seen that while having an idea, even developing a disruptive, innovative technology through research is an important step, but translating that idea, that disruptive technology, into industry-level mass adoption for maximum impact implies challenges due to organization- and system-level resistance to technology adoption. Barriers to technology adoption inherent in these cases include concern of reduced earning margins, risk adversity, and lack of receptiveness to integrating new ideas into old practices. At the same time, progress in technology adoption is evident through CiCLO's growing market share, the patenting of the RECLÉM process, and increased collaboration between fashion industry professionals, scientists, and engineers seeking to innovate in textiles and fibers.

Each of the creativity types discussed, forward incrementation, redirection, and reinitiation, gives a qualitatively different approach to innovation for the fields of textile and apparel. The cases are unified by their eco-positive goals and the common focus on fibers and textiles, yet the innovator's backgrounds show highly heterogenous motivations for initiating the creative process; CiCLO is the result of industry professionals attaining a saturation point with the status quo, RECLÉM evolved from parental influence and is rooted in textile craft, and SMARC bio-fibers are emerging from scientific ambition with an eco-positive drive. It is important to understand that identifying types of creative contributions to a field is neither hierarchical nor implies equality in impact. Each type of creative contribution adds a unique value to helping the field move forward.

REFERENCES

Al-Emran, M., and Griffy-Brown, C. 2023. The role of technology adoption in sustainable development: Overview, opportunities, challenges, and future research agendas. *Technology in Society* 73: 102240.

Black, S. 2011. Sustainable design strategies: Eco chic the fashion paradox. *Journal of the Textile Society* 38: 24–30.

Borneman, J. 2025. JEC World 2025, Textileworld.com. https://www.textileworld.com/textile-world/knitting-apparel/2025/07/ciclo-technology-expands-into-new-markets-and-announces-additional-brand-collaborations-ahead-of-functional-fabric-fair-nyc/ [Accessed 10 December 2025].

Braungart, M., McDonough, W. and Bollinger, A. 2007. Cradle-to-cradle design: Creating healthy emissions: A strategy for eco-effective product and system design. *Journal of Cleaner Production* 15(13–14): 1337–1348.

Carroll, L. S. L. 2017. A comprehensive definition of technology from an ethological perspective. *Social Sciences* 6(4): 126.

Chen, H. L. and Burns, L. D. 2006. Environmental analysis of textile products. *Clothing and Textiles Research Journal* 24(3): 248–261.

Dai, X., Ye, X., Shi, L., Yu, S., Wang, X. and Zhong, B. 2024. High mechanical property silk produced by transgenic silkworms expressing the Drosophila Dumpy. *Frontiers in Bioengineering and Biotechnology* 12: 1359587.

Dastane, D. O. 2020. The impact of technology adoption on organizational productivity. *Journal of Industrial Distribution and Business* 11(4): 7–18.

Egan, J. and Salmon, S. 2022. Strategies and progress in synthetic textile fiber biodegradability. *SN Applied Sciences* 4(1): 22.

Fletcher, K. 2008. *Sustainable Fashion and Clothing*. Design Journeys, Earthscan, London.

Grand View Research. 2024. Synthetic fibers market (2023–2030). https://www.grandviewresearch.com/industry-analysis/synthetic-fibers-market [Accessed 16 November 2025].

Hooks, D., Davis, Z., Agrawal, V. and Li, Z. 2022. Exploring factors influencing technology adoption rate at the macro level: A predictive model. *Technology in Society* 68:101826.

Hughes, D. J., Lee, A., Tian, A. W., Newman, A. and Legood, A. 2018. Leadership, creativity, and innovation: A critical review and practical recommendations. *The Leadership Quarterly* 29(5): 549–569.

Lakhwani, M., Dastane, O., Mohd Satar, N. S. and Johari, Z. 2020. The impact of technology adoption on organizational productivity. *Journal of Industrial Distribution & Business* 11(4): 7–18.

Nova, A., Keten, S., Pugno, N., Redaelli, A. and Buehler, M. 2010. Molecular and nanostructural mechanisms of deformation, strength and toughness of spider silk fibrils. *Nano Letters* 10(7): 2626–2634.

Periyasamy, A. P. 2023. Microfiber emissions from functionalized textiles: Potential threat for human health and environmental risks. *Toxics* 11(5): 406.

Pfaffenberger, B. 1992. Social anthropology of technology. *Annual Review of Anthropology* 21: 491–516.

Ruppert-Stroescu, M. and Hawley, J. M. 2014. A typology of creativity in fashion design and development. *Fashion Practice* 6(1): 9–35.

Ruppert-Stroescu, M. S. 2018. Redirecting the fashion industry towards eco-effective practices. In *Eco-friendly and Fair* (Eds. Heuer, M. and Becker-Leifhold, C.), Routledge, London, 144–152.

Spiber Inc. 2025. More than materials. https://spiber.inc/en [Accessed 27 March 2026].

Sternberg, R. J. 1999. A propulsion model of types of creative contributions. *Review of General Psychology* 3(2): 83–100.

Sternberg, R. J., Kaufman, J. C. and Pretz, J. E. 2001. The propulsion model of creative contributions applied to the arts and letters. *The Journal of Creative Behavior* 35(2): 75–101.

Sternberg, R. J., Kaufman, J. C. and Pretz, J. E. 2013. *The Creativity Conundrum: A Propulsion Model of Kinds of Creative Contributions*. Psychology Press, New York.

Stoneman, P. and Battisti, G. 2010. The diffusion of new technology. In *Handbook of the Economics of Innovation, vol.* 2, (Eds. Hall, B. H. and Rosenberg, N.), North-Holland, Amsterdam, 733–760.

Stornelli, A., Ozcan, S. and Simms, C. 2021. Advanced manufacturing technology adoption and innovation: A systematic literature review on barriers, enablers, and innovation types. *Research Policy* 50(6): 104229.

Traylor, S. D., Granek, E. F., Duncan, M. and Brander, S. M. 2024. From the ocean to our kitchen table: Anthropogenic particles in the edible tissue of US West Coast seafood species. *Frontiers in Toxicology* 6: 1469995.

Vergragt, P., Akenji, L. and Dewick, P. 2014. Sustainable production, consumption, and livelihoods: Global and regional research perspectives. *Journal of Cleaner Production* 63: 1–12.

Yusuf, S. 2009. From creativity to innovation. *Technology in Society* 31(1): 1–8.

10 Fashion and Textile Design Thinking

Arul Jothi Biniya, Harsha Rani, Kakoli Das, and Arumugam Natarajan Monica

10.1 UNDERSTANDING DESIGN THINKING

Design thinking has emerged as a widely accepted methodology for "out-of-the-box" thinking among designers and innovators to foster innovation and problem-solving in areas such as product design, service design, and social innovation. The methodology has been widely applied across industries ranging from healthcare to fashion. As a human-centered and iterative approach, design thinking is typically described as a five-step process, as represented in Figure 10.1, beginning with developing a comprehensive understanding of user requirements, defining problems, and generating innovative solutions through prototyping and testing (Brown 2008; Razzouk and Shute 2012; Dam and Siang 2021; Bertola et al. 2026). The chapter further discusses the design thinking stages in detail, followed by the application of methodology in the field of fashion and design.

10.1.1 EMPATHIZE

Design thinking begins with a deep understanding of users, including their behaviors, motivations, wants, and needs. The goal is to develop empathy for users by understanding their perspectives and the context in which the problem exists. This stage ensures that the design process is grounded in real human experiences rather than assumptions. This understanding helps organizations frame problems more effectively and identify opportunities for innovation. Methods commonly used in this stage include interviews, observations, ethnographic research, user journey mapping, and developing personas (ElShishtawy et al. 2022; Pan et al. 2015; Buchanan 1992).

10.1.2 DEFINE

In the define stage, the insights gathered during the empathy phase are analyzed and synthesized to clearly articulate the design problem. Designers identify patterns and key issues emerging from user research and translate them into a clear problem statement. This stage helps frame the challenge in a way that focuses on the user's needs. A well-defined problem statement guides the rest of the design process and ensures that design solutions remain user-centered (D'Itria and Vacca 2024; Aus et al. 2021; Fletcher and Grose 2012).

 DOI: 10.1201/9781003689225-10

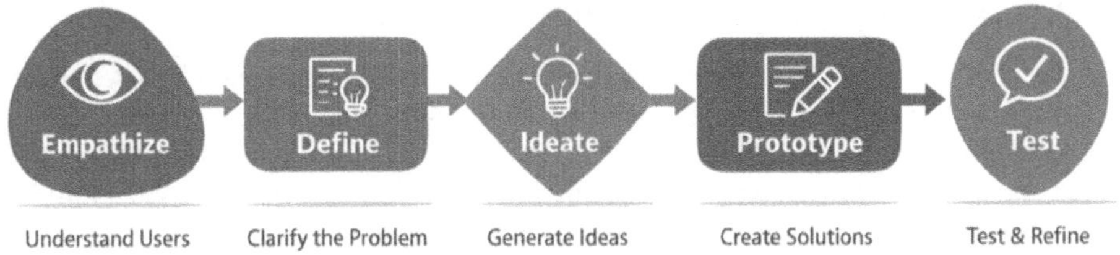

FIGURE 10.1 Design thinking process.

10.1.3 Ideate

The ideation stage focuses on generating a wide range of potential solutions to the defined problem. Designers explore creative possibilities without immediately evaluating or restricting ideas. Brainstorming, mind mapping, sketching, and collaborative workshops are commonly used methods in this phase. The goal is to encourage innovative thinking and explore multiple perspectives before narrowing down to promising concepts (D'Itria and Vacca 2024; Pan et al. 2015; Plattner et al. 2011).

10.1.4 Prototype

Prototyping involves developing tangible representations of ideas so that they can be explored and evaluated. Prototypes may take many forms, such as sketches, digital models, physical mock-ups, or interactive simulations. These prototypes do not need to be fully functional; rather, they serve as experimental tools that allow designers to test ideas quickly and identify potential improvements (D'Itria and Vacca 2024; Razzouk and Shute 2012).

10.1.5 Testing

In the testing stage, prototypes are evaluated with users to gather feedback on how well the proposed solution addresses their needs. User feedback helps designers understand the strengths and weaknesses of the design and identify areas for improvement. The testing stage often leads to further iterations of the design process, where designers refine prototypes, revisit problem definitions, or generate new ideas (Bertola et al. 2026; Pan et al. 2015).

10.2 ITERATIVE NATURE OF DESIGN THINKING

Although the design thinking process is often presented as a sequence of stages, it is not strictly linear. Designers frequently move back and forth between stages as new insights emerge. For example, feedback obtained during testing may lead to redefining the problem or generating new ideas. This iterative process allows designers to continuously refine solutions and respond to evolving user needs (Brown and Katz 2011).

10.2.1 Importance of the Design Thinking Process

The design thinking process is widely valued because it integrates creativity, user understanding, and experimentation. By prioritizing empathy and iterative testing, design thinking helps develop solutions that are both innovative and meaningful for users. As a result, the approach has become increasingly important in areas such as product innovation, service design, business strategy, and social innovation (Simon 1996).

10.2.2 The Application of Design Thinking in Fashion and Textile Design

Design thinking is being applied in fashion and textile design as a human-centered approach that supports innovation, sustainability, and responsiveness to changing consumer needs. In the textile and fashion industry, the design thinking process begins with empathizing with users, where designers study consumer lifestyles, preferences, and challenges through research and observation. The define stage helps identify key issues such as sustainability concerns, ethical production practices, or changing fashion demands. During the ideation stage, designers generate creative solutions for garments, materials, and production methods, exploring possibilities such as sustainable fabrics, improved supply chains, and inclusive design approaches. These ideas are then developed into prototypes, such as sample garments, textile swatches, or process models, which allow designers to experiment with new concepts. Finally, in the testing stage, prototypes are evaluated through user feedback to refine the design and ensure that the final product is desirable, feasible, and viable. By integrating empathy, ideation, prototyping, and testing, design thinking enables fashion and textile companies to develop innovative products while addressing broader challenges such as sustainability, inclusivity, and ethical production practices (Dam and Siang 2021; Valentine et al. 2017; Liedtka 2018).

10.2.3 Design Thinking for Sustainable and Circular Fashion and Textile Design

As the fashion industry faces increasing pressure to mitigate its environmental footprint, design thinking emerges as a critical methodology to reconcile creative innovation with ecological responsibility. Furthermore, this methodology prioritizes interdisciplinary cooperation and radical collaboration to address systemic overconsumption, shifting the industry focus from high-volume output to value-driven, sustainable innovation (Bothner and Grundmeier 2023; Bertola et al. 2026). In order to achieve the goal of sustainability, it is important that the application is carried out throughout the entire process of fashion and textile design, right from the conceptualization phase of fiber selection and material sourcing, extending through patternmaking and manufacturing to the final stages of the garment's life cycle (Hameed et al. 2023; Sayem 2022). The stages in fashion and textile design are represented in Figure 10.2.

In continuation of the stages, Table 10.1 maps different activities that can be carried out in each of the stages from the sustainability perspective. It clearly shows how we can incorporate the principles of conscious design by carefully choosing eco-friendly materials and processes which contribute to minimizing waste or increasing the end of life of the product (Sayem 2022; Shcherbak et al. 2026).

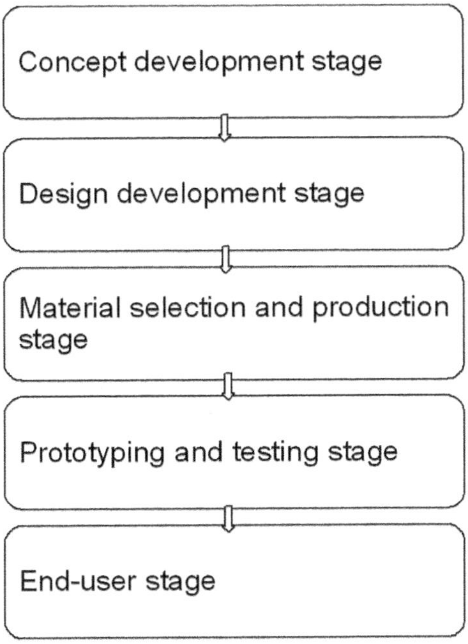

FIGURE 10.2 Fashion and textile design stages.

TABLE 10.1
Mapping Sustainability-Related Activities to Each Stage

Stage	Sustainability-Oriented Activities
Concept development stage	Identifying sustainability goals; researching environmental and social impacts of fashion; studying sustainable fashion trends; understanding user needs and ethical consumption behaviors; developing eco-conscious design concepts; incorporating circular design thinking; defining life cycle considerations.
Design development stage	Creating designs that minimize waste; developing durable and timeless styles; designing modular or multifunctional garments; integrating repairability and recyclability into design; using digital design tools to reduce sampling waste; optimizing pattern layouts for material efficiency.
Material selection and production stage	Selecting eco-friendly or certified materials (organic cotton, recycled fibers, biodegradable textiles); evaluating material life cycle impacts; choosing low-impact dyeing and finishing methods; sourcing materials from ethical and transparent supply chains; implementing energy-efficient and low-waste manufacturing processes.
Prototyping and testing stage	Developing samples using minimal resources; testing garment durability and longevity; evaluating comfort, functionality, and performance; assessing environmental impact during sampling; gathering user feedback for sustainable improvements; refining designs to reduce production waste.
End-user stage	Encouraging responsible consumption; promoting garment care practices that extend product life; enabling repair, reuse, and upcycling; designing for recyclability and circular economy models; providing consumer education on sustainability; facilitating take-back or recycling programs.

10.2.4 Achieving Sustainability through Design Thinking

Empathy, the first step, is very important in determining avenues for sustainability. Designers need to know not only what people want, but also what they do that leads to unsustainable practices, like discarding clothes periodically or choosing affordable fast fashion. Studies reveal that an emotional connection to clothing substantially impacts the lifespan of the garment (Niinimäki and Armstrong 2013; Shcherbak et al. 2026). Designers can empathize with the users; gain an insight into the emotional and utilitarian value they place on clothes and then create garments that people will use for longer and dispose of less. By integrating iterative, human-centered processes, designers can effectively transform textile production waste into high-value garments, thereby fostering a more circular ecosystem (Koca and Koç 2020).

Empathy is not limited to the consumers, but it additionally includes workers, artisans, and other people in the supply chain. For sustainable fashion to become a practice, one needs to know the social and economic conditions under the apparel are manufactured (Bertola et al. 2026). For example, collaborating with artisans or craft groups might lead to creative solutions that help local economies while retaining traditional textile knowledge. Such collaborative approaches are consistent with the participatory principles of design thinking and can promote more equitable production systems.

The define stage involves reframing sustainability challenges into actionable design problems. Instead of approaching sustainability solely as an environmental issue, designers may define problems in terms of user behavior, emotional engagement, or product life cycle. For instance, a designer might frame the challenge as "How might we design garments that adapt to changing user needs over time?" This reframing opens opportunities for solutions such as modular clothing, repairable garments, or adaptable designs that extend product lifespan (Aus et al. 2021).

The ideation stage encourages designers to explore a wide range of sustainable solutions. Ideas may include the use of alternative materials such as biodegradable fibers, innovative textile structures that reduce waste, or new business models such as clothing rental and resale platforms. Designers can act as agents that facilitate a shift toward design for disassembly, enabling practitioners to navigate the complexities of material blends while prioritizing recyclability (Forst 2022; Glogar et al. 2025). Creative ideation often benefits from interdisciplinary collaboration, bringing together designers, engineers, environmental scientists, and social researchers to develop holistic solutions.

In the prototyping stage, ideas are translated into experimental products or systems that can be evaluated. In sustainable fashion, prototypes may include garments made from recycled materials, zero-waste pattern cutting techniques, or digitally simulated garments that reduce physical sampling. Advances in digital technologies such as 3D garment visualization allow designers to test and refine designs without producing multiple physical samples, thereby reducing material waste during the design process (Glogar et al. 2025; Jung and Suh 2026; Lee 2026).

Finally, the testing stage evaluates the effectiveness of sustainable solutions in real-world contexts. User testing may reveal whether consumers are willing to adopt new behaviors such as garment repair, clothing sharing, or extended product use. Feedback from stakeholders can also highlight practical challenges related to manufacturing, cost, or scalability (Zou et al. 2022; Wang et al. 2022).

These five stages of design thinking aids designers in graduating from incremental advances and investigating concepts that pursue sustainability on a systemic scale. Through the integration of empathy, experimentation, and interdisciplinary collaboration, design thinking fosters the creation of fashion systems that emphasize environmental responsibility, social equality, and enduring value. The following sections discuss few case studies that have effectively addressed problems through various design thinking methods.

10.3 CASE STUDIES DEMONSTRATING DESIGN THINKING IN SUSTAINABLE FASHION

The fashion industry has increasingly adopted design thinking methodologies to address sustainability challenges such as resource consumption, textile waste, and ethical production. Design thinking, with its emphasis on user-centered problem-solving, experimentation, and iterative development, enables designers and organizations to create innovative solutions that integrate environmental, social, and economic sustainability. The following case studies illustrate how design thinking has been applied across different areas of the fashion ecosystem, including product innovation, production processes, material development, and circular consumption models.

10.3.1 Adaptive Clothing Innovation: Petit Pli

Petit Pli represents a notable example of design thinking applied to sustainable fashion through product innovation. Developed by Ryan Mario Yasin, the concept emerged from observing the problem that children rapidly outgrow their clothes, resulting in frequent replacement and increased textile waste. Through a user-centered design approach, the designer investigated the needs of both children and parents and explored alternative textile structures that could adapt to growth (Petit Pli 2025).

The resulting innovation uses origami-inspired pleating and advanced textile engineering, allowing garments to expand in both length and width as the child grows. These adaptive garments can accommodate multiple sizes over several years, significantly extending the product life cycle. By focusing on durability and adaptability, Petit Pli demonstrates how design thinking can address sustainability challenges through innovative material engineering and life cycle-oriented design strategies.

10.3.2 Social Sustainability and Craft Preservation: Amounee

Amounee provides an example of how design thinking can support social sustainability and cultural preservation within the fashion ecosystem. The platform connects rural artisans and traditional craft communities directly with consumers through a digital marketplace. By studying the challenges faced by artisan communities—such as limited market access and the loss of traditional crafts—the platform developed a solution that bridges the gap between artisans and contemporary markets (Amounee 2025).

Through the integration of technology and traditional craft knowledge, Amounee enables artisans to showcase handloom textiles, embroidery, and regional craft traditions while receiving fair compensation for their work. The platform also promotes transparency in the supply chain and encourages consumers to support ethically produced fashion. This approach demonstrates how design thinking principles such as empathy, collaboration, and stakeholder engagement can help create inclusive and sustainable fashion systems.

10.3.3 Process Innovation for Resource Efficiency: Levi Strauss & Co.

Large-scale fashion brands have also applied design thinking to improve sustainability in manufacturing processes. Levi's Water Less™ initiative is a prominent example of innovation aimed at reducing water consumption in denim production. Traditional denim finishing processes require large volumes of water, particularly during washing and dyeing stages (Levi Strauss & Co 2025). Through experimentation and collaboration with manufacturing partners, Levi's developed new finishing techniques that significantly reduce water usage while maintaining product quality and aesthetics. This initiative involved iterative testing, process redesign, and technological innovation, reflecting core principles of design thinking. By rethinking conventional production methods, the company was able to improve resource efficiency and reduce the environmental impact of denim manufacturing.

10.3.4 Circular Fashion and Consumer Engagement: Patagonia—Worn Wear Initiative

The Worn Wear initiative by Patagonia illustrates how design thinking can promote circular fashion models by encouraging repair, reuse, and extended product life. Recognizing that consumer behavior plays a significant role in the environmental impact of fashion, Patagonia focused on understanding how customers interact with their clothing and why garments are often discarded prematurely (Patagonia 2025).

The Worn Wear program encourages customers to repair and reuse their garments rather than replace them. Through repair services, resale platforms, and consumer education, the initiative promotes a shift toward responsible consumption and longer garment life cycles. By engaging users in sustainable practices, Patagonia demonstrates how design thinking can influence both product design and consumer behavior to support circular economy principles.

10.3.5 Material Innovation for Environmental Impact: Adidas–Parley Ocean Plastic Collaboration

Another example of design thinking applied to sustainability is the collaboration between Adidas and Parley for the Oceans. This initiative addresses the growing problem of marine plastic pollution by transforming recovered ocean plastic into performance sportswear materials. Through collaboration with environmental organizations and material scientists, Adidas explored innovative ways to integrate recycled plastic waste into footwear and apparel design (Adidas 2025).

The project involved extensive prototyping and testing to ensure that recycled materials met the functional and performance requirements of sportswear products. As a result, Adidas successfully developed shoes and apparel made partially from recycled ocean plastic, demonstrating how design thinking can drive material innovation and environmental responsibility within product development.

10.3.5.1 Findings and Analysis

These case studies highlight the diverse ways in which design thinking can contribute to sustainability in the fashion industry. From adaptive clothing and craft preservation to manufacturing innovation, circular consumption models, and recycled materials, design thinking provides a framework for addressing complex environmental and social challenges (Table 10.2).

By integrating user-centered research, experimentation, collaboration, and iterative development, fashion designers and organizations can create products and systems that balance creativity with sustainability. The examples discussed demonstrate that design thinking is not limited to product aesthetics but can influence the entire fashion value chain—from concept development and production to consumer engagement and product life cycle management.

TABLE 10.2
Comparative Findings from the Cases

Case Study/ Initiative	Brand/ Organization	Design Thinking Application	Sustainability Impact
Petit Pli adaptive clothing	Petit Pli	User-centered innovation inspired by children quickly outgrowing clothes; experimentation with origami-inspired textile structures.	Garments expand as children grow, extending product life cycle and reducing textile waste.
Amounee artisan platform	Amounee	Platform designed to connect rural artisans directly with consumers while preserving traditional textile crafts.	Promotes ethical fashion, supports artisan livelihoods, and sustains traditional craft practices.
Water Less™ manufacturing process	Levi Strauss & Co.	Reimagining denim finishing techniques through experimentation and iterative process improvement.	Significant reduction in water usage during denim production.
Worn Wear initiative	Patagonia	Focused on understanding user behavior around clothing longevity, repair, and reuse.	Encourages repair, reuse, and circular consumption to reduce textile waste.
Parley ocean plastic collaboration	Adidas	Collaborative design thinking with environmental organizations to rethink material sourcing and product development.	Development of footwear and apparel using recycled ocean plastic, reducing marine pollution.

10.4 OPPORTUNITIES AND CHALLENGES IN APPLYING DESIGN THINKING TO SUSTAINABLE FASHION

The integration of design thinking into sustainable fashion presents significant opportunities for innovation and systemic transformation within the fashion industry. At the same time, designers and organizations face several challenges in implementing sustainable practices effectively. Understanding these opportunities and challenges is essential for developing strategies that enable sustainable innovation across the fashion value chain.

10.4.1 Innovation in Sustainable Materials

Design thinking encourages experimentation and prototyping, which can lead to the development of innovative materials such as recycled fibers, biodegradable textiles, and bio-based fabrics. These materials can significantly reduce the environmental footprint of fashion products while enabling designers to explore new aesthetic and functional possibilities (Aus et al. 2021).

10.4.2 Circular Fashion Models

One of the key opportunities provided by design thinking is the development of circular fashion systems. Designers can create garments that are durable, repairable, recyclable, or adaptable to extend product life cycles. Initiatives such as clothing repair programs, resale platforms, and modular garments demonstrate how design thinking can support circular economic principles in fashion (Lee 2025).

10.4.3 User-Centered Sustainable Design

Design thinking emphasizes empathy and a deep understanding of user behavior. This approach enables designers to create products that align with consumers' sustainability values, encouraging responsible consumption practices such as garment care, reuse, and recycling.

10.4.4 Collaboration and Interdisciplinary Innovation

Design thinking promotes collaboration among designers, engineers, material scientists, manufacturers, and consumers. Such interdisciplinary collaboration can accelerate innovation in sustainable materials, production processes, and business models.

10.4.5 Preservation of Traditional Craft and Social Sustainability

By engaging with local communities and artisans, design thinking can help preserve traditional textile crafts while creating new market opportunities. Platforms that connect artisans with consumers contribute to economic empowerment and cultural sustainability.

10.5 CHALLENGES

10.5.1 High Cost of Sustainable Materials

Many sustainable materials and production technologies remain more expensive than conventional alternatives. This can limit their adoption, particularly for small fashion businesses or cost-sensitive markets.

10.5.2 Technological and Manufacturing Limitations

Although innovative sustainable materials are emerging, scaling them for mass production can be challenging. Manufacturing infrastructure may not always support new materials or processes.

10.5.3 Consumer Awareness and Behavior

Despite increasing awareness of sustainability, many consumers still prioritize price and trends over environmental considerations. Changing consumer behavior requires education, awareness campaigns, and accessible sustainable products.

10.5.4 Complexity of the Fashion Supply Chain

The global fashion supply chain involves multiple stakeholders, making it difficult to ensure transparency and sustainability at every stage. Implementing sustainable practices often requires coordination across different suppliers and production networks.

10.5.5 Balancing Sustainability with Fashion Trends

Fashion trends change rapidly, often encouraging overconsumption. Designers must balance the industry's demand for novelty with the need to promote long-lasting and sustainable products.

10.6 CONCLUSION

While the application of design thinking offers promising opportunities to transform the fashion industry toward sustainability, several barriers must be addressed to achieve widespread adoption. Overcoming these challenges requires collaboration among designers, industry stakeholders, policymakers, and consumers. By leveraging the innovative and user-centered nature of design thinking, the fashion industry can move toward more responsible and sustainable production and consumption practices.

REFERENCES

Adidas. 2025. ADIDAS x PARLEY. https://parley.tv/initiatives/adidasxparley [Accessed 24 March 2026].

Aus, R., Moora, H., Vihma, M., Unt, R., Kiisa, M. and Kapur, S. 2021. Designing for circular fashion: Integrating upcycling into conventional garment manufacturing processes. *Fashion and Textiles* 8: 34.

Amounee. 2025. From one home to home. https://amounee.com/ [Accessed 24 March 2026].

Bertola, P., Colombi, C., Celi, M. and Rodriguez Schön, V. 2026. Fashion futures as design scenarios for the triple transition framework. *Platforms* 4(2): 5.

Bothner, S. and Grundmeier, A.-M. 2023. Education for sustainable development through design thinking. *AIP Conference Proceedings* 2889(1): 040010.

Brown, T. 2008. Design thinking. *Harvard Business Review* 86(6): 84–92.

Brown, T. and Katz, B. 2011. Change by design. *Journal of Product Innovation Management* 28(3): 381–383.

Buchanan, R. 1992. Wicked problems in design thinking. *Design Issues* 8(2): 5–21.

Dam, R. F. and Siang, T. Y. 2021. What is design thinking and why is it so popular? Interaction Design Foundation. https://www.interaction-design.org/literature/article/what-is-design-thinking-and-why-is-it-so-popular [Accessed 28 March 2026].

D'Itria, E. and Vacca, F. 2024. Shaping sustainable solutions in fashion through design-led strategies, approaches, and practices. *Discover Sustainability* 5: 409.

ElShishtawy, N., Sinha, P. and Bennell, J. A. 2022. A comparative review of zero-waste fashion design thinking and operational research on cutting and packing optimization. *International Journal of Fashion Design, Technology and Education* 15(2):187–199.

Fletcher, K. and Grose, L. 2012. *Fashion and Sustainability: Design for Change*. Laurence King Publishing, London.

Forst, L. 2022. Textile thinking in practice: Creative textile design methods as research in a circular economy. *DRS Biennial Conference Series, DRS 2022*: Bilbao, 25 June to 3 July, Spain.

Glogar, M., Petrak, S. and Mahnić Naglić, M. 2025. Digital technologies in the sustainable design and development of textiles and clothing: A literature review. *Sustainability* 17(4): 1371.

Hameed, U., Zaheer, S., Amin, N. and Hussain, S. 2023. Exploring low-waste patternmaking techniques for sustainable solutions in fashion industry. *Journal of Design and Textiles* 2(2): 41–64.

Jung, D. and Suh, S. 2026. AI as a creative partner in fashion textile design: Pedagogical case study of AI literacy practices. *International Journal of Fashion Design, Technology and Education*, 1–15. https://doi.org/10.1080/17543266.2026.2630264

Koca, E. and Koç, F. 2020. Example of iterative process in upcycled clothing design: Unused neckties and upholstery scraps. *The Research Journal of the Costume Culture* 28(6): 890–911.

Lee, H. 2025. A user-centered recycling fashion design process: A sustainable approach through emotional awareness and engagement with clothing waste. *Fashion and Textiles* 12: 14.

Lee, Y. 2026. Digital fashion ideology: Towards a critical public sphere. *International Journal of Cultural Studies* 29(1): 75–99.

Levi Strauss & Co. 2025. How Levi's® is saving water. https://www.levistrauss.com/2019/03/25/world-water-day-2019-saving-h2o/ [Accessed 20 March 2026].

Liedtka, J. 2018. Why design thinking works. *Harvard Business Review* 96(5): 72–79.

Niinimäki, K. and Armstrong, C. 2013. From pleasure in use to preservation of meaningful memories: A closer look at the sustainability of clothing via longevity and attachment. *International Journal of Fashion Design, Technology and Education* 6(3): 190–199.

Pan, Y., Roedl, D., Blevis, E. and Thomas, J. C. 2015. Fashion thinking: Fashion practices and sustainable interaction design. *International Journal of Design* 9(1): 53–66.

Patagonia – Worn Wear Initiative. 2025. https://wornwear.patagonia.com/ [Accessed 24 March 2026].

Petit Pli. 2025. Clothes that grow with your child. https://shop.petitpli.com/?srsltid=AfmBOorh7CfKXi-8STUPthzOka-Zoas53jpIEB0nsjxj8E329mzaWI2y [Accessed 28 March 2026].

Plattner, H., Meinel, C. and Leifer, L. 2011. *Design Thinking: Understand – Improve – Apply*. Springer, Berlin, Heidelberg.

Razzouk, R. and Shute, V. 2012. What is design thinking and why is it important? *Review of Educational Research* 82(3): 330–348.

Sayem, A. S. M. 2022. Digital fashion innovations for the real world and metaverse. *International Journal of Fashion Design, Technology and Education* 15(2): 139–141.

Shcherbak, V., Dorokhov, O., Riashchenko, V., Storozhuk, M., Bertoncelj, A. and Meško, M. 2026. From virtual prototyping to digital fashion: How emerging technologies are setting new standards for sustainability in the creative industries. *Sustainability* 18(7): 3281.

Simon, H. A. 1996. *The Sciences of Artificial* (3rd Ed.). MIT Press, Cambridge, MA.

Valentine, L., Ballie, J., Bletcher, J., Robertson, S. and Stevenson, F. 2017. Design thinking for textiles: Let's make it meaningful. *The Design Journal* 20: S964–S976.

Wang, M., Kim, E. and Du, B. 2022. Promoting emotional durability and sustainable fashion consumption through art derivatives design methods. *The Design Journal* 25(5): 789–806.

Zou, Y., Pintong, S., Shen, T. and Luh, D. B. 2022. Evaluation and trend of fashion design research: Visualization analysis based on CiteSpace. *Fashion and Textiles* 9(1): 45.

11 Digital Technologies and Sustainability in the Fashion, Textile, and Clothing Industries

Patrick Hui and Qilong Feng

11.1 INTRODUCTION

The design, fashion, and textile industries are facing increasing pressure to adopt sustainable practices, driven by growing environmental concerns and a rising demand for ethical production. Fashion sustainability, or 'eco fashion', constitutes a multifaceted concept focusing on the system maintainability encompassing design, manufacturing, and consumption philosophy that creates regularity in terms of minimizing human impact on the environment while upholding social responsibility (Kozlowski et al. 2019; Woodside and Fine 2019). Therefore, it addresses three interrelated dimensions: the reduction of environmental impact, the promotion of social responsibility, and the assurance of economic viability (Daukantiene 2022). Digital innovation is emerging as a pivotal force in transforming this sector by improving efficiency, minimizing waste, and enabling circular economy frameworks. Digital technologies, such as artificial intelligence (AI), 3D printing, blockchain, and the Internet of Things (IoT), are revolutionizing processes of design, manufacturing, and supply chain transparency. AI-driven predictive analytics optimize material utilization, while virtual prototyping minimizes physical waste. Blockchain technology enhances traceability, thereby supporting ethical sourcing practices, and IoT technologies improve resource management through comprehensive data acquisition. Additionally, digital platforms foster sustainable consumer behavior through resale and rental markets. This chapter examines the role of digital technologies in accelerating sustainability efforts in the fashion and textile industries, emphasizing key technological advancements, challenges, and prospective opportunities for a more eco-conscious industry.

11.2 DIGITAL TECHNOLOGIES IN SUSTAINABILITY, TRACEABILITY, AND SMART MANUFACTURING

11.2.1 Artificial Intelligence and Machine Learning

The integration of AI in the fashion design process successfully assists designers by automating repetitive design tasks, allowing them to concentrate on the

 DOI: 10.1201/9781003689225-11

decision-making aspects. Machine learning algorithms, by forecasting fashion trends and design elements, align design choices more closely with consumer preferences, which in turn reduces errors in production planning and minimizes surplus inventory. Technological practices enable retailing to move closer to the design phase of the supply chain, directly generating styles that drive more sales (Giri et al. 2019). In the context of the rapid expansion of e-commerce, the vast volume of data reflecting the consumer preferences and fashion tastes – derived from retail transactions and social media platforms – in turn nurtured the iterative update of the AI techniques to facilitate the creative part of the fashion industry (Kotouza et al. 2020). Techniques such as deep learning, generative adversarial networks (GANs), machine learning, and natural language processing (NLP) are integrated into interactive algorithms that analyzes online data and provides recommendations (Liu et al. 2019; Kim and Cho 2000; Shushi and Abdulazeez 2024; Wu and Li 2024; Imtiaz et al. 2024). Such innovation enhances industry primarily in three areas: predictive analytics for forecasting fashion trends and managing inventory, generative design for personalized recommendations, and real-time demand forecasting to optimize production processes.

When the utilization of AI and machine learning initially emerged as an approach to enhance the appeal to customers, concurrently, supply chains have already incorporated AI-driven systems to optimize production activities. Due to constraints such as limited financial resources, inadequate technical expertise, and restricted access to high-quality data, the adoption of AI varies considerably between small and medium-sized enterprises (SMEs) and large enterprises. These differences manifest across several dimensions (Ayinaddis 2025), including (1) scale and budget allocation, (2) data resource and technical capacity, (3) impact on workforce or operational procedures, (4) degree of integration and innovation, and (5) areas of application focus.

Large enterprises implement AI extensively to achieve scalability, automation, and innovation throughout the entire supply chain. In contrast, SMEs employ AI tactically for immediate operational improvements, focusing on specific, practical applications. Large manufacturers significantly invest in AI technologies including robotics, machine learning, and advanced algorithms. With their extensive access to global supply chain data, the large enterprises can coordinate comprehensively across multiple departments such as design, production, logistics, and marketing. Routine tasks are replaced with AI-driven processes, including material selection, automated stitching, product tracking, and repetitive operations in product lines and quality inspection. This integration allows for seamless communication across different stages of production and distribution, thereby ensuring smooth and efficient operations across the board. Furthermore, large enterprises can experiment with cutting-edge innovations, such as predictive analytics for fashion trends, virtual reality fashion shows, and 3D garment design, enabling them to rapidly adapt to evolving consumer preferences.

On the contrary, SMEs tend to concentrate on inventory management and quality control, where the return on investment is more immediately visible. This makes SMEs selective in adopting AI technologies, often limiting their application to smaller-scale operations such as customer services, automated pattern generation, or prioritizing customization for personalized batches targeting niche markets.

The implementation of AI in SMEs is predominantly about augmenting human labor rather than replacing it, which allows employees to engage in more creative and specialized tasks. Data resource for SMEs is not sufficient, mainly from internal productions and local sales analytics, which limits them to make sophisticated AI-driven predictions or optimize the supply chains as efficiently as the large enterprises.

AI technologies are implemented by large enterprises and SMEs to promote sustainability, but in different projects. Large enterprises employ AI-driven systems to analyze production processes, facilitating waste reduction, intelligent fabric selection, and the optimization of resource management. Conversely, SMEs typically utilize AI to enhance material usage and minimize waste on a smaller scale. For example, they adopt AI technology in cutting operations to reduce fabric waste, or in predicting production demands to avoid overstocking. No matter how the AI technologies are employed in the effort, the entire industry ultimately benefited from the prompt analysis of technological change.

11.2.2 Internet of Things and Smart Factories

As the fashion industry advances into the 4.0 era, digital technologies are being adopted to enhance the supply chain processes, encompassing design, production, marketing, and sales. The IoT is characterized as 'the novel paradigm that connects a variety of things or objects, such as sensors, tags, actuators, etc., and programs to interact with each other or cooperate with the neighbors through the Internet or wireless communication networks' (Atzori et al. 2010; Gubbi et al. 2013). IoT devices comprised of NFC (near-field communication), RFID (radio-frequency identification), and Bluetooth technologies facilitate traceability technologies using built-in technologies that have been standardized across electronic components (Zhang et al. 2014), therefore assisting the business operational management by capturing the object traceability and integrating visibility procedures, supply chain, monitoring, manufacture, and associated human resource activities in the industry (Akram et al. 2022). Moreover, the network links physical commercial products to cloud-based data systems, enabling real-time monitoring through information sharing via cloud platforms.

Wireless communication technologies are utilized in a wide array of applications, with their specific implementations varying according to their inherent characteristics. RFID uses radio waves to identify and track tags affixed to objects. The operational range of RFID spans from centimeters to several meters and can be utilized either passively or actively. Passive RFID system does not require internal power sources, whereas active tags are battery-powered. NFC, a subset of RFID, operates passively over very short distances, typically up to 10 cm, enabling bidirectional interactive communication. Bluetooth technology facilitates short- to medium-range radio communication, with a range up to 100 m, supporting two-way data exchange. Table 11.1 provides a generalized summary of the principal differences and applications of these major IoT technologies. By comparing power consumption and material usage, the table also highlights their respective sustainability impacts.

The integration of IoT wireless communication technologies in fashion supply chains significantly advance sustainability initiatives. RFID technology is widely

TABLE 11.1
Summary of the Principal Differences between Major IoT Technologies

	RFID	NFC	Bluetooth
Definition	Radio-frequency identification technology	Near-field communication technology	A global standard for short-range wireless interconnections
Bandwidth	Low frequency: 125 kHz High frequency: 13.56 MHz Ultrahigh frequency: 865–960 MHz	13.56 MHz	2.4 GHz
Range	Centimeters to meters	Up to 10 cm	Up to 100 m
Coverage area	Up to 100 m	From 0.01 to 0.1 m	Between 100 and 400 m
Communication model	One-way communication	Two-way communication	Two-way communication
Power consumption	No battery/battery	No battery	Battery-powered
Data transfer rate	Less than 10 kbps -640 kbps (according to frequency)	From 106 to 424 kbps	1 Mbps
Material usage	Minimal: small tags, some e-waste risk	Minimal: small tags, limited recycling	Higher: complex devices with electronics and batteries
Application	Inventory and supply chain management; transportation and logistics; warehouse and yard management	Mobile payments; asset tracking and management; access control	Smartwatches and wearables; access control
Environmental footprint impact	Low (passive) Moderate (active)	Low	Moderate
Role in sustainability	Improves inventory accuracy and reduces waste	Avoids counterfeit and enhances consumer engagement	Enables smart wearable functionalities

used for inventory tracking and management purposes. NFC is predominantly applied in secure data exchanges, including payment processing, authentication, and smart tagging. Bluetooth Low Energy, characterized by its low power consumption, is commonly employed in smart wearable devices and real time tracking applications, supporting more sophisticated features. The incorporation of NFC tags and Bluetooth into personal devices has enhanced consumers' engagement in mitigating accessory overconsumption and has also promoted the development of fabric waste recycling practices. Among the various applications of IoT technologies, the transformation of manufacturing facilities into smart factories represents the most impactful advancement. By integrating tags and readers within management systems, production processes become more transparent and precise, reducing waste and minimizing reliance on manual operations. The integration facilitates seamless coordination among robotic-assisted manufacturing, automated storage solutions, and predictive maintenance systems that enable the realization of 'smart' factory. Smart factories play a central role in the industrial revolution, addressing critical challenges associated with the traditional 'push supply chain' model (Christopher et al. 2004; Fisher 1997; Jin and Shin 2020). The challenges include the demand–supply mismatches from forecasting discrepancy, oversupply driven by technological advancements during the Fashion Industry 3.0 era (Bertola and Teunissen 2018), and inefficiencies in accurately fulfilling consumer demand (Jin and Shin 2021). By utilizing AI-powered automated systems within manufacturing processes, smart factories are evolving to tackle challenges related to environmental sustainability, hyper-personalization, and efficient productivity issues that are closer to consumers (Jin and Shin 2021; Priyadharshini 2023).

As an increasing number of brands are accelerating the network transparency within smart factories in the fashion industry, the application of AI technology to assist the supply chain has become evident: it provides real-time monitoring solutions over the whole supply chain (McKeegan 2024), including the products tracking through IoT sensors, water consumption, and machine efficiency. The collected data are subsequently analyzed to inform predictive maintenance strategy, minimize downtime automatically, and extend machinery lifespan. By maintaining transparency at all organizational levels, the supply chain achieves greater visibility, so it ensures ethical sourcing practices and mitigates the risk of counterfeit products.

11.2.3 3D Printing and Digital Textile Manufacturing

The manufacturing processes extend beyond traditional mass production paradigms. Alternative solutions appear toward on-demand production, alongside the integration of semi-personalized services within e-commerce platforms. Digital technologies have emerged as the optimal tool to facilitate agile and personalized production workflows. An initial application of digital technology was digital textile printing, particularly on polyester fabrics, targeting large-scale sectors such as sportswear, fast fashion, and home textiles (Digital Textile Printing Market 2025). This innovation aimed to reduce water consumption during fabric processing and mitigate environmental pollution. Still, with the integration of additional technological advancements, the trend among manufacturers is to offer one-step digital textile treatment (McKeegan 2023) from concept to retail marketing. 3D printing technology offers efficient and flexible production capabilities, enabling rapid prototyping of complex structures. When combined with recycled materials and advanced fused deposition modeling (FDM) techniques to precisely control the properties of the final product, experimental applications within fashion manufacturing have the potential to diversify significantly (Glogar et al. 2025), and most importantly, contribute to cost reduction in customization.

The shift from conventional manufacturing to 3D printing technology has significantly decreased the consumption of chemicals and water, and reduced environment pollution. Nonetheless, the use of polymeric materials in 3D printing introduces new sustainability concerns. Firstly, 3D printing processes typically generate more energy per unit volume than traditional manufacturing methods like injection molding (Hopkins et al. 2021), which is largely attributed to the heat for the extruder and the layer-by-layer fabrication approach. Take PLA (polylactic acid) material as example; conventional plastic injection molding consumed about 38.2% less energy than FDM process (Enemuoh et al. 2021). In addition, the recyclability of polymer materials such as PLA, polyethylene terephthalate (PET), polypropylene, and polystyrene presents challenges in technological innovation. Most 3D-printed plastics currently end up in landfills, although PLA is biodegradable and compostable under industrial conditions. While PLA can be recycled separately from other plastics, its mechanical properties degrade with recycling, maintaining usable strength for about 2–3 recycling cycles, with significant loss after 3–5 cycles (Hidalgo-Carvajal et al. 2023). Despite this, 3D printing is widely recognized for its potential to reduce overall environmental impact by minimizing waste and enabling on-demand manufacturing, shortening supply chains, and fabricating complex components with less material usage. The predominance of 3D printing effectively prevents overproduction. Therefore, the sustainability of 3D printing hinges on addressing recycling challenges and enhancing energy efficiency, which in turn drives ongoing research and technological advancements in this field.

11.2.4 Blockchain for Ethical Fashion

Blockchain technology is a digital and distributed ledger that exhibits characteristics such as decentralization, persistence, anonymity, and auditability (Zheng et al. 2018). Therefore, not merely in cryptocurrencies, it can be applied to a wide array of applications for the next-generation Internet interaction systems, including but not limited to smart contracts (Kosba et al. 2016), public services (Akins et al. 2014), IoT (Zhang and Wen 2015), reputation management systems (Sharples and Domingue 2016), and security services (Noyes 2016). With promising potential of enhancing traceability and transparency, blockchain technology has yet to be fully leveraged as a holistic solution for revolutionizing the fashion and textile industry (Badhwar et al. 2023). Given the heterogeneous nature of supply chains across various businesses, blockchain technology currently supports fashion industry primarily in three critical areas: provenance tracking of raw materials, anti-counterfeiting in luxury fashion, and the circular economy of garment recycling and reselling.

11.2.4.1 Provenance Tracking

In the mass production of fashion and textiles, a diverse array of materials from various countries is transferred to processors and manufacturers who operate within a complex network comprising multiple partners distributed across extensive geographical spread of the supply chain. This complexity engenders a lack of transparency in tracking the whole process, thereby enabling upstream suppliers to deviate from established standards to feather their nest, or causing intermediate segments to produce quantities exceeding actual demand, resulting in unnecessary costs (Chen et al. 2019; Agrawal et al. 2021). The implementation of private blockchain networks with class-based permissions systems for different stakeholders (Agrawal et al. 2021), combined with integration of to the IoT technologies such as sensors, RFID tags, and application programming interfaces, facilitates the collection of raw material data (Pal and Ansar-UI-Haque 2020; Malik et al. 2018). This data can be traced throughout the supply chain – from fiber producers, spinners, fabric makers, garment makers, and ultimately retailers, thus ensuring the transparency in raw material sourcing within the industry.

11.2.4.2 Anti-Counterfeiting

In the high-end luxury fashion sector, blockchain technology is being advanced to facilitate digital connectivity through non-fungible tokens (NFTs), ensuring the authenticity of luxury products. The NFT-based authentication system is unique: it is minted through the digital ledger of blockchain, where every transaction is immutably recorded (Alexander and Bellandi 2022), thereby preserving evidence of proprietorship, authenticity, and the price of luxury goods (Gallagher 2021). NFTs have established an indispensable role in augmenting value for luxury firms by effectively preventing the circulation of counterfeit goods during transactions. The impact on enhancing the unique consumer experience and fostering long-term relationships with clients, especially in e-commerce (Joy et al. 2022), has subsequently generated new revenue streams for luxury brands (Sung et al. 2023).

11.2.4.3 Circular Economy

When everything, whether tangible or intangible, can be traced virtually in the ledger of blockchain technology, real-time monitoring of industry processes becomes possible, and the circular economy becomes inevitable. The tracking mechanism is also applied to the circular fashion industry to quantify greenhouse gas emissions, carbon footprints, the recycling of second-hand products, and production forecasting, all aimed at mitigating the excessive consumption of raw materials. The concept of a 'material passport' has been introduced in the circular supply chain framework (Ellen MacArthur Foundation 2021), where blockchain technology serves as an optimal solution for creating digital identifiers for materials or products and documenting their transactional history, thus prolonging the life cycle of fashion items. Regardless of the integration of IoT technology, fashion products are registered with intelligent labeling and governed by smart contracts on blockchain platforms for traceability purposes (Ghoreishi and Happonen 2021). Assets can be examined through interactive graphs and data visualizations, alongside on-demand records of provenance, transaction histories, warranty details, and compliance with industry standards (Alves et al. 2022; Chen 2024). By maintaining transparent traceability records, the traditional fashion industry holds potential to transform its negative reputation into a major contributor to the sustainable environment, particularly concerning water usage, chemical pollution, carbon dioxide emissions, and substantial waste generation (Vercalsteren et al. 2019; Niinimäki et al. 2020).

The development of blockchain technology in the fashion industry enhanced transparency and traceability throughout the entire life cycle of fashion products. The interlinks among the three domains are close. Provenance tracking is fundamental to using digital solutions to securely capture data related to a fashion item, spanning from raw materials to the final commercial product delivered to consumers. The tamper-resistant nature of the data ensures the labeled product is verifiable, and this feature strengthens the anti-counterfeiting protection by guaranteeing authenticity.

The traceable provenance data further facilitates open access to information for all stakeholders involved in the supply chain, enhancing auditability during the redistribution of the recycled materials. In this way, digital technology is integrated into the management system of physical products and robustly supports the sustainability efforts in the complex fashion supply chain.

11.3 CASE STUDY

Digital technology is increasingly being adopted in the fashion and textile industry to enhance transparency, traceability, and sustainability. The following section presents key case studies illustrating the application of digital technologies across various sectors.

11.3.1 Application of AI in Fashion Design and Product Development

11.3.1.1 Stitch Fix: AI Stylist for Online Shopping

Designers initially used AI-assisted design to alleviate the burden of creating numerous sketches, but now, the solution of using AI to replace designers has become possible (Guo et al. 2023). The online fashion retailing brand Stitch Fix utilizes recommendation algorithms to generate customized fashion advice, complete with a complete set of personalized outfits (Lake 2018; iSpot 2025).

Machine learning technology is specifically developed for the learning process of creating a latent style map, utilizing a series of individual profiles, including measurements, scenarios, budgets, and other relevant data, to produce a comprehensive outcome for users (Wahba 2019). The membership system requires consumers to pay the subscription fee. Such a cost is for the consultancy service rather than buying mass-produced fashion items, thereby reducing the stress of decision-making for members and further enhancing the reliance of consumers, which in turn creates a user experience of intimacy toward the professional AI stylist (Bhatt et al. 2021).

The AI fashion stylist proved to be successful, effectively increasing customer retention by 25% through hyper-personalization. As an early attempt of using AI-driven interactive recommendation agents (Kim et al. 2022) in the curated retailing, the pivotal challenge of using AI to stimulate the fashion consumption on curated shopping subscription is how to understand whether the consumers' shopping value obtained from the brand is good enough for them to choose outsourced service as their decision-making (Hasan and Liu 2020). The expansion of Stitch Fix presents the possibility of using an AI solution to engage consumers with hedonic goals and ultimately influences the users' intention to meet their expectations (Kim et al. 2022). Furthermore, the machine learning algorithm also assists in decision-making process for the consumers by analyzing available data.

Recognized as one of the world's top '50 Most Innovative Companies' in 2019 (Smily 2019), Stitch Fix is presently undergoing a restructuring process by layoffs and shutting down some facilities of international operations. The trajectory of this fashion innovation enterprise underscores the necessity for strategical realignment and operational modifications to effectively cooperate with emerging technology, particularly in areas such as inventory management and revenue optimization.

11.3.1.2 Zara – AI for Trend Forecasting

For fast fashion brands, the speed of seizing the fashion trend almost determines the success of a whole season's sales. Zara has integrated AI to utilize predictive consumer analytics, adopting a data-driven approach to personalized marketing strategies. This approach replaces reliance on human intuition and experience in determining production and distribution strategies. By partnering with Jetlore, an AI-powered prediction platform, Zara developed a Just-In-Telligent supply chain system using real-time data to ensure timely and optimized product delivery (IFAB Media 2024). It uses computer vision and NLP technologies to 'see' and 'understand' the attributes such as fabric textures, garment styles, design details from social media, runway shows, and sales data (Shi et al. 2021; Shi and Lewis 2020). Big data analytics are applied to evaluate the attributes retrieved from resources and generate predictions of the following significant trends.

Unlike Stitch Fix's using AI to serve as personal stylists for individual customers, Zara employs AI-driven style forecasting primarily for the improvement of supply chain management to avoid overproduction, therefore providing precise designs that align to the target market's preferences. In the context of trend forecast, data regarding fabric textures and styles are aggregated to generalize user persona with specific preferences, which could reflect the popular designs trends in society rather than unique personal styles of individual client. This emerging AI forecasting solution serves undoubtedly as a powerful tool to strengthen the competitive position of fashion brands by accurately fathoming consumers' tastes. From an inventory management standpoint, it reduces overproduction by synchronizing stock levels with real-time demand. Empirical evidence indicates that the reduced unsold inventory is 10%–15%, which further contributes to improved sustainability.

11.3.2 Application of AI in Retail and Customer Experience

11.3.2.1 H&M – AI-Powered Chatbots and Virtual Fitting

Collaborating with Nuance Communications, Inc., H&M launched virtual assistant and live chat agents to users in 2018 (TyN Magazine 2021). Developed based on NLP technology, the AI chatbot is designed to improve the online shopping experience by analyzing browsing behavior, purchase history, and customer preferences to provide tailored, personalized services. The AI agents not only play a role in customer service by tracking orders but also act as styling consultants, giving advice to online shoppers.

Collaborating with NeXR, a technology company spanning the reality value chain, H&M introduced a virtual fitting solution in Thailand in 2023 (Fibre2Fashion 2022). The solution adopted an augmented reality (AR) visual try-on system, integrating the customer's avatar with the cloud mobile app, allowing users to compare the recommended size and eliminating the need for physical fitting rooms. The 3D scanner technology is also developed to enable customers to create a digital twin of themselves with exact body measurements. The avatar is not merely for the online shopping experience but also connects to the physical store. Additionally, the AR virtual fitting can be used when clients scan garments in retail stores, allowing them to select the proper size without queuing in fitting rooms.

The strategy of adopting AI technologies in the pre-sale phase has alleviated the operational pressures faced by physical retailing stores. AI-powered chatbots function as personal consultants, providing customers with tailored recommendations that help avoid excessive purchase and return behaviors, which in turn optimize sales by minimizing redundant logistics processes. The deployment of virtual try-on systems expands the scope of traditional fitting room experience beyond the confines of physical stores, thereby reducing time and labor expenditures for both consumers and brands.

11.3.2.2 Amazon – AI Fashion Recommendations (Echo Look)

Powered by Alexa, Amazon unveiled a voice-controlled selfie camera specifically for the fashion industry. The camera uses the depth information to create 'computer vision-based' image analysis to process the best of users in full-length selfies or videos (Gibbs 2017). The deep learning algorithm programmed for outfit matching enables Echo Look to serve as a personal stylist, suggesting outfits based on the user's preferences. It can also analyze users' past purchases and, combined with fashion trends, assign ratings to different choices (Thompson 2018).

Being the closest role to the customers, Echo Look can provide professional advice while fostering a more intimate relationship with users. Human–machine interactions add entertainment value to the customers, and potential applications are possible when connected with other roles in the fashion industry (Horst and Mohammid 2022). Statistics show that the style recommendations of Echo Look are practical, increasing the cross-selling of fashion items on e-commerce platforms.

11.3.3 Application of IoT in Supply Chain and Inventory Management

11.3.3.1 Nike – Digital Thread and AI-Driven Supply Chain

The digital transformation of Nike's strategy begins with the 'Nike Digital Thread', which integrates digital technologies and data across its entire business, from design and manufacturing to marketing and sales, to create a more seamless and personalized experience for consumers. During the COVID-19 pandemic, Nike utilized the time to transition its supply chain to digitalized management, leveraging IoT to affix RFID labels on textile products (GlobalData 2024). Furthermore, toward consumers, Nike also launched mobile apps, including Run Club, Training Club, and a shopping-centric retail app, especially SNKRS, which serves as an online sneakerhead community, to expand digital sales close to shoppers (Bain 2022). Digital retail is also supported by an AI-driven supply chain, which initially uses data collected from shoppers to create new products.

The products are also equipped with RFID tags to enable real-time tracking across the supply chain. By keeping track of all products, AI-driven supply chain management can predict demand and reduce overproduction, thereby avoiding excessive waste. From the perspective of producing hyper-personalized sales, Nike adopted digital twins to facilitate sustainability by enabling virtual prototyping and testing, utilizing digital models to reduce material waste and consumption, rather than relying on physical sampling (Murugesan et al. 2024).

The transformation throughout the entire supply chain proved to be a success even in the bankruptcy period of the traditional fashion industry. The AI-driven supply chain has reduced the lead time from design to end-users and improved sustainability.

11.3.3.2 Uniqlo – Automated Warehouse and AI Inventory (Japan)

The transformation of Uniqlo, utilizing AI to facilitate business operations, is highlighted by the 100% automation of supply chain management. AI streamlines support production processes, supplier selection, and inventory management, thereby enhancing efficiency and minimizing waste while reducing production costs (Seles 2019; Zhang 2023). Uniqlo has collaborated with Daifuku, a material handling system provider, to implement AI robotics in the manufacturing processes and the automated storage and retrieval system (AS/RS) in their warehouses (Jimbo 2018; Wightman-Stone 2018). AS/RS systems are designed for the automated storage and retrieval of parts and items in manufacturing, distribution, retail, wholesale, and institutional settings (Exotec 2023). Since the transformation, the AI-powered inventory system has contributed to more accurate demand predictions and automated restocking, improving efficiency and speed. This has resulted in a 90% reduction in human labor for specific tasks. Excluded from the limitations of human labor and errors, the AI-driven management system has achieved faster order fulfillment and reduced overstock. RFID tags facilitate the inventory management system by integrating IoT technology for standardized monitoring of the supply chain. IoT technology is also the foundation of Uniqlo's business, enabling it to extend smart payment worldwide, reduce labor costs, and increase efficiency and customer satisfaction at checkout.

11.3.4 Application of AI and IoT in Smart Manufacturing

11.3.4.1 Adidas – Speedfactory (Germany/USA)

In 2015, Adidas built a speed factory in Germany and the US, featuring robotic arms, multilayered particle machines, laser cutting robots, and 3D printers. The revolutionary design of the factory is to produce shoes more efficiently and closer to the consumer. The automation experiment of the automated factory is developed based on additive manufacturing and thoughtful planning to create high-performance footwear, particularly notable for its use of 3D technology to enable a creative digital mock-up and rapid prototyping. The system not only reduced production time from months to days, but it also cut down the number of laborers by more than 70% (Bain 2018). Not limited to the concept from the literature, Adidas took a step closer to realizing the possibility of enabling mass customization with personalized shoe designs using 3D printing technology in the fashion industry.

The productivity of customization using 3D printing technology is obviously high and imminent; however, the cost of adopting new technologies in the industry remains a challenge (Peels 2019). The high initial investment led to the shutdown of the two facilities (Tabansi 2025; Hernández 2020), prompting practitioners to reflect on how to properly balance investment in new technology for sustainability with the risks of losing profits. The proper solution is still awaited to be announced, especially when similar experimental speed factory technologies are deployed to the Asian suppliers (Herzogenaurach 2019).

11.3.4.2 Levi's – Laser Finishing for Custom Jeans

The digital technology used by Levi Strauss and Co., 'Future Finish', boosts the laser-power personalization technology to replace the finishing process of jeans at the end of the retailing stage (LEVI'S® 2019; Unzipped staff 2019). The finishing process is typically conducted before the jeans are delivered to retail stores, involving a series of treatments that include tinting, stone-washing, bleaching, dyeing, and finishing. This process utilizes chemicals and requires a substantial amount of water, resulting in environmental pollution (Periyasamy and Periyasami 2023). Levi's developed the AI-driven laser cutting technology to replace the manual distressing procedure with AI-guided lasers for precision design. IoT sensors are located as a reference framework to monitor machine efficiency in real-time treatment.

The state-of-the-art advancement utilizing laser-powered technology was developed through Project F.L.X. to digitize design and development, enabling online customization business to gain valuable experience (Unzipped staff 2018). The project not only proved to be effective and exceeded environmental protection expectations, but also was classified into tint colors, patterns, and texture conditions to a simplified standard. The finishing system is further promoted as a customization solution to provide agile manufacturing in response to demand, thereby reducing overproduction and shortening processing time to minutes, rather than the previous hours in conventional factories. By using the new technology, Levi's has cut down its water usage by 96% in the finishing processes and avoided releasing hazardous chemicals into the environment during production.

11.3.4.3 Shima Seiki – Wholegarment Knitting (Japan)

Shima Seiki's Wholegarment is the world's first seam-free knitting machine, offering one-step production of knitwear as a single, three-dimensional piece directly on the knitting machine. The revolutionary business is supported by three different technologies: (1) the modification of traditional knitwear design from separate panels and sleeves into a one-piece structure design, (2) AI-driven design tools to convert the concept into virtual sampling and computer-aided design (CAD)/ computer-aided manufacturing (CAM) plan, and (3) 3D knitting machines and an AI-driven supply chain management system.

The integration of IoT-enabled production could minimize the waste of raw materials to zero and optimize the inventory for quick response to market changes. The computerized automation of knitting garments requires no additional labor in post-production procedures and eliminates the need for skilled sewing labor to further reduce costs. Similar technologies for producing knitwear are also adopted by fast fashion brands such as Zara and H&M, using seamless design and customization for on-demand production. As a result, the optimized design and 3D knitting solution reduce fabric waste by up to 30% compared to the traditional industry standard.

11.3.5 Application of Blockchain in Sustainable Tracking and Authentication

11.3.5.1 VeChain: Blockchain Serves Fashion Brands for Sustainable Tracking

VeChain is a blockchain platform that aims to enhance supply chain transparency and efficiency through innovative technology. Its own cryptocurrency, VeChainThor blockchain, uses a dual-token economic model (VET and VTHO) to distinguish between utility and incentive/payment activities, and can be integrated with IoT devices. The blockchain platform enables the integration of IoT sensors on products, updating manufacturing processes along the supply chain. VeChain has partnered with H&M, which utilizes blockchain technology to track organic cotton and recycled materials throughout its supply chain. The transparency of production processes is not only for the management but is also displayed to the consumers via the QR code on the garments. Consumers can scan the QR codes to view a product's sustainability journey, from the origin of the raw materials (such as cotton and wool) to the final product in retail stores. The integration of digitalized technology ensures ethical sourcing of the brand by recording every step of the entire fashion industry, say, the 'farm → factory → store' course (VeChain official 2021). More importantly, the exposure of practices in the circular economy boosts consumers' trust in H&M's conscious collection.

11.3.5.2 TextileGenesis (From Lenzing Group) Produced for Fiber Tracking

TextileGenesis is a platform that uses blockchain technology to enhance fiber tracking in the textile and fashion industry. It collaborated with Lenzing Group from Australia to track the wool material production and further follow up with the textile products, specifically for its TENCEL™ Lyocell (eco-friendly fiber) from pulp to final garment (Ahmed and MacCarthy, 2021). TextileGenesis developed a Hyperledger platform, Fibercoin™ technology, to convert tangible assets into tokenized fiber credits (Fibree 2023). Tokenization enables faster, safer, and more efficient payments between parties, even without a transaction via banks. Except for the self-owned fiber brand, fashion brands such as Patagonia and Eileen Fisher also utilize it to verify their sustainable sourcing practices. By adopting the Hyperledger platform, Lenzing Group can prevent greenwashing by providing immutable proof of eco-friendly materials, ensuring it adheres strictly to standards throughout its supply chain.

11.3.5.3 Arianee Digital Product Passport for Luxury Brands Tracking

Arianee provides solutions to create a digital product passport for luxury goods using NFT-based digital IDs, which track the trading history of the items on its blockchain, built using the Ethereum or Polygon platform. Together with the digital product passport of the physical luxury item, it also creates a digital twin to store the information about the product's origin, materials, ownership history, and transaction records (Fashionbi 2024). Partners of Arianee include Breitling, which launches digital product passports to verify the authenticity of products and provide additional services using a warranty card bearing a QR code for the watch. This QR code connects customers with Breitling NFTs, allowing them to access information about initiating and tracking the repair process. Additional services could be subscribed for further insurance in case of loss. Balmain and Moncler also collaborated with Arianee to launch digital product passports for their products, utilizing unique NFT certificates that incorporate exclusive features, including asynchronous transfer and minted information on the blockchain, thereby proving authenticity and ownership. The strategies announced the march of high-end luxury fashion brands into Web3 projects, which will reduce counterfeiting by allowing customers to verify product history and further enable the possibility of resale tracking in the secondary market.

11.3.5.4 LVMH X Aura: Blockchain Consortium Luxury Authentication

Founded in 2021, the Aura Blockchain Consortium is a collaborative initiative among luxury brands aimed at creating a blockchain solution tailored to the luxury goods industry. It seeks to establish a shared, decentralized platform that enables brands to track their products and share their stories with consumers. LVMH is a key founding member of the Aura Blockchain Consortium, alongside a group of luxury brands, including Prada, Cartier, Bulgari, Dior, Maison Martin Margiela, and Tod's. Developed based on the consortium blockchain, which is Quorum by J.P. Morgan, Aura assists LVMH brands (Louis Vuitton, Dior, Prada) in tracking their high-end products. In the consortium blockchain, each item has a digital birth certificate that records its manufacturing and ownership history, making the entire supply chain, from manufacturing to retail, transparent to consumers and establishing an industry-wide standard. The impact of this step forward is essential for the high-end fashion industry, especially for jewelry brands, as it effectively combats counterfeit luxury goods that have caused long-term problems for the market, worth over $500 billion.

11.3.6 Application of AI in Circular Fashion

11.3.6.1 Fashion for Good x CircularID™ in Circular Fashion Economy

Fashion for Good is a platform that addresses problems in the fashion industry and accelerates collaborative innovations, offering hands-on support, access to capital, and building a network of experts (Fashion for Good 2025). Toward the circular economy and environmental sustainability, one of their projects is to create a circular fashion system in which the brands could track their garments and footwear for recycling and resale processes. H&M, Microsoft, PVH Corp., and Adidas were partners in the project, collaborating with CircularID™, which was initially designed to establish a global standard for managing physical goods throughout the rental, resale, and recycling processes (McDowell 2019; Tommy and Hilfiger 2025). Built on the Ethereum platform, blockchain technology is specifically applied to track the entire life cycle of physical goods, utilizing digital certificates for product IDs. The project provides access to information such as material, origin, authenticity, price, and style, as a support to encourage the growth of second-hand fashion re-commerce on a larger scale, and to inspire the development of new business models for fashion brands to extend their retailing trade into the resale of goods. Compared to reducing the consumption of raw materials for new mass production, the tracking solution to manage physical ready-to-wear garments could directly solve the problem of massive textile waste storage and reduce waste by extending the garment life cycle.

11.3.6.2 Patagonia – AI for Circular Fashion

Patagonia, being the leading fashion brand in outdoor clothing and gear, adopted blockchain technology to fulfill its ethical mission in the fashion industry and enhance its impact on the environmental and social footprint. The brand developed a provenance blockchain based on the Ethereum platform to track the origin of their materials, including organic cotton, wool, down, and other accessories, directly from the farm. The advanced traceability of fabrics enables the recycling of postindustrial waste and postconsumer products to replace virgin materials in new manufacturing processes (Patagonia 2025a, 2025b). With the precise calculation of material reuse strategies, Patagonia aims to achieve a goal of 99% of its line collections being made with recycled materials, including waxed coating, nylon, down, thermoplastic polyurethane, spandex, polyester, and denim (Patagonia 2025a). Another technology using an AI solution is the development of a trading platform for the resale of used products, such as Worn Wear, which enables the sale of goods in good condition as second-hand products rather than sending them to landfills. The AI-assisted grading tools are used to determine the quality of clothing for resale, and IoT technology tracks the garment to evaluate its life cycle for further recycling. In this way, Patagonia can precisely extend the lifespan of garments while reducing emissions from waste material destruction. The new standard for managing the supply chain of the fashion industry will set an example for how to address the carbon issue in the age of Industry 4.0.

11.4 OPPORTUNITIES AND CHALLENGES

11.4.1 Opportunities

With the advent of Industry 4.0, the fashion industry can expand in multiple directions. This evolution involves the transformation of the fashion sector into an innovative system underpinned by advanced technologies such as AI, IoT, and blockchain, with which to create a digitalized world characterized by enhanced traceability, transparency, customization, and optimization to refine production in alignment with consumer demands. IoT technology can benefit almost every stage of the fashion value chain (Manglani et al. 2019). Notably, fashion design and trend forecasting are increasingly driven by AI algorithms that analyze vast amounts of consumer data, rather than relying solely on traditional designer input. This shift fosters greater consumer involvement in the design phases, making interactions between brands and customers closer. Furthermore, data sources such as social media platforms, online fashion news, and user feedback collected from the chatbots serve as training datasets for GANs and NLP algorithms, presenting valuable opportunities for developers to further improve their AI performance through the analysis of diverse and heterogeneous data repositories.

Robotics, combined with AI-driven analytics and IoT network, holds substantial potential to advance automation within manufacturing processes. Compared to conventional manufacturing methods, digital technologies lessen dependence on intensive human labor in the supply chain. Opportunities arise in reshaping supply chain management by enabling enhanced connectivity among various sectors involved in production. Based on precise computational algorithms, AI-powered robotic arms are capable of executing tasks throughout the entire production line, including design, patternmaking, cutting, material picking, sewing, dyeing, quality inspection, and final packaging for shipment. The smart factory operates and undergoes a process of digitalization grounded in a modular and decentralized structure, facilitating standardized process monitoring, as well as optimizing inventory and logistics management. Furthermore, robotic systems employed in storage and transportation demonstrate superior accuracy compared to manual labor. In the unmanned warehouse of raw materials and finished goods, robots can supplant human workers by autonomously locating specific items through IoT sensor data and retrieving them for subsequent stages.

During Industry 3.0, the just-in-time manufacturing and lean supply chain management shortened the operational lead time within the global market. Advancing to Industry 4.0, the operational efficiency and precision are further improved through the integration of robotics into the shipping process. IoT-enabled machines can minimize human error and maintain real-time monitoring of logistics networks on a global scale. Based on a network composed of IoT technologies, automation management is employed within warehouses, AI-driven robots facilitate continuous 24/7 production, and technologies such as digital printing and laser cutting support product customization. Consequently, productivity transcends the limitations of manual labor, relying instead on collaborative efforts between the fashion industry and other industrial sectors to drive digital innovation. Thus, this convergence creates expanded opportunities for emerging enterprises in the digital era.

In the intricate network of blockchain technology, digital twins can be generated in cloud-based data environments to represent physical products. The provision of real-time product information animates garments and accessories by conferring upon them distinct identities, verifiable authenticity, defined lifespans, and documented trajectories throughout their existence. This paradigm shift has fundamentally altered our perspective on commercial goods and ethical standards in the fashion industry, both pre- and post-purchase. The data utilized for product tracking and tracing provides opportunity to improve predictive maintenance, reuse, and recycling processes. Such transparency necessitates closer collaboration among industry stakeholders, consumers, and policymakers to establish precise regulations on the use of digital data for product maintenance and consumer behavior.

11.4.2 Challenges

Despite the benefits, several barriers hinder widespread implementation. Firstly, digital technology in the current fashion industry is still often associated with substantial initial capital expenditures. Advanced technological infrastructure often necessitates significant financial investment, which may exceed the fiscal capacities of certain fashion enterprises. This could lead to heterogeneous rates of business development among brands, with those unable to adopt new technologies facing competitive disadvantages in the market. Another challenge is the workforce reskilling. Although digital tools free human labor from the manufacturing processes, the supply chain still demands employees with interdisciplinary competencies. For instance, the personnel in factories and warehouses should be proficient in managing mechanical devices, robots, and automated systems, while also coordinating with technical teams to handle maintenance tasks. In addition, as the fashion and textile sectors transit into the digital era, competitive dynamics are shifting from traditional management paradigms toward the development of AI algorithms and network-based projects. More talents with skills in network engineering and algorithm development will be attracted to the industry (Caldarelli et al. 2021), and such situation intensifies the need for investment in human capital. The establishment of comprehensive network to support AI and IoT technologies, which enhances the supply chain efficiency and the diversification of talents in the industry, constitutes a pivotal factor influencing long-term investment outcomes. The transformation necessitates extensive employee training in novel digital tools, further amplifying the financial commitment to talent development. Lastly, the proliferation of digitalization introduces heightened data security risks. Increased digitization raises cybersecurity concerns and will likely prompt the upgrade of network system guardians (Dragomir et al. 2016). Finally, with the entire fashion industry progressively evolving toward a circular economy, the supply chain is reconfigured from a linear to a cyclical model, wherein waste materials are repurposed as raw material. This shift exemplifies the utilization of textile waste and second-hand products, challenges traditional consumption patterns, and elicits diverse responses from different parties. Such complexities frequently pose governance challenges, particularly in relation to compliance with evolving sustainability standards and laws.

Successful advancement of sustainability in the fashion industry requires the tacit cooperation of stakeholders from diverse backgrounds. The changes in cost investment throughout the entire supply chain indicate that fashion companies are recalibrating their business goals, prioritizing not only financial performance but also the positive impact of business profits by mitigating risks of longer development. The elongated life cycle of fabrics in fashion circulation contributes to the reduction of carbon emissions, and technological innovations help cutting down pollution of the planet, thereby enhancing the corporate social responsibility. Further, with the involvement of a broad spectrum of professional technical experts and consumers involved in sustainable fashion, conventional social networks are also transformed within the business. This transformation demands collaborative efforts from all stakeholders, fostering a community characterized by increased diversity and potential. Technologies allow the fashion business to develop toward sustainability, while challenges appear for us to pay attention to economic, environmental, and social elements, as illustrated in the triple bottom line framework. When we embrace sustainability in the digital era of fashion industry, we are also reshaping the way of making business decisions, even broader lifestyle practices.

11.5 FUTURE TRENDS

In future, digital fashion is projected to remain promising, extending far beyond the current scope of understanding and application. One noteworthy advancement in this field is the development of digital twins, which act as virtual replicas of physical products.

These digital counterparts are instrumental in enabling sustainable prototyping processes by eliminating the need for physical products during the evaluation stage.

In pursuit of eco-sustainability within the fashion industry, nonrecyclable materials are likely to be phased out. Instead, technology innovations are paving the way for the adoption of bio-fabrication techniques, including the lab-grown textiles as environmentally friendly alternatives to synthetic fibers, thereby holding promise for mitigating environmental pollution (D'Itria and Colombi 2022; Absharina et al. 2025; Jensen et al. 2023; WWF-Hong Kong 2023; Textile Exchange 2023). Additionally, AI-powered recycling is inevitably integral due to the existing foundation of digital platforms. The IoT-facilitated technologies generate substantial data of fabric tracking, offering rich opportunities for utilizing AI in the automated sorting and upcycling of textile waste (Nisa et al. 2025; H&M Foundation 2024; Faghih et al. 2025; Buenaventura 2024). Finally, the blockchain-enabled circular fashion is considered the ideal benchmark in achieving a sustainable fashion ecosystem. While still in its nascent stages, this paradigm is expected to evolve and holds significant promise for shaping a more sustainable and circular growth stage.

11.6 CONCLUSION

Digital innovation and smart manufacturing are transforming the fashion and textile industry, providing a path toward sustainability without compromising efficiency. While challenges remain, the integration of AI, IoT, 3D printing, and blockchain technologies can significantly reduce environmental disruption while meeting consumer expectations. This study examines recent advancements in the adoption of digital technologies to enhance the fashion supply chain, alongside the growing influence of sustainability in fashion consumption. It includes case studies that showcase the integration of these technologies into existing business models and how supply chains have evolved into smart manufacturing systems. By dialectically exploring the implementation of digital technologies, the study aims to provide practical insights for industry professionals to make informed decisions regarding technology adoption. Findings confirm that digital transformation is unavoidable, although further efforts are necessary to foster its growth and development. Limitations of integrating AI and IoT technologies in the fashion sector exist. Primarily, the legacy system employed in manufacturing and logistics for shipments management presents significant challenges in terms of compatibility with more intellectualized systems. The standardization and regulation of devices and platforms are fraught with complexity during the transition. Additionally, regional policies governing the use of cloud data should also be carefully considered within the system's framework to ensure compliance. One critical issue is the limited availability of high-quality data required for training AI algorithms, because traditional fashion supply chain and its customer feedback may not generate sufficient volume of data for engineers to develop precise AI-driven predictions. The classification of the high-quality, labeled data from the extensive and heterogeneous datasets within the supply chain remains a vital component for the long-term success of such initiatives.

As advanced technologies are at the forefront of the industry, AI and IoT require the collective progress of all elements in the supply chain. In other words, the coordination of different sectors of the supply chain highly relies on the connectivity and sci-tech foundations. Manufacturing regions characterized by limited connectivity may face challenges in integration into IoT systems. Similarly, stakeholders employing conventional production techniques, particularly those located in raw material source areas, may also encounter difficulties in such integration. The vast investment in the sci-tech transformation may further polarize fashion brands at different stages of business development, and the differentiation between data-driven management and craft-driven management could lead to disagreements and cultural and organizational conflicts within the industry. In advancing the co-development of the global fashion supply chain, the disparities among diversified sectors become a barrier, and it is essential to consider strategies that motivate different sectors to embrace digital transformation proactively. Regardless of the stage of technological advancement, the integration of digitalization by fashion suppliers eventually supports sustainability efforts, particularly in alignment with the United Nations' Sustainable Development Goals (SDGs). Implementing Industry 4.0 technologies optimizes manufacturing processes, enhancing productivity while minimizing waste. These technologies facilitate the improvement in infrastructure planning and management through data analytics and real-time monitoring, promoting more efficient resource allocation and upkeep, which is directly tied to SDG 9. Moreover, digital platforms allow logistic transparency, providing consumers with better insight into sustainable practices. This not only encourages more responsible consumption choices but also enables businesses and governments to make well-informed decisions that foster sustainable production and consumption patterns, and this is in accordance with SDG 12. Policymakers, businesses, and technologists must work together to accelerate this transformation, ensuring a more sustainable future for the fashion industry.

REFERENCES

Absharina, D., Padri, M., Veres, C. and Vágvölgyi, C. 2025. Bacterial cellulose: From biofabrication to applications in sustainable fashion and vegan leather. *Fermentation* 11(1): 23.

Agrawal, T. K., Kumar, V., Pal, R., Wang, L. and Chen, Y. 2021. Blockchain-based framework for supply chain traceability: A case example of textile and clothing industry. *Computers and Industrial Engineering* 154: 107130.

Ahmed, W. A. H. and MacCarthy, B. L. 2021. Blockchain-enabled supply chain traceability in the textile and apparel supply chain: A case study of the fiber producer, Lenzing. *Sustainability* 13(19): 10496.

Akins, B. W., Chapman, J. L. and Gordon, J. M. 2014. A whole new world: Income tax considerations of the Bitcoin economy. *Pittsburgh Tax Review* 12(1): 25.

Akram, S. V., Malik, P. K., Singh, R., Gehlot, A., Juyal, A., Ghafoor, K. Z. and Shrestha, S. 2022. Implementation of digitalized technologies for fashion industry 4.0: Opportunities and challenges. *Scientific Programming* 2022: Article ID 7523246. https://doi.org/10.1155/2022/7523246

Alexander, B. and Bellandi, N. 2022. Limited or limitless? Exploring the potential of NFTs on value creation in luxury fashion. *Fashion Practice* 14(3): 376–400.

Alves, L., Cruz, E. F., Lopes, S. I., Faria, P. M. and da Cruz, A. M. R. 2022. Towards circular economy in the textiles and clothing value chain through blockchain technology and IoT: A review. *Waste Management and Research* 40(1): 3–23.

Atzori, L., Iera, A. and Morabito, G. 2010. The Internet of Things: A survey. *Computer Networks* 54(15): 2787–2805.

Ayinaddis, S. G. 2025. Artificial intelligence adoption dynamics and knowledge in SMEs and large firms: A systematic review and bibliometric analysis. *Journal of Innovation & Knowledge* 10(3): 100682.

Badhwar, A., Islam, S. and Tan, S. L. C. 2023. Exploring the potential of blockchain technology within the fashion and textile supply chain with a focus on traceability, transparency, and product authenticity: A systematic review. *Frontiers in Blockchain* 6: 1044723.

Bain, M. 2018. A German company built a "Speedfactory" to produce sneakers in the most efficient way. *Quartz.com.* https://classic.qz.com/perfect-company-2/1145012/a-german-company-built-a-speedfactory-to-produce-sneakers-in-the-most-efficient-way/ [Accessed 5 November 2025].

Bain, M. 2022. Nike's digital strategy is to treat everyone the way it treats sneakerheads. *Quartz.com.* https://qz.com/quartzy/1747382/how-nikes-snkrs-app-community-inspired-its-digital-strategy [Accessed 21 July 2025].

Bertola, P. and Teunissen, J. 2018. Fashion 4.0. Innovating fashion industry through digital transformation. *Research Journal of Textile and Apparel* 22(4): 352–369.

Bhatt, D., Kim, H.S. and Bhatt, S. 2021. Shopping motivations of fashion subscription service consumers. *The International Review of Retail, Distribution and Consumer Research* 31(5): 549–565.

Buenaventura, M. I. 2024. Sustainable AI-driven fashion tech: A new era for Chile's Atacama Desert. ReVista*: Harvard Review of Latin America.* https://revista.drclas.harvard.edu/sustainable-ai-driven-fashion-tech-a-new-era-for-chiles-atacama-desert/ [Accessed 10 November 2025].

Caldarelli, G., Zardini, A. and Rossignoli, C. 2021. Blockchain adoption in the fashion sustainable supply chain: Pragmatically addressing barriers. *Journal of Organizational Change Management* 34(2): 507–524.

Chen, C., Gu, T., Cai, Y. and Yang, Y. 2019. Impact of supply chain information sharing on performance of fashion enterprises: An empirical study using SEM. *Journal of Enterprise Information Management* 32(6): 913–935.

Chen, Y. 2024. How blockchain adoption affects supply chain sustainability in the fashion industry: A systematic review and case studies. *International Transactions in Operational Research* 31(6): 3592–3620.

Christopher, M., Lowson, R. and Peck, H. 2004. Creating agile supply chains in the fashion industry. *International Journal of Retail & Distribution Management* 32(8): 367–376.

Daukantiene, V. 2022. Analysis of the sustainability aspects of fashion: A literature review. *Textile Research Journal* 93(3–4): 991–1002.

Digital Textile Printing Market. 2025. Digital textile printing market outlook: Size, share and innovations 2025 to 2035. *Futuremarketinsights.com.* https://www.futuremarketinsights.com/reports/digital-textile-printing-market#:~:text=The%20digital%20textile%20printing%20market,38.2%%20market%20share%20by%202025 [Accessed 30 June 2025].

D'Itria, E. and Colombi, C. 2022. Biobased innovation as a fashion and textile design must: A European perspective. *Sustainability* 14(1): 570.

Dragomir, D., Gheorghe, L., Costea, S. and Radovici, A. 2016. A survey on secure communication protocols for IoT systems. In *2016 International Workshop on Secure Internet of Things (*SIoT*)*, 26–30 September 2016, Heraklion, Greece: 47–62, IEEE.

Ellen MacArthur Foundation. 2021. How policymakers can pave the way to a low-carbon and prosperous future. *Ellenmacarthurfoundation.org.* https://www.ellenmacarthurfoundation.org/covid-policymakers [Accessed 16 August 2025].

Enemuoh, E. U., Menta, V. G., Abutunis, A., O'brien, S., Kaya, L. I. and Rapinac, J. 2021. Energy and eco-impact evaluation of fused deposition modeling and injection molding of polylactic acid. *Sustainability* 13(4): 1875.

Exotec. 2023. The ASRS revolution in the warehouse. *Exotec.com.* https://www.exotec.com/en-gb/insights/revolution-asrs-warehouse/ [Accessed 23 May 2025].

Faghih, E., Saki, Z. and Moore, M. 2025. A systematic literature review: AI-enabled textile waste sorting. *Sustainability* 17(10): 4264.

Fashion For Good. 2025. Journey to scale: TextileGenesis™. *fashionforgood.com.* https://www.fashionforgood.com/our_news/meet-the-innovator-textilegenesis/ [Accessed 17 July 2025].

Fashionbi. 2024. How luxury brands are embracing the digital product passports? *Fashionbi.com.* https://www.fashionbi.com/insights/how-luxury-brands-are-embracing-the-digital-product-passports [Accessed 30 May 2025].

Fibre2Fashion. 2022. NeXR implements virtual fitting solution for H&M Thailand. https://www.fibre2fashion.com/news/retail-industry/nexr-implements-virtual-fitting-solution-for-h-m-thailand-283060-newsdetails.htm [Accessed 14 September 2025].

Fibree. 2023. Tokenization: How it's transforming business and technology. *fibree.org.* https://fibree.org/tokenization-how-its-transforming-business-and-technology/ [Accessed 24 May 2025].

Fisher, M. 1997. What is the right supply chain for your product? *Harvard Business Review.* https://hbr.org/1997/03/what-is-the-right-supply-chain-for-your-product [Accessed 10 July 2025].

Gallagher, J. 2021. NFTs are the biggest internet craze. Do they work for sneakers? *The Wall Street Journal.* https://www.wsj.com/style/fashion/nfts-and-fashion-collectors-pay-big-money-for-virtual-sneakers-11615829266 [Accessed 15 March 2025].

Gibbs, S. 2017. Amazon unveils Echo Look, a selfie camera to help you choose what to wear. *Theguardian.com.* https://www.theguardian.com/technology/2017/apr/26/amazon-echo-look-webcam-choose-fashion-outfits-alexa-smart-selfie-camera [Accessed 26 April 2025].

Giri, C., Jain, S., Zeng, X. and Bruniaux, P. 2019. A detailed review of artificial intelligence applied in the fashion and apparel industry. *IEEE Access* 7: 95376–95396.

GlobalData. 2024. NIKE: Digital transformation strategies. *globaldata.com.* https://www.globaldata.com/store/report/nike-enterprise-tech-analysis/ [Accessed 01 April 2025].

Glogar, M., Petrak, S. and Naglić, M. M. 2025. Digital technologies in the sustainable design and development of textiles and clothing: A literature review. *Sustainability* 17(4): 1371–1397.

Ghoreishi, M., and Happonen, A. 2021. The case of fabric and textile industry: The emerging role of digitalization, Internet-of-Things and industry 4.0 for circularity. *Proceedings of Sixth International Congress on Information and Communication Technology: ICICT 2021* London 3: 189–200. Singapore: Springer Singapore.

Gubbi, J., Buyya, R., Marusic, S. and Palaniswami, M. 2013. Internet of Things (IoT): A vision, architectural elements, and future directions. *Future Generation Computer Systems* 29(7): 1645–1660.

Guo, Z., Zhu, Z., Li, Y., Cao, S., Chen, H. and Wang, G. 2023. AI assisted fashion design: A review. *IEEE Access* 11: 88403–88415.

Hasan, R. and Liu, C. 2020. Consumer shopping value in using curated fashion subscription: The case of Stitchfix.com. *International Textile and Apparel Association Annual Conference Proceedings* 77(1). Iowa State University Digital Press.

H&M Foundation. 2024. AI-powered smart garment sorting system automating textile recycling. https://hmfoundation.com/project/ai-powered-smart-garment-sorting-system-automating-textile-recycling/ [Accessed 10 September 2025].

Hernández, A. 2020. Learning from Adidas' Speedfactory blunder. *Supplychaindive.com.* https://www.supplychaindive.com/news/adidas-speedfactory-blunder-distributed-operations/571678/ [Accessed 04 February 2025].

Herzogenaurach. 2019. Adidas deploys Speedfactory technology at Asian suppliers by end of 2019. *Adidas Media.* https://www.adidas-group.com/en/media/press-releases/adidas-deploys-speedfactory-technology-at-asian-suppliers-by-end-of-2019 [Accessed 5 November 2025].

Hidalgo-Carvajal, D., Muñoz, Á. H., Garrido-González, J. J., Carrasco-Gallego, R. and Alcázar Montero, V. 2023. Recycled PLA for 3D printing: A comparison of recycled PLA filaments from waste of different origins after repeated cycles of extrusion. *Polymers* 15(17): 3651.

Hopkins, N., Jiang, L. and Brooks, H. 2021. Energy consumption of common desktop additive manufacturing technologies. *Cleaner Engineering and Technology* 2: 100068.

Horst, H. A. and Mohammid, S. 2022. Framing fashion: Human-machine learning and the Amazon Echo Look. In Pink, S., Berg, M., Lupton, D. and Ruckenstein, M. (Eds.), *Everyday Automation*: 102–113. London: Routledge.

IFAB Media. 2024. Mapping fashion trends through AI: Revolutionizing the fashion industry. *InFashionBusiness.com.* https://infashionbusiness.com/home/news_details/3863/13 [Accessed 7 September 2025].

Imtiaz, A., Pathirana, N., Saheel, S., Karunanayaka, K. and Trenado, C. 2024. A review on the Influence of deep learning and generative AI in the fashion industry. *Journal of Future Artificial Intelligence and Technologies* 1(3): 201–216.

iSpot. 2025. Stitch Fix tv spot, 'So Easy'. *iSpot.tv.* https://www.ispot.tv/ad/Tt47/stitch-fix-so-easy [Accessed 16 February 2025].

Jensen, B., Stoneburner, M., Martinez-Pardo, C., Wilkinson, J., Khodabocus, H., Durbiano, L., Faist, M., Frehland, S., Meister, P. and Vargas-Gonzalez, M. 2023. Sustainable raw materials will drive profitability for fashion and apparel brands. *BCG and Textile Exchange.* https://web-assets.bcg.com/2f/a3/8bf925184dfe969547a484ddc5d4/bcg-sustainable-raw-materials-will-drive-profitability-oct-23.pdf [Accessed 16 September 2025].

Jimbo, T. 2018. Strategic global partnership with Daifuku. *FastRetailing.com.* https://www.fastretailing.com/eng/ir/library/pdf/presen181009_FR-Daifuku-ConcludeStrategicGlobalPartnership_eng.pdf [Accessed 16 August 2025].

Jin, B. E. and Shin, D. C. 2020. Changing the game to compete: Innovations in the fashion retail industry from the disruptive business model. *Business Horizons* 63(3): 301–311.

Jin, B. E. and Shin, D. C. 2021. The power of 4th industrial revolution in the fashion industry: What, why, and how has the industry changed? *Fashion and Textiles* 8(1): 31.

Joy, A., Zhu, Y., Peña, C. and Brouard, M. 2022. Digital future of luxury brands: Metaverse, digital fashion, and non-fungible tokens. *Strategic Change* 31(3): 337–343.

Kim, H. S. and Cho, S.B. 2000. Application of interactive genetic algorithm to fashion design. *Engineering Applications of Artificial Intelligence* 13(6): 635–644.

Kim, J., Kang, S. and Bae, J. 2022. The effects of customer consumption goals on artificial intelligence driven recommendation agents: Evidence from Stitch Fix. *International Journal of Advertising* 41(6): 997–1016.

Kotouza, M. T., Tsarouchis, S. F., Kyprianidis, A. C., Chrysopoulos, A. C. and Mitkas, P. A. 2020. Towards fashion recommendation: An AI system for clothing data retrieval and analysis. In Maglogiannis, I., Iliadis, L. and Pimenidis, E. (Eds.), *IFIP International Conference on Artificial Intelligence Applications and Innovations:* 433–444. Cham: Springer International Publishing.

Kosba, A., Miller, A., Shi, E., Wen, Z. and Papamanthou, C. 2016. Hawk: The blockchain model of cryptography and privacy-preserving smart contracts. In *2016 IEEE Symposium on Security and Privacy (SP)*, 18 August 2016, San Jose, CA: 839–858.

Kozlowski, A., Bardecki, M. and Searcy. C. 2019. Tools for sustainable fashion design: An analysis of their fitness for purpose. *Sustainability* 11(13): 3581.

Lake, K. 2018. Stitch Fix's CEO on selling personal style to the mass market. *Harvard Business Review* 96(3): 35–40.

Levi's®. 2019. LEVI'S® presents future finish. *levi.com.* https://www.levi.com/US/en_US/blog/article/levis-presents-future-finish?srsltid=AfmBOopYfAeybd2eQmhjOwgL23tzVIWuA-knIyXxb-06WZsfsZ7qr7MP [Accessed 05 August 2025].

Liu, L., Zhang, H., Ji, Y. and Wu, Q. M. J. 2019. Toward AI fashion design: An attribute-GAN model for clothing match. *Neurocomputing* 341: 156–167.

Liu, Z., Luo, P., Qiu, S., Wang, X. and Tang, X. 2016. Deepfashion: Powering robust clothes recognition and retrieval with rich annotations. In *Proceedings of the IEEE conference on Computer Vision and Pattern Recognition*, 12 December 2016, Las Vegas, NV: 1096–1104.

Malik, S., Kanhere, S. S. and Jurdak, R. 2018. ProductChain: Scalable blockchain framework to support provenance in supply chains. In *2018 IEEE 17th International Symposium on Network Computing and Applications (NCA)*, 29 November 2018, Cambridge, MA: 1–10.

Manglani, H., Hodge, G. L. and Oxenham, W. 2019. Application of the internet of things in the textile industry. *Textile Progress* 51(3): 225–297.

McDowell, M. 2019. H&M, Microsoft, PVH Corp collaborate in circular fashion initiative. *voguebusiness.com.* https://www.voguebusiness.com/technology/circular-id-eon-sustainability-blockchain [Accessed 16 July 2025].

Mckeegan, D. 2023. Innovations driving the growth of digital textile printing. *Texintel.com.* https://www.texintel.com/blog/innovations-driving-the-growth-of-digital-textile-printing-we-dive-into-new-tech-and-evolving-trends [Accessed 20 July 2025].

McKeegan, D. 2024. How smart manufacturing will change fashion, the textile industry, and the future of retail. *WhatTheyThink.com.* https://whattheythink.com/articles/121207-how-smart-manufacturing-will-change-fashion-textile-industry-future-retail/ [Accessed 19 September 2025].

Murugesan, B., Jayanthi, K. B. and Karthikeyan, G. 2024. Integrating digital twins and 3D technologies in fashion: Advancing sustainability and consumer engagement. In Raj, P., Rocha, A., Dutta, P.K., Fiorini, M. and Prakash, C. (Eds.), *Illustrating Digital Innovations towards Intelligent Fashion: Leveraging Information System Engineering and Digital Twins for Efficient Design of Next-Generation Fashion*: 1–88. Cham: Springer Nature Switzerland.

Niinimäki, K., Peters, G., Dahlbo, H., Perry, P., Rissanen, T. and Gwilt, A. 2020. The environmental price of fast fashion. *Nature Reviews Earth & Environment* 1(4): 189–200.

Nisa, H., Van Amber, R., English, J., Islam, S., McCorkill, G. and Alavi, A. 2025. A systematic review of reimagining fashion and textiles sustainability with AI: A circular economy approach. *Applied Sciences* 15(10): 5691.

Noyes, C. 2016. Bitav: Fast anti-malware by distributed blockchain consensus and feedforward scanning. *arXiv preprint arXiv*:1601.01405.

Pal, K. and Ansar-Ul-Haque, Y. 2020. Internet of things and blockchain technology in apparel manufacturing supply chain data management. *Procedia Computer Science* 170: 450–457.

Patagonia. 2025a. How clean are your clothes? The hidden cost of the clothes we're buying. *patagonia.com.* https://www.patagonia.com/hidden-cost-of-clothes/ [Accessed 10 September 2025].

Patagonia. 2025b. Our environmental impact programs. *patagonia.com.* https://www.patagonia.com/environmental-responsibility-materials/ [Accessed 20 September 2025].

Peels, J. 2019. Adidas kills the Speedfactory: Industrie 4.0 funeral date not yet set. *3dprint.com.* https://3dprint.com/259437/adidas-kills-the-speedfactory-industrie-4-0-funeral-date-not-yet-set/ [Accessed 12 November 2025].

Periyasamy, A. P. and Periyasami, S. 2023. Critical review on sustainability in denim: A step toward sustainable production and consumption of denim. *ACS omega* 8(5): 4472–4490.

Priyadharshini, K. 2023. Garment production: Towards smart manufacturing. *Fibre2Fashion.com.* https://www.fibre2fashion.com/industry-article/9750/garment-production-towards-smart-manufacturing [Accessed 19 August 2025].

Seles. 2019. Uniqlo, [AI Artificial Intelligence] Will Uniqlo's warehouses be 100% automated? Garment folding and packaging will all be done by machines! https://businessfocus.io/article/125469/ai人工智能-uniqlo倉庫將100-自動化-服裝摺疊-包裝全由機械代工 [Accessed 20 August 2025].

Sharples, M. and Domingue, J. 2016. The blockchain and kudos: A distributed system for educational records, reputation and reward. In Verbert, K., Sharples, M. and Klobučar, T. (Eds.), *European Conference on Technology Enhanced Learning*: 490–496. Cham: Springer International Publishing.

Shi, M., Chussid, C., Yang, P., Jia, M., Lewis, V. D. and Cao, W. 2021. The exploration of artificial intelligence application in fashion trend forecasting. *Textile Research Journal* 91(19–20): 2357–2386.

Shi, M. and Lewis, V. D. 2020. Using artificial intelligence to analyze fashion trends. *arXiv preprint arXiv*:2005.00986.

Shushi, A. and Abdulazeez, A. M. 2024. Fashion design classification based on machine learning and deep learning algorithms: A review. *The Indonesian Journal of Computer Science* 13(3): 3622–3639.

Smily, L. 2019. Stitch Fix's radical data-driven way to sell clothes-$1.2 billion last year-is reinventing retail. *FastCompany.com.* https://www.fastcompany.com/90298900/stitch-fix-most-innovative-companies-2019 [Accessed 25 August 2025].

Sung, E., Kwon, O. and Sohn, K. 2023. NFT luxury brand marketing in the metaverse: Leveraging blockchain - Certified NFTs to drive consumer behavior. *Psychology & Marketing* 40(11): 2306–2325.

Tabansi, O. 2025. Lessons from the failed Adidas speed factory experiment. *SupplyChainNuggets.com.* https://supplychainnuggets.com/why-the-adidas-speed-factory-experiment-failed/ [Accessed 8 June 2025].

Textile Exchange. 2023. Materials market report. https://textileexchange.org/app/uploads/2023/11/Materials-Market-Report-2023.pdf [Accessed 29 August 2025].

Thompson, A. C. 2018. Amazon Echo Look review: Amazon's Echo Look camera turns Alexa into a fashion stylist. *Cnet.com.* https://www.cnet.com/reviews/amazon-echo-look-review/ [Accessed 06 June 2025].

Tommy & Hilfiger. 2025. Waste nothing: Tommy for tomorrow, responsibility. *pvh.com.* https://responsibility.pvh.com/waste-nothing/ [Accessed 30 August 2025].

TyN Magazine. 2021. H&M extends nuance virtual assistant and live chat to Google's business messages. https://tynmagazine.com/hm-extends-nuance-virtual-assistant-and-live-chat-to-googles-business-messages/ [Accessed 21 February 2025].

Unzipped Staff. 2018. Project F.L.X. Redefines the future of how jeans are designed, made and sold. *levistrauss.com.* https://www.levistrauss.com/2018/02/27/project-f-l-x-redefines-future-jeans-designed-made-sold/ [Accessed 27 February 2025].

Unzipped Staff. 2019. Future finish, A new way to customize your denim. *levistrauss.com.* https://www.levistrauss.com/2019/07/24/future-finish-a-new-way-to-customize-your-denim/ [Accessed 24 July 2025].

Vechain official, 2021. VeChainThor is one of the most eco-friendly public blockchains worldwide, CTI Verified. *medium.com.* https://medium.com/vechain-foundation/vechainthor-is-one-of-the-most-eco-friendly-public-blockchains-worldwide-cti-verified-66d7aa2c5ccd [Accessed 27 September 2025].

Vercalsteren, A., Nicolau, M. and Lafond, E. 2019. Textiles and the environment in a circular economy. *ETC/WMGE*: Copenhagen, Denmark.

Wahba, P. 2019. Stitch Fix is diving deep into data for a more tailored experience. *Fortune.com.* https://fortune.com/longform/stitch-fix-data-algorithm-growth-katrina-lake/ [Accessed 30 October 2025].

Wightman-Stone. D. 2018. Fast retailing signs deal with Daifuku for warehouse automation. *FashionUnited.com.* https://fashionunited.uk/news/retail/fast-retailing-signs-deal-with-daifuku-for-warehouse-automation/2018100939343 [Accessed 09 October 2025].

Woodside, A. G. and Fine, M. B. 2019. Sustainable fashion themes in luxury brand storytelling: The sustainability fashion research grid. *Journal of Global Fashion Marketing* 10(2): 111–128.

Wu, X. and Li, L. 2024. An application of generative AI for knitted textile design in fashion. *The Design Journal* 27(2): 270–290. [Accessed 15 October 2025].

WWF-Hong Kong. 2023. Making zero impact fashionable. https://www.wwf.org.hk/en/cities/fashion/clothing/ [Accessed 15 October 2025].

Zebra. 2025. What is the difference between RFID and Bluetooth technology? https://www.zebra.com/ap/en/resource-library/faq/what-is-the-difference-between-rfid-and-bluetooth-technology.html [Accessed 08 November 2025].

Zhang, D., Yang, L. T., Chen, M., Zhao, S., Guo, M. and Zhang, Y. 2014. Real-time locating systems using active RFID for Internet of Things. *IEEE Systems journal* 10(3): 1226–1235.

Zhang, J. 2023. UNIQLO: Visual revolution with AI and technology. *WordPress.com.* https://wearit24.wordpress.com/2023/10/19/uniqlo-visual-revolution-with-ai-and-technology/ [Accessed 19 October 2025].

Zhang, Y. and Wen, J. 2015. An IoT electric business model based on the protocol of bitcoin. In *2015, 18th International Conference on Intelligence in Next Generation Networks*, 2 April 2015, Paris, France: 184–191.

Zheng, Z., Xie, S., Dai, H. N., Chen, X. and Wang, H. 2018. Blockchain challenges and opportunities: A survey. *International Journal of Web and Grid Services* 14(4): 352–375.

12 Sustainable Chemical Processing Techniques

Rasiah Ladchumananandasivam,
Iris Oliveira da Silva, Miguel Higino Neto,
and Christiane Siqueira de Azevedo Sá

12.1 INTRODUCTION

The textile industry has historically been one of the most resource-intensive and environmentally challenging sectors worldwide. From fiber preparation to finishing, chemical processes have enabled the development of a vast range of fabrics with desirable aesthetic, functional, and performance characteristics. Pretreatment operations such as scouring and bleaching, dyeing and printing with synthetic colorants, and functional finishing treatments have long formed the backbone of textile manufacturing. However, these conventional chemical processes are also responsible for high water consumption, toxic effluent discharge, substantial energy demand, and the emission of greenhouse gases.

According to global assessments, textile wet processing contributes to nearly 20% of all industrial water pollution, and the dyeing and finishing segments alone account for up to 17%–20% of global wastewater generation (UNEP 2020; World Bank 2019; Ellen MacArthur Foundation 2017; McKinsey and Company 2020). In regions where environmental regulations are weak, untreated effluents laden with dyes, salts, and finishing chemicals continue to pollute rivers and groundwater, with serious consequences for ecosystems and human health. In parallel, energy-intensive thermal processes, reliance on synthetic auxiliaries derived from petrochemicals, and high carbon footprints have raised urgent concerns regarding climate change and resource depletion.

Over the past three decades, mounting consumer awareness, governmental policies, and industry-led initiatives such as the Zero Discharge of Hazardous Chemicals (ZDHC) program, Registration, Evaluation, Authorization and Restriction of Chemicals (REACH) in the European Union (EU), and the Sustainable Apparel Coalition's Higg Index have accelerated the push toward environmentally responsible textile production. Researchers and manufacturers alike are now compelled to explore sustainable chemical processing techniques that reduce ecological burdens while maintaining or even enhancing fabric quality.

In this context, the emergence of green chemistry principles, enzymatic bioprocesses, plasma and ozone-based surface modifications, supercritical carbon dioxide ($scCO_2$) dyeing, natural dyes and bio-based auxiliaries, and nanotechnology-driven functional finishes represents not only technological innovation but also a paradigm

DOI: 10.1201/9781003689225-12

shifts toward cleaner production models. At the same time, integrated strategies such as wastewater recycling, zero liquid discharge systems, and life cycle assessment (LCA) frameworks are being applied to ensure holistic sustainability across the textile value chain.

This chapter, therefore, explores "Sustainable Chemical Processing Techniques" by

1. reviewing traditional chemical processing methods and their limitations;
2. presenting recent developments in sustainable pretreatment, dyeing, printing, and finishing;
3. highlighting case studies from industry and academia that showcase successful applications; and
4. analyzing the opportunities, challenges, and future directions for achieving large-scale transformation into textile processing.

This contribution aims to provide a comprehensive perspective on how the textile industry can transition toward practices that align with the principles of the circular economy and responsible production, while simultaneously meeting consumer expectations regarding performance, safety, and environmental sustainability.

12.2 PRETREATMENT PROCESSES

Pretreatment represents the first and one of the most critical stages of textile wet processing. The purpose of these operations is to remove natural impurities such as waxes, pectins, proteins, and seed fragments from cotton; sericin from silk; and lubricants, oils, or synthetic additives from man-made fibers. Conventional methods typically include desizing, scouring, bleaching, and mercerization, which prepare fabrics for uniform dye uptake and finishing.

12.2.1 Conventional Pretreatment and Environmental Concerns

Traditionally, scouring is performed with concentrated sodium hydroxide (NaOH) solutions at high temperatures, resulting in large volumes of alkaline effluents with a high chemical oxygen demand (COD) and biological oxygen demand (BOD). Similarly, hydrogen peroxide (H_2O_2) bleaching requires stabilizers, optical brighteners, and energy-intensive heating, while mercerization consumes considerable quantities of NaOH (Table 12.1). These practices collectively contribute to intense water and energy consumption, fiber damage, and chemical residues in effluents (Dejene et al. 2025).

12.2.2 Enzymatic Pretreatment

Enzymes such as pectinases, cellulases, lipases, and catalases have been increasingly applied to replace or complement traditional scouring and bleaching. Enzymatic desizing and bio-scouring reduce the need for high-temperature alkaline treatments, lower water consumption by 30%–50%, and minimize fiber weight loss. Catalase enzymes further enable the decomposition of residual peroxide into oxygen and water, eliminating the need for extensive rinsing steps (Khan 2025; Tzanov et al.

TABLE 12.1
Comparison of Conventional vs. Sustainable Pretreatment

Process	Conventional Method	Environmental Issues	Sustainable Alternatives
Desizing	Enzymatic/amylase + hot wash	High water use, residual starch	Cold-pad batch enzymatic desizing
Scouring	NaOH boil-off	High COD, fiber damage	Enzymatic scouring (pectinase, lipase), biosurfactants
Bleaching	H_2O_2 + stabilizers	Alkaline effluents, energy intensive	Low-temperature peroxide, enzymatic bleaching
Mercerization	Concentrated NaOH	Caustic recovery challenges	Plasma, liquid ammonia, enzyme-assisted

Source: Created by the authors based on established textile processing literature and industry practices.

2001). Research on natural fiber modification provides valuable insights into how biological alternatives can preserve fiber integrity while reducing chemical load (Nascimento et al. 2015; Silva et al. 2019).

12.2.3 Emerging Physical and Hybrid Methods

Besides enzymatic approaches, novel physical methods such as plasma, ozone, and ultrasonic pretreatments are being explored. Plasma pretreatment modifies fiber surfaces at low temperatures without wet chemicals, improving wettability and reducing subsequent dyeing demand (Militky and Mishra 2019). Ozone bleaching, carried out in aqueous or gaseous phases, offers energy savings and effluent reductions, though fiber yellowing and equipment costs remain challenges. Ultrasonics can enhance scouring efficiency by improving fiber penetration of mild chemical agents. Hybrid processes, where enzymatic and physical treatments are combined, represent the next step toward industrial adoption. For example, enzymatic scouring followed by low-temperature peroxide bleaching achieves a balance between whiteness index and ecological footprint.

12.2.4 Sustainability Perspectives

Sustainable pretreatment is not only about process substitution but also about integrated water and energy management. Adoption of countercurrent washing, low-liquor ratio machinery, and chemical recovery systems further reduces the footprint of preparation. Comparative LCA consistently show that enzymatic and low-temperature bleaching have a 30%–60% lower environmental impact compared to conventional methods (Saleem and Zaidi 2020; Luo et al. 2026). The combination of biotechnology, nanotechnology, and process optimization thus positions sustainable pretreatment as a cornerstone of greener textile production. As industry shifts toward closed-loop and circular systems, these innovations demonstrate that performance, cost-effectiveness, and environmental protection can be achieved simultaneously.

12.3 SUSTAINABLE DYEING TECHNIQUES

Dyeing is central to textile processing, imparting color, aesthetic appeal, and cultural identity to fabrics. Yet it is also one of the most polluting and resource-intensive operations, consuming an estimated 80–100L of water/kg of fabric and discharging effluents loaded with unfixed dyes, salts, and auxiliaries (Holkar et al. 2016). Conventional dyeing with reactive, disperse, or vat dye typically requires high salt concentrations, alkali, and thermal energy, leading to serious environmental challenges (Figure 12.1). The push toward sustainable dyeing has led to innovations in biological processes, natural dyes, $scCO_2$ systems, low-liquor ratio equipment, digital technologies, and nanotechnology-assisted methods (Table 12.2).

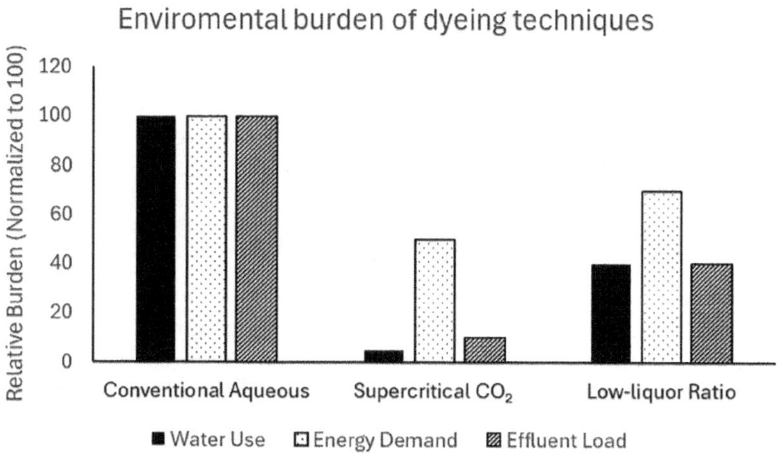

FIGURE 12.1 Environmental burden of conventional vs. sustainable dyeing techniques.

TABLE 12.2
Environmental Comparison of Conventional Aqueous Dyeing and $scCO_2$ Dyeing (de Oliveira et al. 2024; Park et al. 2021; Eren and Özyurt 2024; Yiğit 2026)

Parameters	Conventional Aqueous Dyeing	$scCO_2$ Dyeing
Water consumption	80–100 L/kg fabric	0 L/kg fabric
Salt use	50–80 g/L	0
Energy consumption	High (heating water)	Moderate (CO_2 compression)
Effluent load	High COD/BOD, color, salts	Minimal
Equipment investment	Low–moderate	High

12.3.1 Natural and Bio-Based Dyes

Natural dyes derived from plants, insects, and microorganisms represent the oldest known coloring agents but lost prominence with the rise of synthetic dyes in the 19th century. Renewed interest has emerged due to their biodegradability, low toxicity, and cultural heritage value. Recent research has focused on optimizing natural dye extraction, standardizing mordanting, and improving fastness properties. For example, bio-mordants such as tannins, chitosan, and alum reduce reliance on heavy-metal mordants. Advanced techniques like enzymatic extraction and microwave-assisted dyeing enhance color yield while lowering chemical load.

12.3.2 Enzymatic Dyeing and Biocatalysis

Enzymes are increasingly used in dyeing to replace harsh oxidizing or reducing agents. Laccases and peroxidases can catalyze in-situ dye fixation, improve color yield, and reduce unfixed dye in effluents. Biocatalytic dyeing also enables the modification of dye chromophores, generating novel eco-friendly shades (Osugi et al. 2020; Abbate et al. 2024). This approach aligns with green chemistry by reducing auxiliary use, lowering COD/BOD in effluents, and enabling wastewater reuse.

12.3.3 Supercritical Carbon Dioxide Dyeing

One of the most promising innovations is the use of $scCO_2$ as a dyeing medium, eliminating water altogether. In $scCO_2$ dyeing, disperse dyes are solubilized in compressed CO_2 and diffused into hydrophobic fibers such as polyester (Park et al. 2021; Eren and Özyurt 2024; Yiğit 2026).

Key advantages include the following:

- Zero water usage
- Complete elimination of salt and alkali
- Shorter dyeing times and energy savings
- Easy dye recovery and CO_2 recycling.

However, high equipment cost and limited dye classes remain challenges. Recent pilot plants in Europe (e.g., DyeCoo, The Netherlands) have shown industrial feasibility for polyester sportswear.

12.3.4 Digital and Low-Liquor Dyeing Technologies

Technological innovation has also focused on reducing liquor ratios (e.g., airflow dyeing machines, foam dyeing) and adopting digital textile printing (DTP). DTP significantly reduces dye and water use by directly applying pigments/inks to fabrics with computer-controlled precision (Smithers 2022; Athalye and Patil 2025). While challenges remain in terms of color fastness and ink costs, DTP offers dramatic savings in process stages, effluents, and energy.

12.3.5 Nanotechnology-Assisted Dyeing

Nanoparticles and nanostructured carriers are being used to improve dye uptake, reduce the need for high salt concentrations, and enhance functional properties. For example, TiO_2 nanoparticles can act as photocatalysts, improving dye fixation and self-cleaning properties (Felipe et al. 2022a, 2022b). Chitosan-based nanocarriers also enhance dye–fiber affinity, reducing dye wastage. Research group's ongoing work on cellulose nanofibers and nanocrystals from textile waste demonstrates how nanostructured materials can support innovative dyeing auxiliaries (Nascimento et al. 2015; Silva et al. 2019).

12.3.6 Sustainability Perspectives

Sustainable dyeing is not only about reducing water and chemical loads but also about holistic systems integration. For example, LCA studies indicate that $scCO_2$ dyeing reduces overall environmental impacts by 40%–60% compared to conventional aqueous dyeing when considering water, energy, and effluent parameters (de Oliveira et al. 2024). Industry adoption will depend on balancing economic feasibility, regulatory compliance, and consumer acceptance. As global textile supply chains face increasing scrutiny, sustainable dyeing techniques represent one of the most immediate opportunities to reduce the ecological burden of clothing production.

12.4 PRINTING TECHNIQUES

Textile printing, the art of applying color to fabric in defined patterns, is an ancient craft that has evolved into a major industrial process. Unlike dyeing, which imparts uniform color to a fabric, printing requires precision placement of dyes or pigments onto selected areas, often in multiple colors and complex designs. Conventional printing techniques include screen printing, block printing, roller printing, and transfer printing, with rotary screen printing dominating modern mass production (Miles 2003; Smithers 2022; Athalye and Patil 2025). Figure 12.2 shows a comparison of conventional and digital printing in terms of sustainability.

12.4.1 Environmental Challenges in Conventional Printing

Traditional pigment and reactive printing involve large amounts of thickeners, urea, binders, and steaming processes, generating significant wastewater and chemical residues. Reactive printing, while offering bright shades and good wash fastness, demands alkali fixation and extensive post-washing, consuming water and energy (Miles 2003; Smithers 2022). Pigment printing avoids washing but relies heavily on synthetic binders and plasticizers, raising concerns about volatile organic compound (VOC) emissions and fabric handle.

12.4.2 Digital Textile Printing

The most significant innovation in sustainable textile printing is DTP, which uses inkjet technology to apply colorants directly onto fabrics under computer control

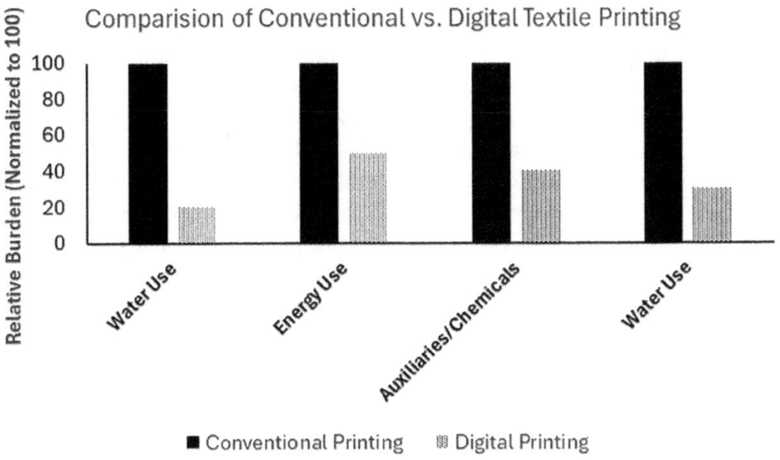

FIGURE 12.2 Comparison of conventional vs. digital textile printing.

TABLE 12.3
Environmental Comparison of Rotary Screen vs. Digital Printing

Parameters	Rotary Screen Printing	Digital Textile Printing
Water use	High (washing, steaming)	Very low
Chemical auxiliaries	High (thickeners, urea, binders)	Minimal
Energy consumption	High (drying, steaming)	Lower
Design flexibility	Limited by screens	Unlimited, digital files
Waste generation	High (screen cleaning, paste residues)	Minimal

(Smithers 2022; Athalye and Patil 2025). DTP eliminates the need for printing screens, drastically reduces paste preparation, and achieves high design flexibility (Table 12.3).

Key sustainability advantages include the following:

- **Water savings**: up to 90% compared to rotary screen printing
- **Reduced chemical use**: elimination of thickeners and urea
- **Energy savings**: fewer steaming and washing steps
- **On-demand production**: reduces overproduction and inventory waste.

12.4.3 Enzymatic and Eco-Friendly Printing Inks

Enzymatic treatments are being explored in pre-printing and post-printing processes, such as using laccases to improve pigment fixation and cellulases for fabric surface preparation (Khan 2025). Eco-friendly inks, including water-based pigment inks and bio-derived binders, further align printing with sustainability goals. Research on nano pigments and functional inks (antimicrobial, ultraviolet [UV]-protective) also promises to integrate functionalization with patterning (Felipe et al. 2022a, 2022b).

12.4.4 Sustainability Perspectives

Digital and enzymatic printing techniques address some of the key sustainability concerns of conventional printing, but wider adoption faces economic and technological barriers. High equipment costs, ink availability, and limitations in printing on certain fiber blends remain challenges. Nonetheless, DTP already accounts for ~7% of global textile printing and is expected to grow rapidly, driven by fast fashion customization and environmental regulations (Smithers 2022). By reducing water, chemicals, and energy, sustainable printing techniques align directly with the broader agenda of green chemistry, resource efficiency, and circular textile production.

12.5 FINISHING PROCESSES

Finishing processes enhance the functional properties of textiles, providing durability, comfort, and performance characteristics beyond basic dyeing and printing. Conventional finishing relies heavily on synthetic resins, formaldehyde-based cross-linkers, and fluorinated compounds, many of which are environmentally persistent and hazardous to human health (Luo et al. 2026; Hasanbeigi and Price 2020; Catarino et al. 2025). The shift toward sustainable finishing focuses on adopting green chemistry, bio-based coatings, nanotechnology-enabled finishes, and eco-friendly functionalization techniques (Figure 12.3).

12.5.1 Conventional Finishing and Its Challenges

Traditional durable-press finishes are typically achieved with formaldehyde-based resins, which release VOCs and raise concerns about carcinogenicity. Similarly, perfluorinated compounds (PFCs) widely used for water and oil repellency are now heavily restricted under international regulations (e.g., EU REACH) due to persistence, bioaccumulation,

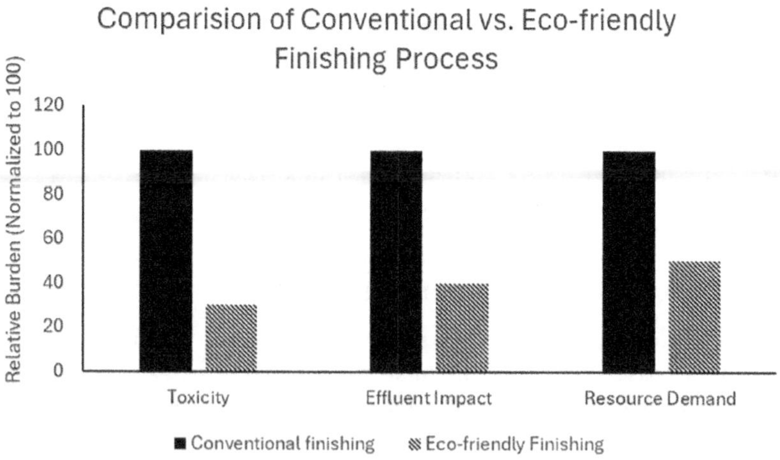

FIGURE 12.3 Comparison of conventional vs. eco-friendly finishing processes.

TABLE 12.4
Comparison of Conventional and Sustainable Textile Finishing Approaches (Militky and Mishra 2019; Silva et al. 2019; Felipe et al. 2022a, 2022b)

Properties	Conventional Finish	Environmental Issues	Sustainable Alternatives
Wrinkle resistance	Formaldehyde-based resins	VOCs, health hazards	Polycarboxylic acids, citric acid
Water repellency	Fluorocarbon (PFCs)	Persistence, bioaccumulation	Wax emulsions, silicones, biopolymers
Antimicrobial	Triclosan, quaternary ammonium	Toxicity, resistance risk	Chitosan, silver/metallic nanoparticles
Flame retardancy	Halogenated flame retardants	Toxic combustion gases	Phosphorus-based finishes, nanocoatings

and toxicity. Flame retardants, antimicrobial agents (quaternary ammonium salts, triclosan), and softeners often contribute to toxic effluents and poor biodegradability (Silva et al. 2019; Saleem and Zaidi 2020; Luo et al. 2026) (Table 12.4).

12.5.2 Biopolymer-Based Finishes

Biopolymers such as chitosan, alginate, and starch derivatives have gained attention as renewable finishing agents. Chitosan, derived from chitin (shellfish waste), is biodegradable, antimicrobial, and film-forming. It has been widely applied for antimicrobial textiles, dye–fiber binding, and wastewater treatment. Silva et al. (2019) reported on the use of multifunctional chitosan/gold nanoparticle coatings for biomedical textiles, and demonstrated enhanced antimicrobial activity and washing durability, establishing a clear pathway for sustainable functional finishes.

12.5.3 Nanotechnology-Enabled Finishes

Nanotechnology plays a pivotal role in next-generation textile finishing. By incorporating nanoparticles or nanostructured coatings, fabrics can achieve durable functionalities with reduced chemical load. Examples include the following:

- **UV-protective finishes**: TiO_2, zinc oxide (ZnO), and graphene quantum dots.
- **Antimicrobial finishes**: Silver, copper, and biopolymer–nanoparticle hybrids.
- **Self-cleaning and hydrophobic finishes**: Nanostructured silica, fluorine-free nanocoatings.

Felipe et al. (2022a) reported nitrogen-doped graphene quantum dots/TiO_2/polyvinyl alcohol (PVA) nanocoatings on cotton fabrics, and their results showed that it was possible to provide UV protection and photoluminescence without relying on hazardous chemicals. This research highlights how nanostructured hybrid systems can combine functionality, durability, and sustainability.

12.5.4 Plasma- and Enzyme-Assisted Finishing

Plasma surface modification offers a dry finishing route that enhances fabric surface reactivity, reducing the need for chemical auxiliaries. Low-temperature plasma can impart water repellency, hydrophilicity, or enhanced dye affinity depending on gas composition. Enzyme-assisted finishing processes, such as laccase-based cross-linking, further replace formaldehyde resins in wrinkle-resistant finishes and enable eco-friendly surface functionalization (Jelil 2015).

12.5.5 Sustainable Flame Retardant and Smart Finishes

Eco-friendly flame retardants based on phosphorus- and nitrogen-containing polymers and nano-clay coatings are replacing halogenated compounds. Meanwhile, "smart finishes" that respond to stimuli (UV, pH, temperature) are being developed using sustainable nanomaterials. Several researchers reported on nanofibers and nanocrystals of cellulose derived from textile waste, which opened the door for bio-based nanostructures in finishing, contributing to both recycling and functionalization (Silva et al. 2019; Felipe et al. 2022a, 2022b; Nascimento et al. 2015).

12.5.6 Sustainability Perspectives

Sustainable finishing is about more than replacing one chemical with another; it requires integration of green chemistry principles, circular economy approaches, and regulatory compliance. The elimination of formaldehyde-based resins and PFCs, adoption of biopolymer- and nanotechnology-based alternatives, and deployment of plasma/enzyme hybrid methods are collectively driving the industry toward a toxic-free, resource-efficient finishing future. Incorporating functional properties while reducing ecological burdens is one of the most promising areas for innovation in textiles, offering new opportunities for performance apparel, medical textiles, and technical fabrics.

12.6 WASTE MANAGEMENT AND EFFLUENT TREATMENT

The textile industry is recognized as a major generator of industrial wastewater, with dyeing and finishing units discharging effluents that contain unfixed dyes, high salt loads, surfactants, heavy metals, and other toxic organic compounds (Table 12.5). It is estimated that up to 200,000 tons of dyes are lost annually to effluents worldwide, causing visible water pollution and severe ecological damage (Holkar et al. 2016).

12.6.1 Characteristics of Textile Effluents

Typical textile wastewater is characterized by the following:

- High COD and BOD due to organic dyes and auxiliaries
- Intense color that reduces sunlight penetration in water bodies, affecting photosynthesis

TABLE 12.5
Comparison of Conventional Effluent Treatment Plants (ETPs) and Zero Liquid Discharge (ZLD) Systems (Holkar et al. 2016; Hasanbeigi and Price 2020)

Parameters	Conventional ETP	ZLD System
COD/BOD removal	Partial	Nearly complete
Color removal	Low	High
Water recovery	<10%	70%–90%
Sludge generation	High	Minimal
Energy consumption	Low	High
Capital cost	Moderate	Very high

- Alkalinity and high total dissolved solids from salts used in reactive dyeing
- Toxic metals from some dye classes, mordants, or finishing agents.

These effluents are resistant to conventional treatment due to the complex aromatic structure of dyes, which makes them nonbiodegradable under normal conditions.

12.6.2 Conventional Treatment Methods and Limitations

Traditional effluent treatment plants (ETPs) rely on the following methods:

- Primary treatment (screening, sedimentation)
- Secondary treatment (activated sludge, trickling filters)
- Tertiary treatment (coagulation–flocculation, adsorption, oxidation).

While these methods can reduce suspended solids and COD, they are ineffective in removing color and recalcitrant organics, and they generate secondary sludge that requires further disposal (Holkar et al. 2016; Hasanbeigi and Price 2020).

12.6.3 Advanced and Sustainable Treatment Methods

To address these limitations, advanced approaches are being implemented (Silva et al. 2019; Nascimento et al. 2015).

- **Adsorption**: Activated carbon and low-cost adsorbents such as agricultural by-products and nanocellulose.
- **Advanced oxidation processes**: Ozone, Fenton, photocatalysis using TiO_2 and ZnO nanoparticles.
- **Membrane technologies**: Ultrafiltration, nanofiltration, reverse osmosis, often coupled with zero liquid discharge (ZLD).
- **Biological methods**: White-rot fungi and enzymatic degradation (laccases, peroxidases) for dye removal.

12.6.4 Chitosan-Based Effluent Treatment

Chitosan has emerged as a promising bio-based flocculant and adsorbent due to its cationic nature, which enables binding to anionic dyes and heavy metals. Ongoing research on the treatment of textile effluents using chitosan spheres modified with mixed oxides demonstrates enhanced dye adsorption capacity and improved reusability (Silva et al. 2019; Nascimento et al. 2015). These systems combine biodegradability, non-toxicity, and efficiency, representing a low-cost, sustainable alternative to synthetic polymers like polyacrylamide (Figure 12.4).

12.6.5 ZLD Systems

ZLD has become a benchmark for sustainable wastewater management in textile hubs such as India, China, and the EU. ZLD integrates multiple technologies (Catarino et al. 2025):

- Membrane concentration
- Multiple-effect evaporators
- Salt recovery and reuse.

Though capital-intensive, ZLD ensures no liquid effluent is released, reducing environmental burdens and enabling circular water use (Table 12.5). Industrial case studies show that ZLD can recover 70%–90% of water for reuse in dyeing, though energy requirements remain high. Comparison of conventional ETP vs. ZLD systems is shown in Table 12.5.

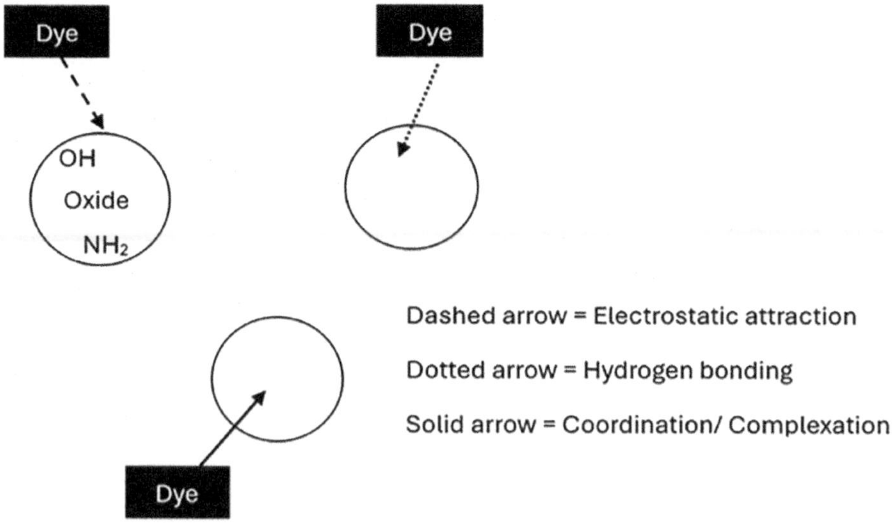

FIGURE 12.4 Mechanism of dye adsorption on chitosan/oxide-modified spheres.

12.6.6 Sustainability Perspectives

Sustainable effluent treatment in the textile sector requires a multi-pronged approach that integrates the following:

- Cleaner production (low-water dyeing, reduced chemical auxiliaries)
- Bio-based adsorbents and nanomaterials (e.g., chitosan, nanocellulose, photocatalysts)
- Closed-loop systems like ZLD to achieve circularity in water use.

Policy frameworks such as EU Best Available Techniques (BAT) Reference Documents (BREF) and India's ZLD mandates are accelerating the adoption of advanced effluent treatment. The integration of bio-based technologies, nanotechnology, and circular water management highlights how waste management itself is evolving into a domain of innovation within sustainable textile processing.

12.7 INNOVATIONS IN GREEN CHEMISTRY

Green chemistry, defined as the design of chemical products and processes that reduce or eliminate the use and generation of hazardous substances, offers a powerful framework for transforming textile wet processing (Table 12.6). Applying its 12 principles to textiles has resulted in breakthroughs in enzymatic bioprocessing, solvent-free dyeing, recyclable auxiliaries, and life cycle-based design (Saleem and Zaidi 2020; Samanta and Agarwal 2009; Saxena and Raja 2014).

12.7.1 Principles of Green Chemistry in Textile Processing

Several principles are particularly relevant:

- **Prevention of waste**: Adoption of low-liquor ratio dyeing and digital printing reduces effluent volumes.

TABLE 12.6
Summary of Green Chemistry Innovations in Textile Processing (Militky and Mishra 2019; Saleem and Zaidi 2020; Luo et al. 2026; Silva et al. 2019)

Areas	Green Chemistry Innovation	Key Benefits	Examples
Pretreatment	Enzymatic scouring/ bleaching	Lower COD/BOD, mild conditions	Pectinase, catalase
Dyeing	$scCO_2$ dyeing	Zero water, recyclable CO_2	Polyester dyeing
Printing	Digital textile printing	90% less water, less waste	Inkjet pigment printing
Finishing	Biopolymer coatings	Biodegradable, nontoxic	Chitosan–AuNP finish
Effluent treatment	Bio-adsorbents, nanocatalysts	High dye removal, reusable	Chitosan/oxide spheres

- **Safer solvents**: Replacement of organic solvents with water, $scCO_2$, or solvent-free systems.
- **Renewable feedstocks**: Use of bio-based dyes, polymers, and auxiliaries.
- **Catalysis**: Enzymes and nanocatalysts enable selective, low-energy transformations.
- **Design for degradation**: Biodegradable finishes and auxiliaries avoid persistent pollutants.

12.7.2 Enzymatic and Biocatalytic Approaches

Biocatalysis is one of the most successful green chemistry applications in textiles. Enzymes are employed not only in pretreatment (desizing, scouring) but also in dyeing and finishing. Examples include the following:

- Laccases for dye fixation and color modification
- Catalases for peroxide removal
- Proteases and cellulases for biopolishing and surface finishing.

These reduce chemical demand, lower wastewater load, and work under milder conditions, aligning with green chemistry principles of energy efficiency and safer chemicals.

12.7.3 Solvent-Free and Waterless Technologies

A central innovation is the move toward solvent-free processing.

- $scCO_2$ dyeing eliminates water and salt, embodying the principle of safer solvents.
- Foam finishing replaces aqueous padding, reducing chemical and water usage by up to 80%.
- Low-liquor ratio machinery achieves high dye fixation at liquor ratios of 1:4–1:6 compared to 1:20 in conventional machines (Figure 12.5).

12.7.4 Renewable and Bio-Based Auxiliaries

Green chemistry encourages the substitution of petrochemical-based auxiliaries with renewable alternatives. Examples include the following:

- Chitosan for antimicrobial finishing and dye fixation (Silva et al. 2019)
- Starch derivatives and alginate as thickeners in printing
- Plant-derived oils and waxes for water repellency
- Nanocellulose from textile waste as a reinforcing or thickening agent (Nascimento et al. 2015).

Nanocellulose extraction from discarded cotton fibers and natural dye extraction from Brazilian plants demonstrate how renewable resources can be reintegrated into textile chemical processing (Nascimento et al. 2015; Samanta and Agarwal 2009; Saxena and Raja 2014).

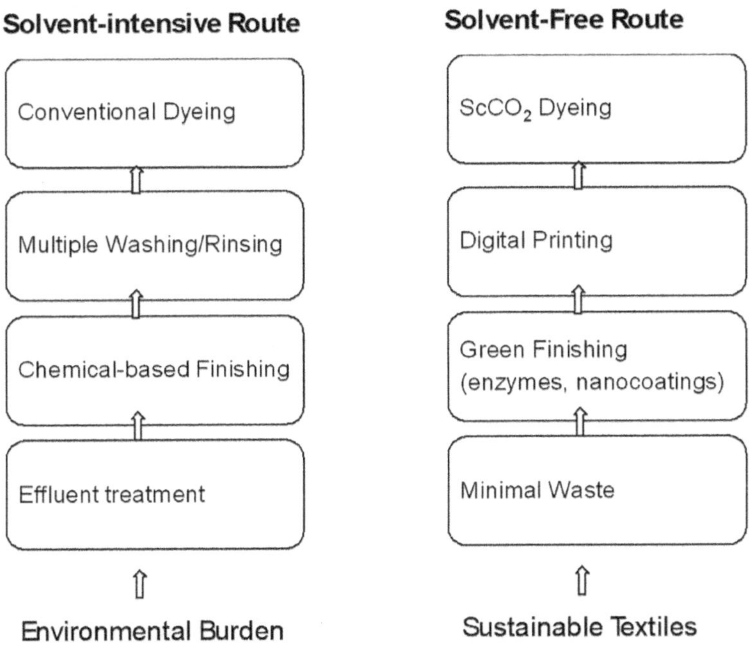

FIGURE 12.5 Comparison of solvent-intensive vs. solvent-free dyeing and finishing routes.

12.7.5 Safer Alternatives and Substitution

Green chemistry emphasizes replacing hazardous substances with safer alternatives.

- Formaldehyde-free crosslinkers (polycarboxylic acids, citric acid) are replacing conventional resins.
- Fluorine-free water repellents based on silicones, paraffin, or dendrimers are gaining traction as PFC bans tighten globally.
- Phosphorus-based flame retardants and nanocoatings are replacing halogenated compounds.

12.7.6 LCA Integration

Green chemistry in textiles extends beyond laboratory substitution to system-wide evaluations. LCA provides a holistic view of impacts from raw materials through production, use, and disposal. Studies have shown that (de Oliveira et al. 2024; Park et al. 2021; Eren and Özyurt 2024; Yiğit 2026)

- switching from conventional scouring to enzymatic scouring reduces global warming potential by ~30%,
- $scCO_2$ dyeing significantly lowers water footprint and eutrophication potential, and
- digital printing reduces resource depletion by minimizing overproduction and waste.

12.7.7 Sustainability Perspectives

The application of green chemistry principles ensures that textile chemical processing evolves into a system that is safer, circular, and resource-efficient. While cost, scalability, and regulatory harmonization remain barriers, innovations in biotechnology, nanotechnology, and solvent-free processes show that sustainable chemical processing is no longer a distant goal but an emerging industrial reality.

12.8 REGULATORY FRAMEWORKS AND CERTIFICATIONS

The transformation of textile chemical processing toward sustainability is strongly shaped by regulatory pressures and voluntary certification schemes. Governments, international bodies, and nongovernmental organizations have introduced a range of frameworks aimed at reducing toxic chemical use, ensuring worker safety, and protecting ecosystems. At the same time, third-party certification provides transparency and credibility for consumers demanding environmentally responsible products.

12.8.1 European Union's REACH Regulation

The REACH is one of the most comprehensive chemical regulations globally (ECHA 2021). Implemented in 2007, REACH requires companies to provide safety data for substances manufactured or imported in volumes exceeding 1 ton/year. In the textile sector, REACH has led to the restriction of the following:

- Azo dyes that release carcinogenic amines
- Formaldehyde and formaldehyde-based resins
- PFCs used in water- and oil-repellent finishes.

By enforcing substitution of hazardous chemicals, REACH has accelerated the adoption of safer alternatives such as fluorine-free repellents, bio-based auxiliaries, and enzymatic processes.

12.8.2 Zero Discharge of Hazardous Chemicals Program

The ZDHC program, launched in 2011, is a multi-stakeholder initiative involving global brands, retailers, and suppliers. Its Manufacturing Restricted Substances List and Wastewater Guidelines set global benchmarks for eliminating hazardous substances from textile production (ZDHC Foundation 2021). ZDHC focuses not only on banned substances but also on supply chain transparency, driving the adoption of cleaner production technologies such as digital printing, waterless dyeing, and enzymatic finishing. Many multinational brands (e.g., H&M, Adidas, Levi's) now require their suppliers to comply with ZDHC standards, making it a de facto global requirement.

12.8.3 OEKO-TEX® Standards

OEKO-TEX® Standard 100 is a widely recognized certification that tests textiles for harmful substances, ensuring consumer safety (OEKO-TEX 2021). Expansions include the following:

- **STeP (sustainable textile production)**: Evaluates factory environmental performance.
- **Made in Green**: Combines product safety with traceability and sustainability credentials. OEKO-TEX® standards encourage chemical management throughout the supply chain, covering dyes, auxiliaries, and finishing agents.

12.8.4 Global Organic Textile Standard (GOTS)

The GOTS certification applies specifically to organic fibers, but its scope extends to chemical processing. It restricts the use of toxic heavy metals, formaldehyde, aromatic solvents, and genetically modified organisms. GOTS also promotes the use of low-impact dyes, enzymatic treatments, and biodegradable finishes, making it one of the most comprehensive eco-labels for textiles (GOTS 2020).

12.8.5 Other International Standards and Guidelines

- **Bluesign®**: Focuses on input stream management, ensuring that only safe and sustainable chemicals enter the production process (Bluesign Technologies AG 2021).
- **ISO 14001**: Provides a framework for environmental management systems applicable to textile mills.
- **Higg Index (sustainable apparel coalition)**: Measures environmental and social performance across the value chain.
- **UN Sustainable Development Goals**: Goal 12 (Responsible Consumption and Production) is particularly relevant to textile chemical sustainability.

12.8.6 Sustainability Perspectives

Regulatory frameworks and certifications are not merely compliance burdens but drivers of innovation. By restricting hazardous substances, mandating safe alternatives, and promoting transparency, they create incentives for industry to invest in enzymatic bioprocesses, plasma treatments, $scCO_2$ dyeing, nanotechnology-based finishes, and wastewater recycling systems (Table 12.7).

12.8.7 Outlook

As consumer awareness and regulatory pressure increase, compliance with these frameworks is becoming a prerequisite for global market access. Future policy trends are expected to include the following:

- Stricter limits on microfiber release from synthetic textiles
- Carbon labeling of textiles to address climate impact
- Expansion of extended producer responsibility for end-of-life textiles.

Together, these frameworks accelerate the textile industry's transition toward chemical sustainability, circular economy practices, and global accountability (Figure 12.6).

TABLE 12.7
Key Regulatory and Certification Frameworks in Textile Chemical Processing

Framework/ Standard	Scope	Key Impact
REACH (EU)	Chemical safety	Ban/restriction of hazardous dyes, auxiliaries, finishes
ZDHC	Supply chain, effluents	Eliminates hazardous substances, enforces wastewater guidelines
OEKO-TEX®	Product and process safety	Harmful substance testing, sustainable production
GOTS	Organic textiles	Restricts toxic chemicals, promotes enzymatic/ eco-friendly processing
Bluesign®	Input stream management	Approves only safe chemicals for production
Higg Index	Value chain assessment	Quantifies environmental and social impacts

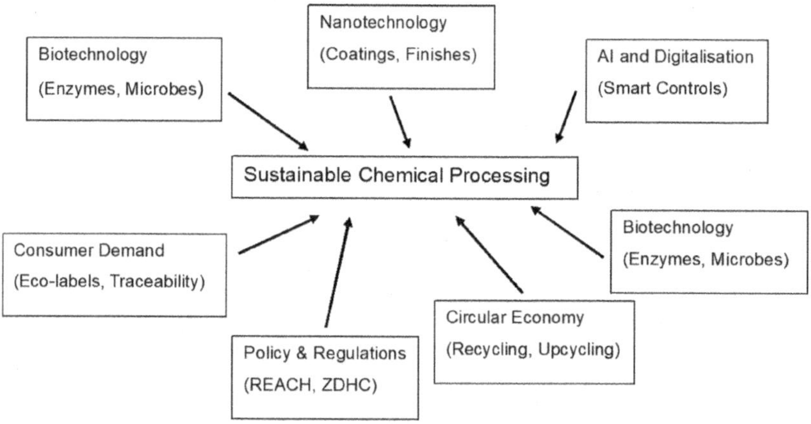

FIGURE 12.6 Future directions in sustainable textile chemical processing.

12.9 CASE STUDIES

Case studies serve as concrete illustrations of how sustainable chemical processing techniques are being developed, tested, and implemented across academia and industry. They demonstrate the feasibility of replacing conventional, resource-intensive practices with innovative methods that align with environmental, economic, and regulatory expectations (Table 12.8).

12.9.1 Case Study 1: Enzymatic Pretreatment for Cotton Fabrics

At the industrial scale, enzymatic scouring has been adopted in textile clusters in India and Bangladesh, replacing conventional NaOH boil-off. Studies showed a 30%–50% reduction in water and energy consumption, improved fabric softness, and reduced fiber weight loss compared to alkaline scouring (Khan 2025).

TABLE 12.8
Overview of Case Studies in Sustainable Chemical Processing (Hasanbeigi and Price 2020; Felipe et al. 2022a; 2022b)

Case Study	Techniques	Key Benefits	Main Challenges
1	Enzymatic scouring	Lower water/energy use, reduced COD	Enzyme cost
2	Natural dyes (Brazilian plants)	Renewable, biodegradable, cultural value	Standardization, color fastness
3	$scCO_2$ dyeing (DyeCoo)	Zero water, high dye uptake, CO_2 recycling	High equipment cost, limited dye classes
4	Nanotechnology UV-protective finishes	Nontoxic, durable, fluorine-free UV protection	Scale-up, nanoparticle safety
5	Chitosan-based effluent treatment	Biodegradable, low-cost dye removal	Industrial efficiency
6	ZLD systems (Tamil Nadu, India)	70%–90% water recovery, pollution prevention	High energy/capital costs

- **Sustainability impact**: Lower COD/BOD in effluents, elimination of harsh alkali, and reduced wastewater load.
- **Scaling challenge**: Higher enzyme costs, although these are decreasing with biotechnology advances.

12.9.2 Case Study 2: Supercritical CO_2 Dyeing (DyeCoo, The Netherlands)

The Dutch company DyeCoo has pioneered $scCO_2$ dyeing systems used commercially by brands such as Nike and Adidas for polyester sportswear. The technology eliminates water and salt entirely, achieving 90% dye uptake and enabling CO_2 recycling.

- **Sustainability impact**: Zero wastewater discharge, reduced dye wastage, and lower overall carbon footprint.
- **Scaling challenge**: High equipment costs and limited dye classes (mainly disperse dyes for polyester).

12.9.3 Case Study 3: Nanotechnology-Enabled UV-Protective Finishes

Felipe et al. (2022a) applied nitrogen-doped graphene quantum dots/TiO_2/PVA nanocoatings to cotton fabrics and developed UV-protective and photoluminescent finishes with enhanced durability (Figure 12.7).

- **Sustainability impact**: Fluorine-free, nontoxic nanocoating as an alternative to synthetic UV blockers.
- **Scaling challenge**: Industrial upscaling and ensuring nanoparticle safety in long-term use.

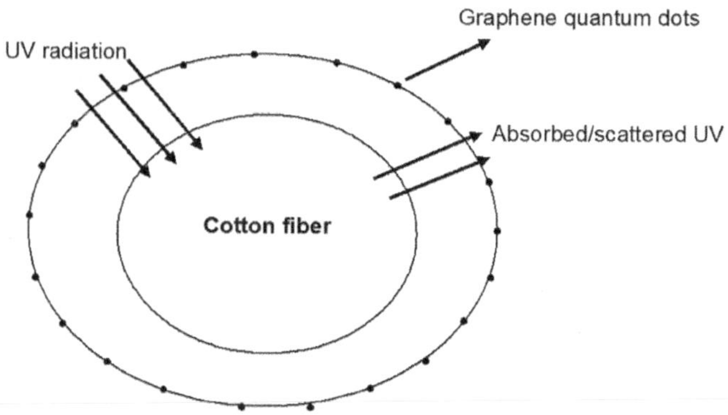

FIGURE 12.7 Mechanism of UV protection in cotton fabrics using graphene quantum dot/ TiO_2 nanocoatings.

12.9.4 Case Study 4: Chitosan-Based Effluent Treatment

Silva et al. (2019) developed chitosan spheres modified with mixed oxides for the adsorption of dyes and heavy metals from textile effluents. Laboratory and pilot trials showed significant removal efficiency, biodegradability, and potential reuse of treated water.

- **Sustainability impact**: Bio-based, low-cost alternative to synthetic flocculants.
- **Scaling challenge**: Maintaining adsorption efficiency under high-volume industrial conditions.

12.9.5 Case Study 5: ZLD in Indian Textile Parks

Several Indian textile parks in Tamil Nadu have implemented ZLD systems, integrating reverse osmosis, multiple-effect evaporators, and salt recovery. These plants recover up to 90% of process water for reuse and reduce dependence on freshwater withdrawals.

- **Sustainability impact**: Virtually eliminates liquid effluent discharge.
- **Scaling challenge**: High energy demand and operational costs.

12.9.6 Sustainability Perspectives

These case studies show that sustainable chemical processing is not only a laboratory concept but also an emerging industrial reality. From enzymatic scouring to nanotechnology-enabled finishes, the transition is underway. However, cost, scalability, and regulatory frameworks remain critical determinants of widespread adoption.

12.10 CHALLENGES AND FUTURE DIRECTIONS

Despite significant progress in developing sustainable chemical processing techniques, their adoption in the textile industry remains uneven and limited. The transition from conventional, resource-intensive methods to eco-friendly alternatives faces multiple technical, economic, and social barriers. At the same time, emerging technologies, policy drivers, and consumer demand offer promising pathways for accelerated transformation.

12.10.1 Key Challenges

1. **High cost**: High initial investment. Technologies such as $scCO_2$ dyeing, plasma finishing, and ZLD systems require substantial capital investment. Small- and medium-sized enterprises (SMEs), which constitute most textile mills globally, often lack access to financing for advanced machinery.
2. **Process scalability**: Laboratory innovations (e.g., enzymatic dyeing, nano-coatings, chitosan-based effluent treatment) demonstrate strong potential, but scaling them to continuous, high-volume industrial operations while maintaining efficiency and cost-effectiveness remains difficult.
3. **Performance trade-offs**: Sustainable alternatives sometimes underperform compared to conventional methods in areas such as color fastness, durability, or cost. For example, natural dyes may require additional mordanting, while fluorine-free water repellents may not yet match the long-lasting effect of PFCs.
4. **Supply chain complexity**: Global textile supply chains involve diverse actors with varying levels of technical expertise and regulatory enforcement. Ensuring uniform adoption of sustainable practices across developing and developed countries is a persistent challenge.
5. **Knowledge and skill gaps**: Implementing novel green chemistry approaches demands new skills in enzymology, nanotechnology, and process optimization. Many mills lack trained personnel, slowing down adoption.
6. **Regulatory and consumer pressures**: While certifications (e.g., OEKO-TEX®, GOTS, ZDHC) push for sustainability, compliance costs can be high. Additionally, consumer willingness to pay a premium for "green" textiles remains inconsistent across markets.

12.10.2 Emerging Opportunities and Future Directions

1. **Integration of biotechnology and nanotechnology**: The combined use of enzymes, bio-based polymers, and nanomaterials will create multifunctional finishes, improved dye uptake, and eco-friendly effluent treatments. This aligns directly with the principles of green chemistry and circular production.
2. **Artificial intelligence and digitalization**: Digital twins, process simulation, and AI-driven optimization can minimize energy and chemical use by predicting optimal process parameters. Coupled with Industry 4.0 technologies, smart factories could reduce the footprint of chemical processing.
3. **Circular economy approaches**: Future systems will increasingly adopt closed-loop water management, chemical recovery, and fiber recycling. Innovations such as nanocellulose extraction from textile waste show how waste can become a valuable raw material for new finishing and processing agents.

4. **Policy and regulatory momentum**: Growing alignment of international policies (REACH, ZDHC, GOTS) will ensure global harmonization and speed up substitution of hazardous substances. Stricter controls on microfiber release and carbon footprint labeling are expected.
5. **Consumer-driven innovation**: As awareness grows, especially among younger generations, demand for sustainable textiles will pressure brands and manufacturers to accelerate the adoption of eco-friendly chemical processing techniques. Transparency via digital product passports (DPPs) and blockchain-enabled traceability may further reinforce consumer trust.
6. **Industrial collaborations**: Partnerships between academia, industry, and policymakers will be essential for accelerating the transition toward sustainable textile chemical processing. Collaborative platforms are already demonstrating measurable success—for example, the adoption of ZLD systems across major textile clusters in India and the commercialization of $scCO_2$ dyeing technologies such as DyeCoo in Europe (Hasanbeigi and Price 2020; Holkar et al. 2016; de Oliveira et al. 2024).

Within this translational landscape, our research group's collaborative work on nanocoating and effluent treatment using biopolymer–nanomaterial hybrids provide an illustrative academic–industrial interface. These collaborations have centered on the development of multifunctional textile coatings based on renewable biopolymers such as chitosan, integrated with advanced nanomaterials including graphene derivatives, metal/metal oxide nanoparticles, and quantum dots. The resulting hybrid systems have demonstrated enhanced antimicrobial performance, UV shielding, catalytic self-cleaning behavior, and improved coating durability under simulated industrial laundering and finishing conditions (Silva et al. 2019; Felipe et al. 2022a).

Parallel efforts have focused on wastewater remediation, where biopolymer–nanomaterial composites have been engineered as adsorption and catalytic platforms for dye removal, heavy-metal sequestration, and degradation of recalcitrant textile auxiliaries. Laboratory findings have been evaluated under scaled processing parameters to assess adsorption kinetics, regeneration capacity, sludge minimization potential, and compatibility with existing effluent treatment infrastructure (Nascimento et al. 2015; Saleem and Zaidi 2020).

Importantly, these collaborative initiatives have involved knowledge exchange with textile processors and finishing units, facilitating the alignment of material design with industrial process requirements such as pad–dry–cure application routes, bath exhaustion compatibility, and regulatory compliance frameworks. While large-scale commercialization remains an evolving pathway, these partnerships exemplify how academic nanomaterials research can inform scalable, environmentally responsible solutions for textile surface functionalization and effluent management (Felipe et al. 2022b; Militky and Mishra 2019).

12.10.3 Outlook

The pathway to sustainable chemical processing is neither linear nor immediate. It requires a balance between technological innovation, cost reduction, regulatory enforcement, and consumer engagement. Nonetheless, evidence from case studies and global initiatives demonstrates that the industry is moving in the right direction.

The future lies in integrated systems where green chemistry principles, biotechnology, nanotechnology, and circular economy models converge. Such integration will enable the textile industry to meet environmental goals without compromising fabric quality, functionality, or economic viability.

12.11 CONCLUSION

Chemical processing remains at the heart of textile manufacturing, enabling fabrics to achieve desirable qualities such as whiteness, color brilliance, comfort, and performance. Yet these very processes have historically contributed to some of the most pressing sustainability challenges in the textile industry, including excessive water consumption, toxic effluent discharge, and dependence on petrochemical-derived auxiliaries.

Over the past two decades, a paradigm shift has taken place. Advances in enzymatic bioprocessing, natural dyes, digital and low-liquor ratio technologies, $scCO_2$ dyeing, plasma and ozone treatments, biopolymer-based finishes, and nanotechnology-enabled functionalization have demonstrated that high-quality textiles can be produced with far lower environmental impact. Coupled with effluent treatment innovations, ZLD systems, and LCA frameworks, these developments have set the foundation for a greener and more resilient textile sector.

At the same time, regulatory frameworks and certifications such as REACH, ZDHC, OEKO-TEX®, GOTS, and Bluesign® have accelerated the substitution of hazardous substances and pushed the industry toward greater accountability and transparency. Case studies from both academia and industry that include enzymatic scouring, natural dye applications, graphene quantum dot nanocoatings, chitosan-based effluent treatment, and industrial ZLD systems show that sustainable chemical processing is no longer theoretical but increasingly a practical reality.

Nevertheless, significant challenges remain. High capital investment, scalability issues, and performance trade-offs still limit widespread adoption, particularly in SMEs. Addressing these challenges will require stronger collaborations between industry, academia, and policymakers, as well as continued innovation in biotechnology, nanotechnology, digitalization, and circular economy approaches.

Looking ahead, the future of textile chemical processing lies in integration: combining green chemistry principles with emerging technologies to create processes that are safe, efficient, and circular. As consumer demand for transparency and sustainability grows, adopting these innovations is not only an environmental necessity but also a market advantage. By aligning technological innovation with environmental stewardship, the textile industry has the opportunity to transform itself into a sector that truly embodies the spirit of *Wear the Change*.

REFERENCES

Abbate, S., Centobelli, P., Cerchione, R., Nadeem, S. P. and Riccio, E. 2024. Sustainability trends and gaps in the textile, apparel and fashion industries. *Environment, Development and Sustainability* 26: 2837–2864.

Athalye, A. and Patil, H. 2025. *Digital Printing: Sustainable Solution to Fashion Industry*. Springer, Singapore.

Bluesign Technologies AG. 2021. *Bluesign® System – Sustainable Textile Production*. Bluesign Technologies AG, St. Gallen.

Catarino, M. L., Sampaio, F. and Gonçalves, A. L. 2025. Sustainable wet processing technologies for the textile industry: A comprehensive review. *Sustainability* 17: 3041.

Dejene, B. K., Abtew, M. A. and Pawlos, M. 2025. Eco-friendly flame retardant and antibacterial finishing solutions for cotton textiles: A comprehensive review. *Journal of Industrial Textiles* 55. doi:10.1177/15280837251325779

de Oliveira, C. R. S., de Oliveira, P. V., Pellenz, L., de Aguiar, C. R. L. and da Silva Júnior, A. H. 2024. Supercritical fluid technology as a sustainable alternative method for textile dyeing: An approach on waste, energy, and CO_2 emission reduction. *Journal of Environmental Sciences* 140: 123–145.

ECHA. 2021. *REACH Regulation: Guidance on Requirements*. European Chemicals Agency (ECHA), Helsinki.

Ellen MacArthur Foundation. 2017. A new textiles economy: Redesigning fashion's future. Ellen MacArthur Foundation.

Eren, S. and Özyurt, İ. 2024. Waterless dyeing of polyamide 6.6. *Polymers* 16: 1472.

Felipe, M. J., Silva, I. O. and Ladchumananandasivam, R. 2022a. Nanostructured graphene quantum dots/TiO_2 coatings for multifunctional cotton textiles. *Journal of Materials Research and Technology* 18: 2456–2468.

Felipe, M. J., Silva, I. O., and Ladchumananandasivam, R. 2022b. Nanotechnology-enabled smart textiles for sustainability. *Materials Today: Proceedings* 49: 189–196.

GOTS. 2020. Global Organic Textile Standard (GOTS), Version 6.0. Hamburg, Germany.

Hasanbeigi, A. and Price, L. 2020. Emerging technologies for sustainable textile dyeing and finishing. Lawrence Berkeley National Laboratory.

Holkar, C. R., Jadhav, A. J., Pinjari, D. V., Mahamuni, N. M. and Pandit, A. B. 2016. A critical review on textile wastewater treatments: Possible approaches. *Journal of Environmental Management* 182: 351–366.

Jelil, R. A. 2015. A review of low-temperature plasma treatment of textile materials. *Journal of Materials Science* 50: 5913–5943.

Khan, M. F. 2025. Recent advances in microbial enzyme applications for sustainable textile processing and waste management. *Sci* 7: 46.

Luo, H., Sun, H., Zhang, M., Wen, J., Chen, M., Fang, J. and Sun, Z. 2026. Advances in anti-wrinkle finishing agent for natural fabrics. *Polymers* 18(3): 407.

McKinsey and Company. 2020. The state of fashion 2020. McKinsey Global Fashion Index.

Miles, L. W. C. 2003. Textile printing. Society of Dyers and Colorists.

Militky, J. and Mishra, R. 2019. *Nanotechnology in Textiles: Theory and Practice*. Woodhead Publishing, Cambridge.

Nascimento, J. H. O., Luz, R. F., Galvão, F. M. F., Melo, J. D. D., Oliveira, F. R., Ladchumananandasivam, R. and Zille, A. 2015. Extraction and characterization of cellulosic nanowhisker obtained from discarded cotton fibers. *Materials Today: Proceedings* 2: 1–7.

OEKO-TEX. 2021. *OEKO-TEX® Standard 100: Criteria and Limit Values*. OEKO-TEX® Association, Zurich.

Osugi, M. E., Silva, M. F. and Oliveira, H. P. 2020. Biodegradation of azo dyes using laccases: Industrial perspectives. *Dyes and Pigments* 173: 107936.

Park, G., Kwon, D. E., Kong, W., Park, J. and Lee, Y.-W. 2021. A dissolution kinetic study of disperse dye in supercritical carbon dioxide to design an efficient supercritical dyeing process. *Processes* 9: 977.

Saleem, H. and Zaidi, S. J. 2020. Sustainable use of nanomaterials in textiles and their environmental impact. *Materials* 13(22): 5134.

Samanta, A. K. and Agarwal, P. 2009. Application of natural dyes on textiles. *Indian Journal of Fiber and Textile Research* 34(4): 384–399.

Saxena, S. and Raja, A. S. M. 2014. Natural dyes: Sources, chemistry, application, and sustainability issues. In *Roadmap to Sustainable Textiles and Clothing: Eco-friendly Raw Materials, Technologies, and Processing* (Ed. Muthu, S. S.), Woodhead Publishing, Cambridge, 37–80.

Silva, I. O., Ladchumananandasivam, R., Nascimento, J. H. O., Silva, K. K. O. S., Oliveira, F. R., Souto, A. P., Felgueiras, H. P. and Zille, A. 2019. Multifunctional chitosan/gold nanoparticles coatings for biomedical textiles. *Nanomaterials* 9(8): 1064.

Smithers. 2022. The future of digital textile printing to 2032. Smithers Pira.

Tzanov, T., Calafell, M., Guebitz, G. M. and Cavaco-Paulo, A. 2001. Biopreparation of cotton fabrics. *Enzyme and Microbial Technology* 29(6–7): 357–362.

UNEP. 2020. Sustainability and circularity in the textile value chain. United Nations Environment Program (UNEP). https://www.unep.org/resources/publication/sustainability-and-circularity-textile-value-chain-global-roadmap [Accessed 02 October 2025].

World Bank. 2019. *How Much Do Our Wardrobes Cost to the Environment?* World Bank Group, Washington, DC. https://www.worldbank.org/en/news/feature/2019/09/23/costo-moda-medio-ambiente [Accessed 05 October 2025].

Yiğit, İ. 2026. Production of sustainable textiles using natural dye and eggshell powder on recycled polyester fabric via waterless supercritical CO_2 dyeing. *Polymers* 18: 431.

ZDHC Foundation. 2021. *ZDHC Manufacturing Restricted Substances List (MRSL), Version 2.0*. ZDHC Foundation, Amsterdam.

13 Sustainable Technical Textiles

Huipu Gao and Shafiqul Islam

13.1 INTRODUCTION

Textiles are an integral part of human civilization and can be categorized into three areas based on their applications: fashion, home, and technical textiles, with technical textiles accounting for about 25% of the textile market in 2026 (Fortune Business Insights 2026a, 2026b). The Textile Institute defines technical textiles as "textile materials and products manufactured primarily for their technical performance and functional properties rather than their aesthetic or decorative characteristics" (The Textile Institute 2021). They are engineered to achieve specific performance properties (e.g., fire resistance, strength, antimicrobial, moisture management, durability, and repellency to water or oil), and are found in applications such as filters, textile-reinforced composites, protective clothing, hygiene products, sportswear, medical gowns, implants, house insulation, and geotextiles.

The United Nations defines sustainability as "meeting the needs of the present without compromising the ability of future generations to meet their own needs" (United Nations 2025), encompassing environmental, social, and economic dimensions. Although all three aspects are important, this chapter specifically focuses on the environmental aspect. In the fashion industry, regulations and growing awareness of environmental impacts lead to significant efforts to limit the use of toxic chemicals, reduce energy and water consumption, improve traceability, address microplastic pollution, and increase the use of biodegradable, renewable, and recycled materials. In contrast, sustainability within the field of technical textiles has received far less attention.

Technical textiles are produced by weaving, knitting, nonwoven processes, or using composite structures, often with chemical finishing to enhance performance. The most common fibers used are synthetic (55%), followed by natural fibers and inorganic fibers (KPMG 2021). Technical textiles face many of the same sustainability challenges as the fashion industry, including microfiber pollution, reliance on petroleum-based fibers, chemical use, energy and water consumption, and emissions. This chapter reviews the different types of technical textiles, materials used, environmental impact, and recent sustainability advances.

13.2 APPLICATION SECTORS AND THEIR IMPACTS

Technical textiles have broad applications, as shown in Figure 13.1.

DOI: 10.1201/9781003689225-13

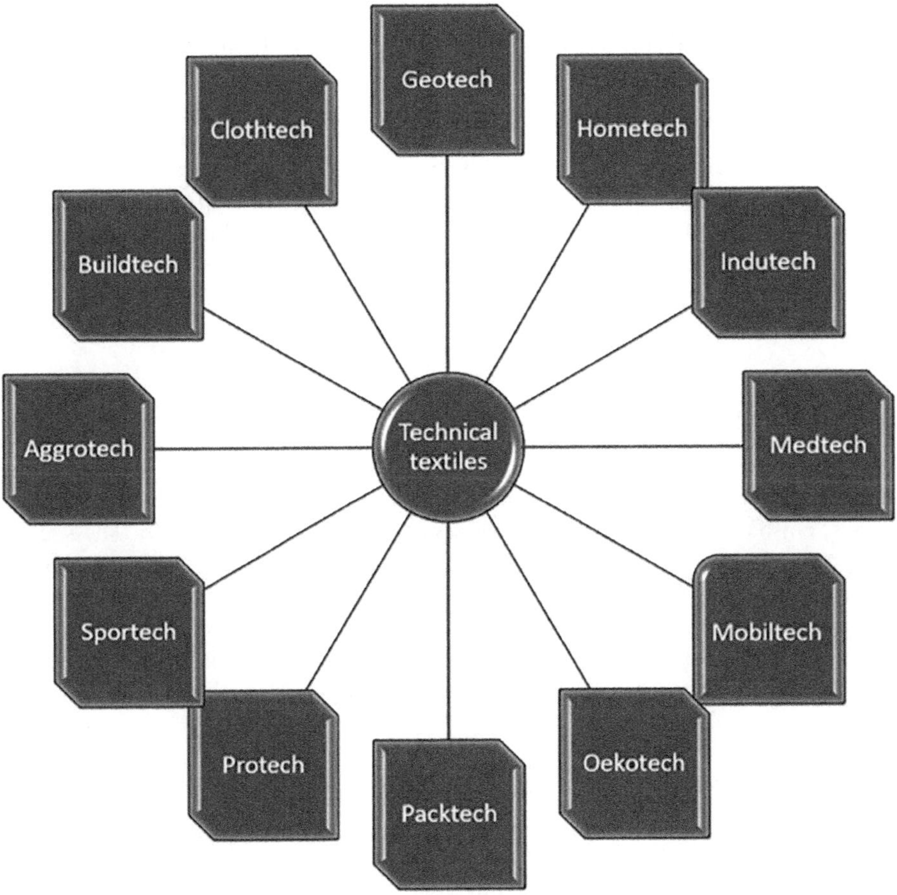

FIGURE 13.1 Application sectors of technical textiles.

Source: **Şahin 2025.**

13.2.1 Agrotech

Agrotech is a category of textile products used to support productivity in crop production, animal husbandry, fisheries, forestry, and horticulture (Bala et al. 2025). Some of their key functions include crop protection from external hazards (crop covers, anti-hail nets, bird protection nets, insect nets, wind protection nets), temperature and moisture regulation (shade and thermal nets, mulch mats, frost protection nets), weed suppression (weed control mats), and aquaculture support (fisher nets, traps) (Azam and Ahmad 2020; Sharma et al. 2022; Grosu et al. 2020).

13.2.2 Buildtech

Buildtech refers to textiles employed in construction and architectural applications for both temporary and permanent structures. Architectural textiles are used to create enclosed or semi-enclosed environments, such as canopies, tents, and façade systems (Tabor and Ghosh 2019). These materials are prepared from glass fiber, nylon, polyester, polypropylene (PP), or polyethylene (PE) fabrics coated with polytetrafluoroethylene, polyvinyl chloride (PVC), polyvinylidene fluoride, silicone, polyurethane (PU), or ethylene tetrafluoroethylene. These fabric-reinforced membranes possess desirable properties such as high tensile strength, low weight, durability, flexibility, resistance to ultraviolet (UV) radiation and moisture, and tunability for thermal performance (Singh 2021). Buildtech textiles are also widely used in building construction, including fiber-reinforced concrete, thermal insulation materials (e.g., mineral wool, glass fiber, recycled textile waste, and natural fibers), and house-wrap fabrics made with flash-spun PE nonwovens.

13.2.3 Clothtech

Clothtech refers to textiles used in clothing and footwear that serve as functional components, such as interfacing fabrics, labels, sewing threads, waterproof membranes, elastic bands, fasteners, and zippers (Yasir and Ahmad 2020). The specific functions of clothtech depend on their end-use. For example, interfacing fabrics enhance the garments' mechanical properties, dimensional stability, and shape retention. Specialized membrane technologies, such as Gore-Tex expanded PE and the North Face's electrospun PU membranes, are designed to provide waterproof protection while maintaining breathability.

13.2.4 Geotech

Geotech refers to textiles used in infrastructure, soil engineering, and environmental applications (Tanasă et al. 2022). They provide important functions including separation, drainage, reinforcement, filtration, barrier, and erosion control (Wu et al. 2020). Due to the stringent requirements for durability, mechanical strength, and resistance to biodegradation in soil environments, synthetic fibers dominate geotextile applications. PP is the most used polymer, followed by polyester and PE (Bérubé and Saunier 2016; Hsuan 2016). For applications where biodegradability is desirable, such as short-term stabilization, natural fibers can be used to provide temporary functional performance (Desai and Kant 2016).

13.2.5 Hometech

Hometech textiles are functional textiles used in household and domestic applications (Rasheed 2020). Common examples include curtains, sleeping bags, upholstery fabrics, carpets, wall coverings, window shades, towels, mattresses, household filters, oven gloves, and mosquito nets. Depending on the specific application, hometech textiles are required to meet a range of performance requirements, including comfort, mechanical strength, UV resistance, flame retardancy, abrasion resistance, absorbency, thermal insulation, soundproofing, moisture regulation, waterproofing, and light-blocking properties.

13.2.6 Indutech

Indutech refers to textile products used for industrial and manufacturing applications (Ahmad et al. 2020). One of indutech's major application areas is filtration. Textile-based filters are widely used for separating air, gases, and liquids. These filtration textiles are mainly produced from nonwoven fabrics, such as needle-punched, meltblown, spunbond, and wet-laid structures. Other indutech applications include conveyor belts, textile-reinforced composites, and industrial hoses and ropes. The selection of fiber type and fabric structure is determined by the specific requirements of the end application. For example, in conveyor belts, where high mechanical strength, dimensional stability, and durability are required, woven textiles of polyester, nylon, para-aramid, glass fibers, or carbon fibers are used as reinforcement layers. These fabrics are typically coated with PVC, PU, or rubber to enhance performance.

13.2.7 Medtech

Medical textiles are fiber-based products used in healthcare and hygiene applications (El-Ghazali et al. 2022). For healthcare applications, medical textiles can be broadly categorized into non-implantable and implantable textiles. Non-implantable textiles are designed for external use and include products such as face masks, surgical gowns, sterilization wraps, wound dressings, compression garments, and bandages (Guru et al. 2022; Molla et al. 2024). Implantable medical textiles are fibrous materials placed inside the body to replace, repair, or reinforce damaged tissues or organs (Guru et al. 2022; Molla et al. 2024). Typical examples include sutures, hernia meshes, scaffolds, and vascular grafts. Hygiene applications include baby diapers, adult incontinence products, and feminine hygiene products. The majority of these products are multilayered and disposable, containing wood pulp, superabsorbent polymers, PP, and PE. A major environmental issue associated with medical textiles is the single-use nature of many non-implantable healthcare and hygiene products, which generate substantial waste. This problem is further exacerbated by the heavy reliance on synthetic plastics and the difficulty of recycling due to biohazard and contamination.

13.2.8 Mobiltech

Mobiltech refers to technical textiles used in transportation applications (Paul 2019). These textiles provide a wide range of functions such as insulation, soundproofing, fire retardancy, comfort, and safety. Common examples include airbags, seat covers, headliners, seat belts, thermal insulation materials, carpets, composites, tires, airline upholstery, NVH (noise, vibration, and harshness) insulation, and sunblinds (Behera 2019). In marine applications, ship furnishing textiles must be durable and flame-resistant (FR) to meet safety and performance standards. Fiber-reinforced composite materials are used to reduce weight and enhance strength. Coated fabrics are commonly employed in life jackets and buoyancy products. For sailcloth and ropes, textiles must exhibit high tensile strength, resistance to seawater, and UV radiation (Fung 2000).

13.2.9 Packtech

Textiles used for packaging applications in industrial, agricultural, consumer goods, and other sectors are referred to as Packtech (Cărpuș et al. 2020). Examples include reusable grocery bags, sacks, tea bags, soft luggage, special envelopes, flexible intermediate bulk containers, and leno nets for packing fruits and vegetables. Some major functions are to protect items from damage, ensure efficient distribution and storage, preserve product quality, and enhance consumer convenience (Samanta et al. 2015). These materials are required to be strong, durable, and non-contaminating to the packaged contents.

13.2.10 Protech

Protech refers to textiles designed to protect wearers from external hazards, including chemical, thermal, impact, ballistic, and abrasion-related risks. Examples include firefighter suits, protective apparel for the oil and gas industry, ballistic vests, cut-resistant garments, chemical protective suits, and radiation-protective clothing. For these products, protective performance takes precedence over other considerations. High-performance fibers are normally used, including para-aramid, meta-aramid, carbon fibers, ultrahigh-molecular-weight PE, polybenzimidazole, and flame-retardant lyocell fibers.

13.2.11 Sporttech

Sporttech refers to textiles used in a wide range of sporting and recreational activities. Some common sporttechs are ski, racing, cycling, and fencing suits; moisture-wicking shirts; swimwear; yoga pants; hiking footwear; headbands; and equipment made with textile-reinforced composite materials. These products are engineered to enhance comfort, performance, and safety. Some key functional requirements of sports textiles include moisture management, compression, protection, thermal comfort, strength, and durability (Zhezhova et al. 2024).

13.3 GLOBAL MARKET AND GROWTH OF TECHNICAL TEXTILES

Due to the high demand for functional textiles across various applications, the global technical textiles market was estimated to reach USD 206 billion in 2024, with a forecasted compound annual growth rate of 4.8% through 2030 (Grand View Research 2025). Based on the Global Technical Textile Market (2020–2027) report from Allied Market Research and KPMG analysis, synthetic fibers accounted for the largest volume in 2020, and were projected to be 29000 kilotons (55%) in 2027, followed by natural fibers at 11000 kilotons (21%) and mineral fibers at 5.600 kilotons (11%) (KPMG 2021). In terms of production processes, woven textiles accounted for the largest market share at 67%, followed by nonwoven (15%), knitted (11%), and other processes (5%) in 2020, with similar projected shares in 2027 (KPMG 2021). The top three application sectors for technical textiles are Mobiltech, Indutech, and Sporttech (KPMG 2021). The automotive industry accounts for the highest portion (15.26% in 2023) of the technical textiles market (Prasad and Tiwarkhede 2025). On average, 30 kg of technical textiles are used per car in 2023, which is 50% higher than in the year 2000 (Prasad and Tiwarkhede 2025). The scale of the technical textiles industry and the heavy reliance on petroleum-based fibers underscore the urgency of addressing their environmental impacts.

13.4 ENVIRONMENTAL CHALLENGES AND THE IMPORTANCE OF SUSTAINABILITY

Technical textiles need to meet stringent end-use requirements, thereby limiting the use of natural fibers. As a result, about 55% of the fibers used in technical textiles are synthetic petrochemical fibers, which are neither renewable nor biodegradable. Technical textiles containing synthetic materials can release microfibers throughout their life cycle. Microfiber fragments are considered as a significant source of environmental pollution and health hazards. Human beings can be exposed to them through digestion, inhalation, or dermal contact (Periyasamy and Tehrani-Bagha 2022). Discarded synthetic technical textiles can accumulate in landfills or the natural environment due to their nonbiodegradable nature. Recycling technical textiles is generally more challenging than recycling conventional fashion textiles because of their composite structures, contamination, and the frequent presence of functional finishes. Other typical environmental impacts include freshwater consumption, water and air pollution, use of toxic chemicals, noise pollution, and carbon emissions (Tripa et al. 2024).

Sustainability is a strategic approach that can help companies enhance brand reputation, reduce waste, lower their carbon footprint, and comply with regulatory requirements (Haider et al. 2025). Possible ways to improve the sustainability of technical textiles include using recycled fibers and bio-based polymers, nonhazardous chemicals, and renewable energy sources; adopting low-temperature and water-efficient wet processing technologies; responsible management of effluents; and recycling of end-of-life products.

13.5 EVALUATING IMPACT – LIFE CYCLE ASSESSMENT (LCA)

LCA is the most widely used methodology for quantifying the environmental impacts of materials and products, as standardized by ISO 14040. This method enables the identification of environmental hotspots across a product's life cycle and supports informed decision-making to improve sustainability. LCA consists of four main phases: goal and scope definition, life cycle inventory (LCI), life cycle impact assessment, and interpretation (Dhiwar and Bedarkar 2025).

During the goal and scope definition phase, the assessment's system boundaries are established, including cradle to grave, cradle to cradle, cradle to gate, and gate to gate (Dhiwar and Bedarkar 2025), as shown in Figure 13.2. The second step of LCI involves collecting primary and secondary data on the inputs (resources consumed) and outputs (emissions and releases). This is the most time-consuming part, and the data quality will determine the accuracy of results. Based on inventory data, individual or comprehensive environmental impacts can be calculated. The final stage would be interpreting the results and making informed recommendations. However, LCA has some limitations, as it does not capture important aspects such as biodiversity, soil health, and animal welfare (Mutambo et al. 2024).

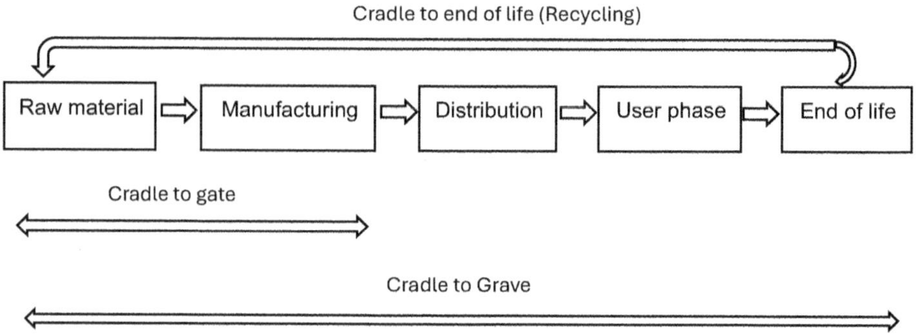

FIGURE 13.2 Scopes of LCA.

13.6 SUSTAINABLE DESIGN IN TECHNICAL TEXTILES

Sustainable design refers to the planning and development of technical textiles with lower environmental impacts. The main goal during designing sustainable technical textiles is to confirm that the developed products consume less water, dyes, chemicals, energy, and raw materials. Replacing synthetic fibers with renewable, bio-based fibers can improve the sustainability of textiles. However, natural fibers may also have adverse environmental impacts to some extent, as their cultivation is associated with large amounts of water and fertilizer. Hence, the use of recycled fibers in the design can further improve sustainability. The use of bio-based chemicals during processing can also improve sustainability.

13.6.1 Bio-Based Fibers

Bio-based fibers refer to fibers derived from biological origins. The use of bio-based fibers for designing sustainable technical textiles has been gradually increasing with the increase of public awareness and imposed regulations (Islam and Bhat 2019).

At present, bio-based fiber-reinforced composite materials are used in several applications, including automobile, aircraft, infrastructure, and other structural applications (Khalid et al. 2021). Bio-based fibers can be classified into natural fibers and bio-based synthetic fibers.

13.6.1.1 Natural Fibers

Natural fibers are the major components used in the development of bio-based sustainable technical textiles (Yashas Gowda et al. 2018). Based on their source, natural fibers can be subdivided into plant fibers (cotton, jute, flax, etc.), animal fibers (wool, silk, etc.), and mineral fibers (glass, rock wool, asbestos, etc.). Among these, plant fibers are mostly used for the fabrication of sustainable technical textiles, followed by animal fibers, whereas mineral-based asbestos fibers have been banned in several countries due to their hazardous nature (Ghalme et al. 2025).

13.6.1.1.1 *Plant Fibers*

The primary element of plant fiber is cellulose. Other elements of plant fibers are hemicellulose, lignin, wax, pectin, oils, and water-soluble substances. Cellulose, a linear-chain polymer, has higher thermal stability than hemicellulose, which has a branched structure and a lower degree of polymerization. Lignin, formed by cross-linking of phenolic precursors, gives structural support to plants. The chemical compositions of some plant fibers are presented in Table 13.1 (Eleutério et al. 2025; Karimah et al. 2021).

The chemical composition varies among plant fiber types, resulting in differences in their physical and mechanical properties. Plant fiber with relatively high cellulose content, like cotton, generally provides higher tensile strength, lower moisture and dye absorption, and better chemical resistance (Karimah et al. 2021). As cellulose is a linear polymer, it can be oriented to form crystalline regions, which provide higher strength and less space for water molecules and other chemicals to penetrate fibers. On the other hand, plant fiber with high hemicellulose content generally provides improved flexibility, higher moisture and dye absorption, and is easily attacked by mild alkali and acid, as hemicellulose is a branched polymer with a low degree of polymerization and forms amorphous regions. Lignin gives lower strength, UV

TABLE 13.1
Chemical Compositions of Some Plant Fibers

Plant Fibers	Cellulose (%)	Hemicellulose (%)	Lignin (%)	Pectin (%)	Wax (%)	Ash (%)
Bamboo	26–43	13–30	21–31	—	—	1.7–5
Flax	64–84	14–20	2–5	2.3	1.5	—
Jute	51–78	12–20	10–15	0.2	0.5	0.5–2
Hemp	67–78	18–22	3–5	0.9	0.8	0.8
Ramie	68–91	5–16	0.6–1	1.9	0.3	—
Banana	60–65	6–19	5–12	—	—	—
Kenaf	45–57	8–13	15–21	0.6	0.8	2–5
Abaca	60–80	15–25	6–14	—	3	3
Sisal	47–78	10–24	7–11	10	2	0.6–1
Pineapple	70–82	16–19	5–12	4	—	—
Cotton	82–96	2–6	0–5	5	0.6	—
Coir	43–46	0.15–0.25	40–45	—	—	—
Kapok	53–64	23–29	20	—	—	—
Rice	30–45	23–33	12–19	—	8–38	15–20
Oil palm	43–63	28–33	17–19	—	4	—
Nettle	77–86	16–20	0.5	0.9–4.8	4	—
Curaua	70–80	9.9	13	—	—	—
Henequen	60–78	10–18	8–13	—	—	—
Agave	60–78	11–17	6.08	—	—	3.29
Sansevieria	80	10.15	7.8	0.45	0.86	1.4

Source: Eleutério et al. (2025) and Karimah et al. (2021).

degradation resistance, and char formation behavior of plant fibers. However, this correlation is not always linear, as many other factors also influence the physical and mechanical properties of plant fibers. The physical and mechanical properties of plant fibers are listed in Table 13.2 (Eleutério et al. 2025; Karimah et al. 2021).

In general, plant fibers have high moisture absorption as water molecules easily bind to the polar groups of cellulose polymers. Furthermore, the hollow spaces within the plant fibers act as capillary tubes that contain water. These hollow spaces also contribute to the lightweight and higher thermal insulation properties of plant fibers. These properties of plant fibers are suitable for the fabrication of several sustainable technical textiles. The applications of plant fibers as sustainable technical textiles are presented in Table 13.3 (Karimah et al. 2021; Rahaman and Moshwan 2026).

However, high moisture absorption results in poor dimensional stability in plant fibers, as they swell in contact with water (Silva et al. 2020). Higher moisture absorption is also responsible for microbial attack on plant fibers (Joseph et al. 2002). Several chemical treatments are conducted to improve the properties of plant fibers. Low-concentration alkali (NaOH) treatment can remove lignin, pectin, and hemicellulose, thereby reducing the amorphous regions and improving the strength of plant fibers (Vijay et al. 2021). The pulping process is useful to remove the lignin from plant fibers.

TABLE 13.2
Physical Properties of Plant Fibers

Fibers	Diameter (μm)	Density (g/cm³)	Moisture Content (%)	Tensile Strength (MPa)	Modulus of Elasticity (GPa)	Elongation (%)
Bamboo	10–20	0.8–1.1		150–220	12–16	
Flax	18–22	1.4–1.5	12	1,500	27.6	3.2
Jute	16–21	1.3–1.5	12	400–900	20–55	1.5–1.8
Hemp	16–23	1.4–1.5	12	350–750	35–65	2–4
Ramie	25–35	1.4–1.6	8.5	400–1,000	44–128	2–3.8
Banana		1.3–1.4		530–914	7.7–33.8	5.3
Kenaf	20–30	1.22–1.4	17	295–1191	22–60	2–17
Abaca	17–22	0.83–1.3	14	650–780	29–32	2–4
Sisal	50–200	1.3–1.5	11	274–855	9–38	2–7
Pineapple	20–80	1.32–1.56		413–1,627	34.5–82.5	0.8–1.6
Cotton	11–22	1.5–1.55	8.5	300–800	6–10	6–8
Coir	12–55	1.15–1.25	13	106–175	6	15–40
Kapok	15–35	0.3–0.4	11	45–95	4	1.2

Source: Eleutério et al. (2025) and Karimah et al. (2021).

TABLE 13.3
Applications of Plant-Based Fibers in Sustainable Technical Textiles (Karimah et al. 2021; Rahaman and Moshwan 2026)

Fibers	Applications
Bamboo	Green composite, medical textiles (surgical masks, bandages).
Flax	Composite (automotive, acoustic insulation), biomedical, and furniture industry.
Jute	Biomedical, geotextile, electrical insulation, and lightweight composites.
Hemp	Lightweight composite (automotive, thermal insulation), tarpaulins, garden mulch, and geotextiles.
Ramie	Industrial canvases, filter fabrics, and composites.
Banana	Art/handicraft, currency note paper, tea bags, cardboard, natural absorbent in colored wastewater, conveyor belts, and lightweight composites.
Kenaf	Ceiling, acoustic and thermal insulator, ballistic armor materials, aerospace, composite, and oil absorbents.
Sisal	Industrial rope, garden mulch, lightweight composites, and geotextiles.
Pineapple	Composites, conveyor belt cord, and geotextiles.
Cotton	Biomedical (gauze, surgical sutures), composite.
Coir	Lightweight composites, geotextiles, and insulation.

13.6.1.1.2 Animal Fibers

Animal fibers are the second most preferred natural fibers for the fabrication of sustainable technical textiles due to their excellent mechanical properties, availability, and non-toxicity. Animal fibers, such as wool and silk, can be obtained from various animal sources, including the hair of sheep, alpaca, goat, and camel, as well as the secretions of silkworms and spiders.

The main chemical composition of animal fibers is protein. Proteins of silk fibers have β-sheet crystalline structure, whereas proteins of wool fibers have α-helical structure, known as α-keratin, which is largely amorphous. Therefore, silk fibers exhibit higher strength, while wool fibers exhibit greater thermal insulation. Animal fibers can be used to fabricate several sustainable technical textiles, including lightweight products and composites for the automobile industry (Ghalme et al. 2025). Silk fibers can also be used in biomedical applications, construction, protective clothing, and nonwoven fabric preparation due to their lustrous appearance, high mechanical strength, and antimicrobial properties (Colomban and Jauzein 2018).

13.6.2 Bio-Based Synthetic Fibers

Bio-based synthetic or regenerated fibers are those fibers that are derived from biological origins like plants (cotton, flax, hemp, wood, etc.), algae, bacteria, and others. Several bio-based synthetic or regenerated fibers (regenerated cellulose, silk fibroin, chitosan, alginates, collagen, polylactic acid [PLA], polyhydroxyalkanoates [PHAs]) can be utilized for developing sustainable technical textiles as their sources are renewable and they have lower environmental impacts.

Regenerated celluloses are extracted from plant-based materials, mainly wood, and provide comfort similar to that of plant fibers while offering enhanced physical and mechanical properties. Waste cotton fibers (e.g., cotton linters and recycled cotton fabrics) can also be used to produce regenerated cellulose, thereby improving the sustainability of products. Cellulose extraction can be done by chemical, biological, or physical processes, each with its own advantages and disadvantages. Biological processes are preferable for environmental issues and biomedical applications. Recently, nanocellulose extraction has been gradually increasing due to its high mechanical strength, surface area, renewability, and their suitability for various biomedical and industrial applications (Luan et al. 2022).

Silk fibroin, obtained by degumming raw silk, exhibits excellent mechanical properties and biodegradability, making it suitable for technical textile applications. Among several degumming processes, sodium carbonate degumming is the most popular (Liu et al. 2024).

Chitosan is a cationic polysaccharide obtained by the deacetylation of chitin, which is commonly found in the exoskeletons of crustaceans and insects, as well as in the cell walls of fungi. This polymer can form electrostatic interactions with negatively charged cell surfaces and exhibits antimicrobial activity. It can be used in tissue engineering, wound dressings, artificial skin, bone repair, hernia mesh, absorbable sutures, antimicrobial applications, drug delivery, catheters, and ophthalmology (Islam et al. 2017).

Alginates, extracted from the cell walls of brown seaweeds, are linear copolymers that can form hydrogels through ionic cross-linking (Ca^{2+}cross-linking). Alginate can be used in several biomedical applications, including tissue engineering, drug delivery, and wound dressing (Hecht and Srebnik 2016).

Collagen is a fibrillar, rod-shaped molecule that can be found in tendons, ligaments, bones, skin, and cornea. Collagens exhibit high strength and melting temperature, and they can be used in wound healing, drug delivery, and tissue engineering.

PLA fibers are derived from microbial fermentation of corn starch, followed by polymerization and melt spinning. PLA is a thermoplastic aliphatic polyester with moderate tensile strength and elasticity, along with biodegradability and biocompatibility. It can be used in agri-textiles, geotextiles, disposable hygiene products, matrix materials for composites, 3D-printed medical implants, and tissue engineering.

PHAs are a class of natural, biodegradable polyester synthesized by microorganisms. The commercially available PHAs are polyhydroxybutyrate (PHB) and polyhydroxybutyrate hydroxyvalerate. Among the various PHAs, PHB has high molecular weight, high crystallinity, and high strength (Jayalath and Alwis 2025). PHAs can be used in marine ropes and nets, geotextiles, composites, hygiene, wound dressing, medical implants, and scaffolds for tissue engineering (Shahi et al. 2024; Philip et al. 2007).

Bio-based polymers obtained from different sources can be converted into fibers using several methods, including wet spinning, electrospinning, melt spinning, and airflow spinning. Wet spinning is the most commonly practiced method for converting bio-based polymers into fibers (Liu et al. 2024). This method consists of several key steps. First, a spinning solution is prepared by dissolving polymer in a suitable solvent. After that, the prepared solution is pumped through a spinneret and then passed through a coagulation bath where polymer solidifies into fibers. Finally, the fiber is collected into a suitable package. This method is suitable for most polymers that can be dissolved in appropriate solvents. Electrospinning is an effective technique for producing nanofibers and is particularly suitable for biomedical applications. Similar to wet spinning, a polymer solution is prepared and pumped through a syringe. A voltage is applied between the syringe needle and the collector, causing the polymer solution at the syringe needle to overcome its surface tension and form nanofibers.

13.6.3 Recycled Fibers

Every year, huge amounts of textile waste are generated, and globally it accounts for about 92 million tons. About 73% of this waste is incinerated or discarded, which causes serious environmental pollution (Biyada and Urbonavičius 2025). Recycling this waste into sustainable technical textiles can minimize adverse environmental effects. Moreover, recycling textile waste can reduce the need for virgin raw materials, thereby providing additional environmental benefits and improving the sustainability of technical textiles. Virgin raw materials, both natural and synthetic, are associated with 4–5 billion tons of CO_2 emissions (10%–12% of total global CO_2 emissions) annually (Abrishami et al. 2024). Hence, textile recycling can contribute to achieving Sustainable Development Goal (SDG) 6 and SDG 14 by promoting the sustainable use of clean water and natural resources (Tadesse et al. 2025). In addition, recycling can give economic and social benefits by creating jobs.

However, textile recycling faces several challenges, including variations in fiber types, colors, contamination, and lack of technological knowledge. Textile recycling process can be classified as upcycling, downcycling, closed-loop recycling, and open-loop recycling. Upcycling is the process of converting into higher-value products (e.g., shredded fabric to acoustic insulation), whereas downcycling is the process of converting into lower-value products (e.g., towel to wiping cloths or industrial rags). Closed-loop recycling is the process of converting nearly identical products (e.g., fiber-to-fiber recycling), whereas open-loop recycling is the process of converting into completely different products (e.g., plastic bottles to polyester fabrics).

Recycling can be done mechanically, chemically, or biologically. In mechanical recycling, used textiles are converted into usable fibers and fabrics without changing basic structures. It is done following several steps, including sorting, shredding, opening, and carding. In chemical recycling, used textiles are converted into polymers or monomers using solvents or chemical reactors, which can be used to produce new materials. In biological recycling, microorganisms or enzymes are used to break down textiles into monomers or polymers.

13.6.4 Green Finishing

A huge amount of water, energy, and chemicals is used during fabrication, specifically during wet processing and finishing of textiles, which contribute to major environmental pollution and are considered as a barrier to the development of sustainable technical textiles. Hence, interest in bio-based finishes and advanced physical processes (like plasma treatment) has been gradually increasing, as these processes can significantly reduce water, energy, and chemical consumption.

13.6.4.1 Enzymatic and Biotechnology-Based Finishes

Enzymatic and biotechnology-based finishes are used at various stages of textile production, including desizing, bioscouring, biofinishing/biopolishing, denim finishing, and functional finishes. Enzymes, obtained from living cells through biochemical reactions, are biocatalysts that can accelerate chemical reactions by activating inorganic catalysts like acids, bases, metals, and metal oxides (Robinson 2015).

13.6.4.2 PFAS-Free Water and Oil Repellents

Per- and polyfluoroalkyl substances (PFAS) have long been used in the textile industry to provide water and oil repellency, stain resistance, and thermal stability to fabrics. However, these synthetic chemicals pose significant environmental and health risks, as they do not break down readily in nature and accumulate in water systems, wildlife, and human bodies (Itumoh et al. 2024). Hence, to develop sustainable technical textiles, it is necessary to use PFAS-free water and oil repellent finishes, like silicone-based finishes (polydimethylsiloxane), bio-based waxes, and plasma treatments.

Silicone-based finishes form a flexible, durable, low-surface-energy barrier on fabric surfaces that can provide excellent repellence against water while maintaining breathability and softness (Chen et al. 2025). Bio-based waxes (carnauba wax, beeswax, soy wax, and palm wax) give water repellency when coated on fabrics due to their inherent hydrophobic nature, but they may negatively impact fabric hand and breathability and are generally less durable than synthetic finishes. The major limitation of these non-PFAS finishes is their inferior performance in repelling oils and other low-surface-energy liquids (Chen et al. 2025). Significant research is ongoing to develop effective non-PFAS chemicals for oil repellency.

13.6.4.3 Plasma Treatments

Plasma, a high concentration of reactive species (electrons, ions, neutrons, radicals, and photons), can reduce the use of water, chemicals, and energy. It can form a polar group on fiber surfaces and enhance wettability, dyeability, and adhesion, resulting in improved performance for biomedical and technical textile applications (Patti 2025).

13.7 CASE STUDIES

13.7.1 Mechanical Recycling of Personal Protective Equipment

The disposal of FR personal protective equipment (PPE) presents a significant sustainability challenge. FR garments are generally made from nonbiodegradable aramid fibers, which release toxic gases when incinerated due to their aromatic, nitrogen-containing polymer chains. Traditionally, end-of-life FR garments have been sent to landfills after decontamination, where they remain indefinitely. This approach not only poses environmental concerns but also represents a substantial loss of material value, as aramid fibers are high-cost materials.

General Recycled Ltd., founded in 2012 in Canada, was established to address this problem. Dave Kasper, Vice President of Sales and Product Development, was interviewed for this chapter on November 21, 2025. The company specializes in recycling inherent FR garments, particularly meta-aramid and para-aramid, by converting them into fibers through shredding. General Recycled has developed a proprietary mechanical recycling process designed specifically for aramids, which can damage conventional textile shredders designed for regular fibers like polyester, cotton, wool, etc.

A distinctive aspect of the company's model is its strict closed-loop system. General Recycled does not accept open-market textile waste. Instead, companies send their used FR garments for recycling and receive the resulting recycled fibers through their preferred value chain, including yarn spinners and fabric manufacturers. This ensures traceability, prevents contamination, and guarantees consistent downstream end-use for the recycled material. The facility has a processing capacity of up to 2.7 million kg/year, although this remains only a small percentage of global aramid waste. This underscores the large, untapped potential for circularity in technical textiles.

The company focuses primarily on meta-aramids and para-aramids, whose high market price (up to USD 40–45/kg for virgin fiber in North America) makes recycling economically attractive. Recycled aramid fiber costs less than half of virgin material. As shredding reduces fiber length by about 50%, recycled fibers are typically blended with virgin aramid to meet required mechanical properties. For example, one of their clients supplied used Nomex IIIA garments (93% meta-aramid, 5% para-aramid, 2% antistatic fiber). After shredding, the recycled fibers were delivered to the client's designated spinner, who incorporated 33% recycled fiber into new yarns blended with virgin fibers. The final yarn composition was 73% meta-aramid, 20% modacrylic, 5% nylon, and 2% antistatic. These yarns were then used to produce new garments and certified for various PPE applications, including flashfire and arc-flash protection.

All protective clothing sent for recycling undergoes cleaning, either dry cleaning or standard laundering, depending on the type of contamination. Firefighting turnout gear, however, remains a major challenge, as they often contain extremely high levels of PFAS chemicals. To avoid introducing PFAS into their facility, General Recycled excludes PFAS-contaminated gears. However, they are actively developing PFAS-removal technologies. This is increasingly urgent, as millions of sets of turnout gear need to be replaced in the coming decade.

FR garments made with recycled aramid fibers can be approximately 10% cheaper than those made with virgin fibers. Disposal currently carries hidden costs, including cleaning, handling, waste management, and landfill fees, that often add several dollars per garment. Redirecting these costs into a recycling system can reduce the overall expense. In PPE manufacturing, protection remains the highest priority, and garments containing recycled aramids have been certified to offer comparable or superior thermal protection.

Although circularity has gained attention in the fashion industry, the technical textile sector has been slower to adopt similar practices. General Recycled emphasizes that broader adoption will require increased education and legislative support.

When recycling remains voluntary, many companies will choose the easier route, i.e., landfilling; however, once the business case is clarified, lower cost with equal or better performance, industrial users often show strong interest. The company's model demonstrates that an economically viable, high-performance circular system for technical textiles is not only feasible but already functioning at scale.

13.7.2 Reusable Medical Gowns

Medical gowns are crucial PPE that protect healthcare workers and patients from infectious agents. Their performance requirements are defined in the AAMI PB70 standard, which classifies gowns from Level 1 to Level 4. Key performance requirements include barriers against fluids and pathogens, durability, flammability, and comfort. Most medical gowns are disposable PP nonwoven products, which are typically landfilled or incinerated due to biohazard concerns. Recognizing the importance of sustainability, many healthcare institutions are exploring the use of reusable gowns to partially or fully replace disposable gowns.

The following are the common structures of medical gowns by protection level:

- **Level 1**: Minimal risk activities (e.g., basic care, standard isolation, visitor gowns) (U.S. Food and Drug Administration 2024). Disposable gowns are typically made from lightweight PP SMS (spunbond-meltblown-spunbond) nonwoven fabrics. Reusable gowns are usually made from polyester or polyester/cotton blend woven fabrics.
- **Level 2**: Low-risk activities (e.g., blood draws, suturing, or use in intensive care units) (U.S. Food and Drug Administration 2024). Disposable gowns are made from higher-weight SMS nonwoven fabrics. Reusable gowns are commonly produced from tightly woven, higher-weight polyester fabrics.
- **Level 3**: Moderate-risk activities (e.g., arterial blood draws, inserting intravenous lines, emergency room procedures, or trauma care) (U.S. Food and Drug Administration 2024). Disposable gowns are made from heavyweight SMS or SMMS fabrics, or nonwovens laminated with microporous or solid PE films. Reusable gowns are typically high-density woven polyester fabrics, reinforced in critical zones with membrane coatings.
- **Level 4**: High-risk activities (e.g., long, fluid-intensive procedures, surgery, and exposure to infectious diseases) (U.S. Food and Drug Administration 2024). Disposable gowns are commonly constructed from PP spunbond fabrics laminated with microporous or solid PE films. Reusable gowns consist of woven fabrics laminated with microporous or monolithic films to provide the highest level of protection.

Most reusable gowns can withstand approximately 70–100 industrial laundering cycles. Replacing disposable medical gowns with reusable alternatives offers significant environmental and economic benefits, including reductions in energy consumption, greenhouse gas emissions, and solid waste generation, as demonstrated by multiple LCA studies (Vozzola et al. 2020; Burguburu et al. 2022). McQuerry et al. tested the performance of medical gowns after industrial laundering cycles, and found that reusable gowns could provide equal or better protection consistently than the disposable gowns (McQuerry et al. 2021).

To improve sustainability and reduce costs, many healthcare institutions have shown interest in adopting reusable medical gowns. In 2012, University of California, Los Angeles (UCLA) Health initiated a pilot program to replace disposable isolation gowns with reusable ones. By July 2022, the program had diverted approximately 14 million single-use gowns from landfills, reduced annual gown consumption from 2.6 million to 1.3 million, and lowered annual gown-related costs from $1.6 million to $500,000 (Kucinskas and Mitry 2024). Similarly, in Canada, Island Health introduced reusable Level 2 isolation gowns equipped with radio-frequency identification tags and designed to withstand up to 100 wash cycles. Early adoption is projected to replace 75% of disposable gowns within one year. Full replacement is anticipated to save approximately $2 million annually while eliminating an estimated 77,000 kg of CO_2 emissions (Island Health 2025).

These cases demonstrate that reusable medical gowns offer sustainable and cost-effective solutions for the healthcare industry. However, some challenges are energy and water consumption during washing and drying, potential performance degradation due to improper care or wear, and the lack of laundering facilities in some healthcare institutions. Thorough inspection after each laundering cycle is essential to ensure garment integrity and continued performance.

13.7.3 Building Insulation Made with Textile Waste

Building insulation can decrease energy consumption by reducing heat exchange with the ambient. The materials are traditionally made from glass fiber, rock wool, and polystyrene foams (Jordeva et al. 2019). Key performance requirements for building insulation include high thermal resistance, fire resistance, and resistance to moisture and mold. Besides traditional insulation materials, mechanically shredded textile waste has emerged as an economical and sustainable alternative for building insulation. A cradle-to-gate LCA found that building insulation made from 100% recycled cotton fibers had an 11-fold lower global warming potential than stone wool insulation (Islam et al. 2024).

One of the most well-known commercial textile-based insulation products is Henry® UltraTouch Recycled Denim Insulation, a collaboration between Cotton Incorporated's Blue Jeans Go Green™ and Bonded Logic, Inc. (acquired by Carlisle Companies Incorporated in 2025). Launched in 2006, the program collects postconsumer denim clothing and converts it into nonwoven insulation materials. By May 2024, the program had recycled approximately 2,630 tons of denim waste, producing nearly 11 million square feet of denim insulation (Cotton Incorporated 2024). This insulation product contains about 80% recycled denim fibers and is treated with borates to enhance flame retardancy and mold resistance (Henry n.d.). The nonwoven insulation offers comparable thermal resistance to that of conventional glass fiber insulation (Henry n.d.).

Several challenges remain in the development of building insulation from textile waste. A major challenge is the consistency of the feedstock quality of postconsumer textile waste, as it varies widely in fiber composition, contamination, and degradation.

In addition, natural fibers generally have lower inherent moisture resistance than inorganic or synthetic insulation materials, increasing the risk of moisture absorption and decreased insulation performance. Also, FR finishing is needed to improve safety.

13.7.4 Natural Fiber-Reinforced Composites

Fiber-reinforced composites are made by embedding fibers, such as carbon, glass, and aramid fibers, into resins to improve the physical and mechanical properties of the composites (Diniţă et al. 2023). Both thermosetting (polyester, epoxy, vinyl ester) and thermoplastic (PE, PP, PVC, polyamide, PLA) resins can be used as matrices. There has been growing interest in using natural fibers, such as hemp, jute, and flax, as reinforcements due to their lower environmental impacts. LCA studies have confirmed the environmental benefits of natural fiber-reinforced composites (Wasti et al. 2024; Skosana et al. 2025).

Due to their environmental benefits, natural fiber-reinforced composites are increasingly adopted across various industrial applications. For example, BMW (Bayerische Motoren Werke AG) announced in 2025 that it would begin using flax-reinforced composites developed in collaboration with Bcomp Ltd. to replace conventional carbon fiber-reinforced composites in future series-production models. Their test concluded that the flax-based composite is suitable for visible exterior, interior components, and even roof structures. According to BMW, using flax-based alternatives can reduce carbon emissions by around 40% (BMW 2025).

Although the use of natural fibers reduced the carbon footprint and the reliance on petrochemicals, there is room for further improvement. While natural fibers are biodegradable, their biodegradability is significantly reduced when encapsulated within nonbiodegradable synthetic polymer matrices. Recyclability also depends on the type of matrix used. Thermosetting resins are difficult to recycle, because thermosets cannot be melted, dissolved, or reshaped due to their covalently cross-linked structure (Diniţă et al. 2023). In contrast, thermoplastic composites can be recycled up to four to six times with minimal degradation in properties using thermomechanical reprocessing methods (Zhao et al. 2022).

13.8 OPPORTUNITIES AND CHALLENGES

There are substantial opportunities to improve the sustainability of technical textiles to benefit both society and environment. The following are some key opportunities identified by the authors:

- **Increase the use of bio-derived, recycled, natural, and biodegradable fibers**: Bio-derived fibers produced from renewable resources, such as castor oil-based nylon, can reduce the environmental impact of petroleum-based alternatives without compromising performance. The use of mechanically or chemically recycled fibers can significantly reduce carbon emissions and energy consumption by decreasing reliance on virgin materials. Natural fibers offer the advantages of biodegradability, renewability, and generally lower environmental impact. Biodegradable materials can reduce end-of-life impacts, such as microplastic pollution and accumulation in ecosystems.

- **Extend service life**: Many technical textile products, such as medical protective clothing and filtration materials, are designed for single-use or short-term applications. Improving reusability can significantly reduce material consumption and overall environmental impact. For example, increasing the dust-holding capacity of filters can extend replacement intervals, and replacing disposable medical textiles with reusable alternatives can reduce waste.
- **Develop low-toxicity chemical finishes**: Technical textiles often rely on chemical finishes to impart functional properties, such as water and oil repellency, flame retardancy, and antimicrobial performance. Many commonly used finishes, particularly PFAS-based chemicals, raise serious environmental concerns. The development of low-toxicity effective finishing chemicals represents a critical opportunity.
- **Increase the adoption of sustainability certifications**: Sustainability certifications offer structured approaches to reducing environmental impacts throughout the supply chain. While performance-based certifications are often mandatory for technical textiles, broader adoption of sustainability-focused certifications can help minimize environmental impacts.
- **Develop lower-impact wet processing technologies**: The environmental hotspot in textile manufacturing is the wet processing stage (dyeing and finishing). Technologies such as solution dyeing, low-temperature, waterless, or water-efficient processes can significantly reduce resource consumption and energy demand.

Despite these opportunities, several challenges remain. Technical textiles are often composed of multi-material structures that include fiber blends, coatings, and chemical finishes, making material separation and recycling difficult. Another challenge is the exposure to hazardous substances during use, including firefighter protective suits, chemical protective suits, and medical textiles. These contaminations pose safety and regulatory barriers to recycling. Technical textiles must meet stringent performance requirements, while sustainable alternatives often compromise performance. For example, compared with PFAS-based repellent finishing, non-PFAS finishes generally provide similar water repellency but inferior oil repellency. This limitation can pose safety concerns in applications such as firefighting and the oil and gas industry. Another example is that natural fiber-reinforced composites typically exhibit lower mechanical performance than carbon- or glass fiber-reinforced composites. Sustainable raw materials, such as bio-derived and recycled fibers, are more expensive than virgin synthetic fibers. Without regulatory requirements or strong market incentives, many industries lack sufficient motivation to adopt these more sustainable alternatives.

13.9 FUTURE OUTLOOK

The field of sustainable technical textiles is expected to grow gradually due to technological advancements, increasingly stringent regulations, shifting consumer preferences toward sustainability, and the need to address global challenges like energy crisis, climate change, and environmental pollution. In the future, along with the automotive sector, the largest sector of technical textiles (about 15%), major growth areas will be healthcare, construction, military and defense, sports and leisure, agriculture, and pharmaceuticals. Technical textiles will play a leading role in shaping the future of these sectors by increasing safety, durability, and performance. It is anticipated that sustainable technical textiles will increasingly be dominated by bio-based and recycled materials. To further improve sustainability, greater emphasis is also being placed on the use of green energy, resource-efficient manufacturing processes, and recycling technologies.

However, future technical textiles will face significant challenges in improving sustainability, including high costs, technological barriers to recycling, and potential compromises in performance. Continuous research and innovation will be required to overcome these limitations. The authors are optimistic that sustainability in the technical textile sector will be improved significantly with ongoing technological advancements and the implementation of stronger environmental regulations.

REFERENCES

Abrishami, S., Shirali, A., Sharples, N., Kartal, G. E., Macintyre, L. and Doustdar, O. 2024. Textile recycling and recovery: an eco-friendly perspective on textile and garment industries challenges. *Textile Research Journal* 94(23–24):2815–2834.

Ahmad, Z., Naeem, M. S., Jabbar, A. and Irfan, M. 2020. Fibers for other technical textiles applications. In *Fibers for Technical Textiles* (Eds. Ahmad, S., Rasheed, A. and Nawab, Y.), Cham, Switzerland: Springer International Publishing, 201–220.

Azam, F. and Ahmad, S. 2020. Fibers for agro textiles. In *Fibers for Technical Textiles* (Eds. Ahmad, S., Rasheed, A. and Nawab, Y.), Cham, Switzerland: Springer International Publishing, 151–168.

Bala, A., Sharma, A. and Bharti, A. 2025. Branches of agriculture. In *Modern Agriculture Science and Sustainability* (Eds. Singh, P. L., Dubey, A., Bharti, A., Devi, M. and Kaur, R.), Uttarakhand, India: String Production, 11–20.

Behera, B. K. 2019. Automotive textiles and composites. In *High Performance Technical Textiles* (Ed. Paul, R.), Hoboken, NJ: Wiley, 353–383.

Bérubé, D. and Saunier, P. 2016. Manufacturing process of geotextiles. In *Geotextiles* (Ed. Koerner, R. M.), Cambridge, UK: Woodhead Publishing, 25–60.

Biyada, S. and Urbonavičius, J. 2025. Circularity in textile waste: challenges and pathways to sustainability. *Cleaner Engineering and Technology* 24:100905.

BMW M GmbH. 2025. From motorsport to series production. https://www.bmw-m.com/en/topics/magazine-article-pool/natural-fiber-composites-series-production.html [Accessed 10 January 2026].

Burguburu, A., Tanné, C., Bosc, K., Laplaud, J., Roth, M. and Czyrnek-Delêtre, M. 2022. Comparative life cycle assessment of reusable and disposable scrub suits used in hospital operating rooms. *Cleaner Environmental Systems* 4:100068.

Cărpuş, E., Stroe, C. and Dorogan, A. 2020. Textile packaging–the connection between science and health. *Industria Textila* 71(2):138–144.

Chen, S., Lai, J., Du, Y. and Hong, Z. 2025. Review on non-fluorinated durable water repellent alternatives for textile and fabric coatings. *Science and Technology of Engineering, Chemistry and Environmental Protection* 1(3):1–12.

Colomban, P. and Jauzein, V. 2018. Silk: Fibers, films, and composites: Types, processing, structure, and mechanics. In *Handbook of Properties of Textile and Technical Fibers* (Ed. Bunsell, A. R.), Cambridge, UK: Woodhead Publishing, 137–183.

Cotton Incorporated. 2024. Cotton incorporated celebrates the benefits of authentic cotton denim for 151st birthday of blue jeans. https://www.cottoninc.com/press-releases/cotton-incorporated-celebrates-the-benefits-of-authentic-cotton-denim-for-151st-birthday-of-blue-jeans/ [Accessed 10 January 2026].

Desai, A. N. and Kant, R. 2016. Geotextiles made from natural fibers. In *Geotextiles* (Ed. Koerner, R. M.), Cambridge, UK: Woodhead Publishing, 61–87.

Dhiwar, K. and Bedarkar, M. 2025. Life cycle assessment in fashion industry: a systematic review. *Discover Sustainability* 6(1):1–19.

Diniță, A., Ripeanu, R. G., Ilincă, C. N., Cursaru, D., Matei, D., Naim, R. I., Tănase, M. and Portoacă, A. I. 2023. Advancements in fiber-reinforced polymer composites: a comprehensive analysis. *Polymers* 16(1):2.

El-Ghazali, S., Khatri, M., Kobayashi, S. and Kim, I. S. 2022. An overview of medical textile materials. In *Medical Textiles from Natural Resources* (Ed. Mondal, M. I. H.), Cambridge, UK: Woodhead Publishing, 3–42.

Eleutério, T., Trota, M. J., Meirelles, M. G. and Vasconcelos, H. C. 2025. A review of natural fibers: classification, composition, extraction, treatments, and applications. *Fibers* 13(9):119.

Fortune Business Insights. 2026a. Technical textiles market size, share & industry analysis 2026–2034. https://www.fortunebusinessinsights.com/technical-textiles-market-102716 [Accessed 11 February 2026].

Fortune Business Insights. 2026b. Textile market size, share & industry analysis 2026–2034. https://www.fortunebusinessinsights.com/textile-market-103879 [Accessed 11 February 2026].

Fung, W. 2000. Textiles in transportation. In *Handbook of Technical Textiles* (Eds. Horrocks, A. R. and Anand, S. C.), Cambridge, UK: Woodhead Publishing, 490–528.

Ghalme, S., Hayat, M. and Harne, M. 2025. A comprehensive review of natural fibers: bio-based constituents for advancing sustainable materials technology. *Journal of Renewable Materials* 13(2):273.

Grand View Research. 2025. Technical textiles market (2025–2030). https://www.grandviewresearch.com/industry-analysis/technical-textiles-market [Accessed 10 January 2026].

Grosu, C., Stroe, C., Sarbu, T., Burnichi, F., Muscalu, A. and Tudora, C. 2020. Design and creativity in the field of agro-textiles. *Agriculture and Food* 8:107–119.

Guru, R., Kumar, A., Grewal, D. and Kumar, R. 2022. Study of the implantable and non-implantable application in medical textile. In *Next-Generation Textiles* (Ed. Ibrahim, H.), London, UK: IntechOpen, 1–14.

Haider, S. Z., Hashim, M. and Iqbal, M. N. 2025. Sustainability in the textile sector: a qualitative analysis of challenges, benefits, and best practices. *International Journal of Management Research and Emerging Sciences* 15(2):38–59.

Hecht, H. and Srebnik, S. 2016. Structural characterization of sodium alginate and calcium alginate. *Biomacromolecules* 17(6):2160–2167.

Henry. n.d. UltraTouch FAQs. https://henryultratouch.com/?page_id=584 [Accessed 10 January 2026].

Hsuan, Y. G. 2016. Geotextile resins and additives. In *Geotextiles* (Ed. Koerner, R. M.), Cambridge, UK: Woodhead Publishing, 17–23.

Islam, S. and Bhat, G. 2019. Environmentally-friendly thermal and acoustic insulation materials from recycled textiles. *Journal of Environmental Management* 251:109536.

Islam, S., Bhat, G. and Mani, S. 2024. Life cycle assessment of thermal insulation materials produced from waste textiles. *Journal of Material Cycles and Waste Management* 26(2):1071–1085.

Islam, S., Bhuiyan, M. A. R. and Islam, M. N. 2017. Chitin and chitosan: structure, properties and applications in biomedical engineering. *Journal of Polymers and the Environment* 25(3):854–866.

Island Health. 2025. An amazing program: reusable gowns initiative launches at Island Health. https://www.islandhealth.ca/news/stories/amazing-program-reusable-gowns-initiative-launches-island-health [Accessed 10 January 2026].

Itumoh, E. J., Data, S., Chen, J. L. Y., Kah, M., Padhye, L. P. and Leitao, E. M. 2024. Addressing the persistence of per-and poly-fluoroalkyl substances (PFAS): current challenges and potential solutions. *RSC Sustainability* 2(11):3183–3201.

Jayalath, S. U. and Alwis, A. P. D. 2025. PHA, the greenest plastic so far: advancing microbial synthesis, recovery, and sustainable applications for circularity. *ACS Omega* 10(30):32564–32586.

Jordeva, S., Golomeova, S., Kertakova, M., Mojsov, K. and Efremov, J. 2019. Textile as a sustainable insulating material for buildings. *Tekstilna Industrija* 67(2):20–28.

Joseph, P. V., Rabello, M. S., Mattoso, L. H. C., Joseph, K. and Thomas, S. 2002. Environmental effects on the degradation behaviour of sisal fiber reinforced polypropylene composites. *Composites Science and Technology* 62(10–11):1357–1372.

Karimah, A., Ridho, M. R., Munawar, S. S., Adi, D. S., Damayanti, R., Subiyanto, B., Fatriasari, W. and Fudholi, A. 2021. A review on natural fibers for development of eco-friendly bio-composite: characteristics, and utilizations. *Journal of Materials Research and Technology* 13:2442–2458.

Khalid, M. Y., Al Rashid, A., Arif, Z. U., Ahmed, W., Arshad, H. and Zaidi, A. A. 2021. Natural fiber reinforced composites: sustainable materials for emerging applications. *Results in Engineering* 11:100263.

KPMG. 2021. Technical textiles: emerging opportunities and investments. https://assets.kpmg.com/content/dam/kpmg/in/pdf/2021/03/technotex-2021-thought-leadership-kpmg-and-ficci.pdf [Accessed 10 January 2026].

Kucinskas, B. S. and Mitry, V. 2024. Prioritizing reduction, reuse, and reprocessing at UCLA Health. *Advancing the Circular Economy in Healthcare.* Webinar recording. Published by Busch Systems. https://www.youtube.com/watch?v=Y4DtWGXqt9g [Accessed 10 February 2026].

Liu, F., Pan, L., Liu, Y., Zhai, G., Sha, Z., Zhang, X., Zhang, Z., Liu, Q., Yu, S., Zhu, L. and Xiang, H. 2024. Biobased fibers from natural to synthetic: processing, manufacturing, and application. *Matter* 7(6):1977–2010.

Luan, P., Zhao, X., Copenhaver, K., Ozcan, S. and Zhu, H. 2022. Turning natural herbaceous fibers into advanced materials for sustainability. *Advanced Fiber Materials* 4(4):736–757.

McQuerry, M., Easter, E. and Cao, A. 2021. Disposable versus reusable medical gowns: a performance comparison. *American Journal of Infection Control* 49(5):563–570.

Molla, S., Abedin, M. M. and Siddique, I. M. 2024. Exploring the versatility of medical textiles: applications in implantable and non-implantable medical textiles. *World Journal of Advanced Research and Reviews* 21(1):603–615.

Mutambo, N., Peirson-Smith, A., KeChi-Okafor, C., Irving-Munro, A., Sheridan, K. J., Prendergast-Miller, M. T., Namdeo, A., Stanton, T., Gallidabino, M. D. and James, A. 2024. Mapping the environmental impact assessment landscape in the fashion and textile industries: critical gaps and challenges. *Sustainability* 16(19):8377.

Patti, A. 2025. Green advances in wet finishing methods and nanoparticles for daily textiles. *Macromolecular Rapid Communications* 46(2):2400636.

Paul, R. 2019. High performance technical textiles: an overview. In *High Performance Technical Textiles* (Ed. Paul, R.), Hoboken, NJ: Wiley, 1–10.

Periyasamy, A. P. and Tehrani-Bagha, A. 2022. A review on microplastic emission from textile materials and its reduction techniques. *Polymer Degradation and Stability* 199:109901.

Philip, S., Keshavarz, T. and Roy, I. 2007. Polyhydroxyalkanoates: biodegradable polymers with a range of applications. *Journal of Chemical Technology & Biotechnology* 82(3):233–247.

Prasad, Y. E. and Tiwarkhede, A. 2025. Technical textile: transforming industries with futuristic fabric solutions. https://www.alliedmarketresearch.com/resource-center/amr-perspectives/materials-and-chemicals/technical-textiles-the-future-of-industrial-fabrics [Accessed 10 January 2026].

Rahaman, M. T. and Moshwan, M. M. 2026. Advancements in natural fibers for sustainable textile manufacturing: importance, sources, applications, and future prospects. *Next Materials* 10:101564.

Rasheed, A. 2020. Classification of technical textiles. In *Fibers for Technical Textiles* (Eds. Ahmad, S., Rasheed, A. and Nawab, Y.), Cham, Switzerland: Springer International Publishing, 49–64.

Robinson, P. K. 2015. Enzymes: principles and biotechnological applications. *Essays in Biochemistry* 59:1.

Şahin, U. K. 2025. Effects of coated knitted polyester fabric properties on strength and color fastness performance of automotive seat cover. *Coatings* 15(12):1494.

Samanta, K. K., Basak, S. and Chattopadhyay, S. K. 2015. Potentials of fibrous and nonfibrous materials in biodegradable packaging. In *Environmental Footprints of Packaging* (Ed. Muthu, S. S.), Singapore: Springer Singapore, 75–113.

Shahi, Z., Mehrizi, M. K. and Firoozabad, M. S. M. 2024. Introduction of polyhydroxyalkanoates (PHAs) and their application in the textile industry. In *Biopolymers in the Textile Industry: Opportunities and Limitations* (Eds. Ahmed, S. and Shabbir, M.), Singapore: Springer Singapore, 29–40.

Sharma, N., Allardyce, B., Rajkhowa, R., Adholeya, A. and Agrawal, R. 2022. A substantial role of agro-textiles in agricultural applications. *Frontiers in Plant Science* 13:895740.

Silva, G., Kim, S., Aguilar, R. and Nakamatsu, J. 2020. Natural fibers as reinforcement additives for geopolymers: a review of potential eco-friendly applications to the construction industry. *Sustainable Materials and Technologies* 23:e00132.

Singh, S. 2021. ArchiTextile: a review on application of textiles in architecture. *Journal of Textile & Apparel Technology & Management (JTATM)* 12(1):1–16.

Skosana, S. J., Khoathane, C. and Malwela, T. 2025. Driving towards sustainability: a review of natural fiber reinforced polymer composites for eco-friendly automotive lightweighting. *Journal of Thermoplastic Composite Materials* 38(2):754–780.

Tabor, J. and Ghosh, T. 2019. Building and construction textiles. In *High Performance Technical Textiles* (Ed. Paul, R.), Hoboken, NJ: Wiley, 319–352.

Tadesse, M. G., Abate, M. T., Felix Lübben, J. and Rauch, M. 2025. Recycling and sustainable design for smart textiles: a review. *Advanced Sustainable Systems* 9:2401072.

Tanasă, F., Nechifor, M., Ignat, M. E. and Teacă, C. A. 2022. Geotextiles: a versatile tool for environmental sensitive applications in geotechnical engineering. *Textiles* 2(2):189–208.

The Textile Institute. 2021. Technical textiles. https://www.textileinstitute.org/sigs/technical-textiles/ [Accessed 10 January 2026].

Tripa, S., Indrie, L., Garcia, P. D. and Mikucioniene, D. 2024. Solutions to reduce the environmental pressure exerted by technical textiles: a review. *Industria Textila* 75(1):66–74.

United Nations. 2025. Sustainability. https://www.un.org/en/academic-impact/sustainability [Accessed 10 January 2026].

U.S. Food and Drug Administration. 2024. Medical gowns. https://www.fda.gov/medical-devices/personal-protective-equipment-infection-control/medical-gowns [Accessed 10 January 2026].

Vijay, R., Vinod, A., Singaravelu, D. L., Sanjay, M. R. and Siengchin, S. 2021. Characterization of chemical treated and untreated natural fibers from Pennisetum orientale grass: a potential reinforcement for lightweight polymeric applications. *International Journal of Lightweight Materials and Manufacture* 4(1):43–49.

Vozzola, E., Overcash, M. and Griffing, E. 2020. An environmental analysis of reusable and disposable surgical gowns. *AORN Journal* 111(3):315–325.

Wasti, S., Kamath, D., Armstrong, K., Clarkson, C., Tekinalp, H., Ozcan, S. and Vaidya, U. 2024. Life cycle assessment of coir fiber-reinforced composites for automotive applications. *Journal of Cleaner Production* 485:144368.

Wu, H., Yao, C., Li, C., Miao, M., Zhong, Y., Lu, Y. and Liu, T. 2020. Review of application and innovation of geotextiles in geotechnical engineering. *Materials* 13(7):1774.

Yashas Gowda, T. G., Sanjay, M. R., Bhat, S., Madhu, K., Senthamaraikannan, P. and Yogesha, B. 2018. Polymer matrix-natural fiber composites: an overview. *Cogent Engineering* 5(1):1446667.

Yasir, N. and Ahmad, S. 2020. Introduction. In *Fibers for Technical Textiles* (Eds. Ahmad, S., Rasheed, A. and Nawab, Y.), Cham, Switzerland: Springer International Publishing, 1–20.

Zhao, X., Copenhaver, K., Wang, L., Korey, M., Gardner, D. J., Li, K., Lamm, M. E., Kishore, V., Bhagia, S., Tajvidi, M. and Tekinalp, H. 2022. Recycling of natural fiber composites: challenges and opportunities. *Resources, Conservation and Recycling* 177:105962.

Zhezhova, S., Jordeva, S., Risteski, S., Golomeova Longurova, S. and Dimitrijeva-Kuzmanoska, V. 2024. Application of technical textiles in sports. Časopis Koža and Obuća 73(2):3–10.

14 Sustainable Biocomposites and Green Composites

Dimakatso Makwakwa-Morajane, Orebotse Joseph Botlhoko, and Sudhakar Muniyasamy

14.1 INTRODUCTION

The increasing environmental impact of petroleum-derived polymers has heightened worldwide interest in sustainable material alternatives. Concerns, including climate change, exhaustion of nonrenewable resources, growing plastic waste, and tightening environmental regulations, have steered both research and industry toward the development of sustainable biocomposites and green composites (Faruk et al. 2012; Mohanty et al. 2002). In general, sustainable biocomposites are engineered materials, which consist of natural fibers or bio-fillers and biodegradable polymers with broader impact to social longevity. This means their end product is recyclable, biodegradable, or compostable. Green biocomposites are typically made from 100% renewable biopolymers reinforced with natural fibers. These materials could reduce their environmental carbon footprint while maintaining, or even enhancing, their functional performance. Biocomposites generally consist of natural or bio-derived reinforcements such as plant fibers or bio-fillers dispersed in polymer matrices that can be bio-based, biodegradable, or conventional (Satyanarayana et al. 2009). Green composites build on this idea by integrating sustainability throughout the full life cycle, encompassing renewable raw materials, resource-efficient manufacturing, lower emissions during use, and environmentally friendly end-of-life solution (La Mantia and Morreale 2011).

The use of biocomposites and green composites has grown quickly in industries including automotive, packaging, construction, biomedical engineering, and additive manufacturing, driven by their low density, advantageous specific mechanical performance, renewability, and lower carbon emissions (Pickering et al. 2016). The global biocomposites market size was estimated at US$ 31.76 billion in 2025 and is projected to reach US$ 79.35 billion by 2033, growing at a compound annual growth rate (CAGR) of 12.1% from 2026 to 2033) (Lee et al. 2021). Despite these advantages, challenges related to moisture sensitivity, interfacial adhesion, long-term durability, and large-scale manufacturing remain key topics of active research. This chapter surveys recent advances in sustainable biocomposites and green composites and presents selected case studies that illustrate their industrial significance and prospects (Ghasemlou et al. 2024).

DOI: 10.1201/9781003689225-14

14.2 RECENT DEVELOPMENTS IN BIOCOMPOSITES AND GREEN COMPOSITES

14.2.1 Advances in Bio-Based Polymeric Materials

The environmental impact of plastic pollution has become the primary reason for moving toward a transition to bio-based plastic materials. The disposal of plastic waste and its incineration and pyrolysis significantly contributed to climate change and greenhouse gas emissions. The postconsumer plastic waste issues have immensely affected the environmental and economic societal factors. Unlike conventional materials, bio-based materials can be developed from various biomass feedstocks like agriculture, forestry, seaweed, and organic waste. This biomass can be potentially utilized to valorize the bio-based chemicals and for conversion into bio-based materials. Through various materials processing techniques, these bio-based materials can be designed to develop various innovative bio-based materials for various industrial applications like automotive, construction, electronics, packaging, medical, consumer goods, and others (Ghasemlou et al. 2024; Gregory 2009; Huerta Lwanga et al. 2016; Islam et al. 2022; Jeyakanthan et al. 2023; Hopewell et al. 2009). The major advantages of these raw materials are made from renewable resources and contribute to the reduction of carbon footprint as compared to nonrenewable fossil-fuel resources. Notable advances have been achieved in the design of bio-based and biodegradable polymer matrices, such as polylactic acid (PLA), polyhydroxyalkanoates, polybutylene succinate, and bio-derived polyamides (Fambri and Migliaresi 2010). These materials are increasingly tailored to deliver mechanical performance, thermal stability, and processing behavior that rival those of traditional petroleum-based plastics. To enhance toughness and heat resistance, various modification approaches, including polymer blending, plasticization, chain extension, and reactive compatibilization, have been implemented (Rasal et al. 2010). Moreover, bio-based thermosetting resins obtained from vegetable oils, lignin, and furan derivatives have emerged as attractive candidates for structural applications (Gandini 2011). For example, Balla and Bikiaris (2025) stated that bio-based and biodegradable polymers have undergone impressive development, with notable progress in environmental, biomedical, packaging, and sustainable processing applications. Several authors discuss how research is increasingly directed toward enhancing material performance, life cycle behavior, and integration into the circular economy, while also recognizing persistent issues related to environmental impact, recyclability, and the industrial-scale deployment of bio-based solutions. Consequently, the authors regard current advances as a pivotal stage in the transition to more sustainable polymer technologies, underscoring both promising opportunities and domains that still demand further innovation (Ghasemlou et al. 2024; Gautam et al. 2007; Gaur et al. 2023).

14.2.2 Natural Fiber Reinforcements and Bio-Fillers

Natural fibers such as flax, hemp, jute, kenaf, sisal, bamboo, and a range of agricultural residues have attracted substantial interest owing to their renewability, biodegradability, low density, and high specific stiffness (Pickering et al. 2016; Bledzki and

Gassan 1999), yet, because these fibers are intrinsically hydrophilic, they typically exhibit poor interfacial compatibility with hydrophobic polymer matrices, which can degrade composite performance. To mitigate this issue, extensive research has centered on modifying fiber surfaces through methods such as alkali treatment, silane coupling, acetylation, and enzymatic processes, all of which have been reported to markedly enhance fiber matrix bonding and the overall properties of the resulting composites (Xie et al. 2010). In addition, using waste biomass sources, including seaweed, rice husk, bagasse, and wood flour, aligns with circular economy concepts and further improves the sustainability profile of composite systems (Mohanty et al. 2002). Collectively, prior studies consistently highlight natural fibers as viable, sustainable substitutes for synthetic reinforcements in polymer composites, primarily due to their environmental benefits and favorable cost (Taj et al. 2025; Ekundayo and Adejuyigbe 2019; Keya et al. 2019). Nonetheless, issues such as sensitivity to moisture, variability in fiber quality, and generally inferior mechanical performance relative to synthetic fibers remain, underscoring the necessity for ongoing progress in surface modification, hybrid reinforcement approaches, and processing optimization to enable wider industrial implementation.

14.2.3 Nanotechnology and Functional Green Composites

Plant fibers are composed of cellulose as one of the major constituents, which is a principal structural component of plant cell walls. Other constituents of plant fibers are lignin, hemicellulose, and extractives (Lu and Hsieh 2012). Cellulose fibers are composed of bundles of microfibrils, of which each fibril, also called cellulose nanofibers (CNFs), comprises crystalline and amorphous domains (Qing et al. 2013). CNFs provides a web-like structured network with high tensile strength and tensile modulus, high aspect ratios, large surface area, and high crystallinity (Qing et al. 2013). The diameters and lengths of CNFs are at nanoscale (less than 100 nm) and microscale, respectively (Qing et al. 2013). The individualized nanofibers are obtained from cellulose by means of mechanical processes. The crystalline regions of nanofibers are called cellulose nanocrystals (CNCs), which are rod-shaped structures, and CNCs remain after removal of the amorphous domain either by sulfuric acid or enzyme hydrolysis (Satyamurthy et al. 2011). In addition, the diameters and lengths of CNCs are in nanoscale (less than 100 nm) and microscale, respectively. CNFs and CNCs are commonly termed as nanocellulose, and cellulose are regarded as their main source. Thereafter, nanocellulose is used as a reinforcement in the development of polymer composites. Research reports have indicated that nanocellulose offers great potential in bio-nanocomposite applications as they provide comparative tensile strength and stiffness to synthetic reinforcements (Lu and Hsieh 2012). Other advantages of nanocellulose include inherent biodegradability and recyclability, renewable resource-based, low cost, low carbon footprint, environmentally benign nature, and low density (Pullawan et al. 2010). With these advantages, the use of cellulose is becoming great potential in developing bio-based products for various applications. The incorporation of nanoscale reinforcements has created new opportunities for improving the performance of biocomposites. Nanocellulose, CNCs, nanoclays, and bio-based graphene have demonstrated the ability to enhance mechanical strength, thermal resistance, and barrier properties even at relatively low filler contents (Saba et al. 2016: Dufresne 2017). More recently, research has focused on multifunctional and intelligent green composites, such as self-healing materials, electrically conductive biocomposites, and stimuli-responsive systems exhibiting shape-memory effects or 4D functionality (Mendez et al. 2011; Tănase and Veres 2025; Moraru et al. 2024; Wang et al. 2024). Collectively, these developments greatly expand the range of applications for sustainable composites in smart structures and biomedical technologies.

14.2.4 Sustainable Processing and Manufacturing Technologies

The scalability of biocomposites has improved owing to progress in processing and manufacturing methods such as extrusion, injection molding, compression molding, resin transfer molding, and filament-based additive manufacturing (Faruk et al. 2012). Notably, additive manufacturing technologies (including 3D and 4D printing) have become highly effective for producing sustainable composites, minimizing material waste while enabling greater design freedom (Ngo et al. 2018; Tănase and Veres 2025). Moreover, the incorporation of life cycle assessment (LCA) and eco-design principles into material selection and processing has further reinforced the environmental performance of green composites (Ashby 2012). Nanocellulose as reinforcement in bionanocomposites for applications in packaging and biomedical fields has attracted considerable interest. There is an increasing demand for developing bio-based materials due to environmental problems caused by conventional materials. Extraction of cellulose and nanocellulose from sugarcane bagasse and maize stalks was chosen as waste biomass, which are abundantly available in South Africa, and the results have indicated the potential for developing advanced composite materials. The CNCs have a rod-like shape, and their diameters and lengths were in nano- and microscale, respectively, while CNFs have a web-like shape and their diameters and lengths were in nano- and microscale, respectively. The reinforcements of polymers with nanocellulose have shown improvement in the properties of the resultant composites and can be used in various applications such as green packaging, biomedical, and automotive parts.

14.3 CASE STUDIES

14.3.1 Automotive Interior Applications

The automotive sector is among the most advanced fields in which biocomposites have been adopted. Natural fiber–reinforced thermoplastics are extensively applied in components such as door panels, dashboards, seat backs, and interior trims, where they provide lower weight and better vibration damping than glass fiber–reinforced composites (Faruk et al. 2012). Figure 14.1 shows that LCA studies have shown that these biocomposites help reduce greenhouse gas emissions and decrease energy requirements during both production and use phases (La Mantia and Morreale 2011; Osman et al. 2024).

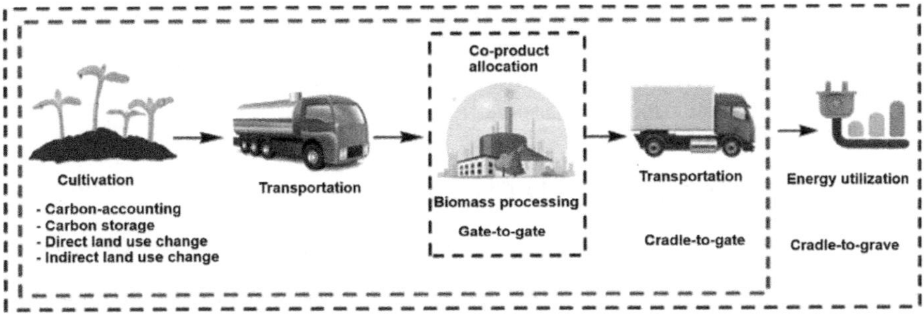

FIGURE 14.1 The life cycle of biofuels and bioproducts from cradle to grave, illustrating the by-products and environmental burden at the processing stage.

Source: **Osman et al. 2024.**

14.3.2 Sustainable Packaging Materials

PLA-, starch-, and cellulose fiber–based biocomposites have been widely studied for use in packaging, especially for food and other consumer products. Adding natural fibers increases stiffness and allows for lower material consumption, while preserving compostability (Auras et al. 2004). In parallel, advances in active and intelligent packaging that employ bio-based fillers and functional additives have further enhanced shelf life and improved product safety (Siracusa et al. 2008; Zhu et al 2023; Asrafali et al. 2026).

14.3.3 Construction and Building Materials

In the construction industry, eco-friendly composites like wood–plastic composites are being adopted more widely for applications such as decking, cladding, and insulation. These composites offer a combination of long service life, resistance to moisture, and lower maintenance requirements, while also relying on sustainably sourced materials (Ashori 2008; Wilczyński et al. 2025). Their integration in buildings enhances energy performance and facilitates adherence to green building certification standards.

14.3.4 Biomedical and Smart Material Applications

Biocomposites are increasingly recognized as highly promising for biomedical uses such as tissue-engineering scaffolds, biodegradable implants, and controlled drug-delivery platforms (Vert et al. 2012). When bio-based polymers are combined with natural fibers or bioactive fillers, they enable adjustable degradation behavior and enhanced biocompatibility. In addition, novel green 4D-printed composites that react to variations in temperature, humidity, or pH exemplify a state-of-the-art trend toward adaptive biomedical devices and smart material systems (Mendez et al. 2011). In a study, Makwakwa et al. (2022) reported that PLA-based 4D-printed green composites, incorporating smart textiles, boehmite alumina, and thermochromic dyes, demonstrated temperature-induced shape memory and adjustable thermomechanical performance. These stimuli-responsive features align the materials with the cutting edge of adaptive smart material technologies and suggest strong promise for future biomedical and other functional applications.

14.4 OPPORTUNITIES AND CHALLENGES

14.4.1 Opportunities in Sustainable Biocomposites and Green Composites

Since the invention of polymers and polymer composites, global consumption of plastics has increased dramatically, as well as manufacturing. Plastic-based composites are a need in today's economy, as they are ubiquitous in practically all industrial and domestic sectors. Plastic-based composites have seen increased opportunities and use in the packaging, automotive, medical, and agricultural industries in recent years, among others. Plastic-based products are widely used and are in high demand due to their low cost, light weight, and good physical and mechanical properties. Unfortunately, conventional plastics and composites are not degradable, nor are they effectively recycled. The majority of plastic waste materials end up in landfills and continue to pollute the environment (Geyer et al. 2017; Zhang et al. 2023). Conventional plastics and composites are made primarily of polyolefin-based plastic, which is generally considered nonbiodegradable under natural environmental conditions. One of the challenges in relation to plastic waste material is the frequent burning of plastic products, which generates CO_2 and toxic carcinogenic fumes. Hence, conventional plastics and composites are not sustainable and harmful to the environment upon disposal.

In this regard, sustainable biocomposites and green composites employing natural reinforcement material and biodegradable polymer matrix offer a promising strategy to develop high-performance, environmentally friendly materials with a lower carbon footprint. This material development strategy could add value to the emerging global regulations on the use of sustainable materials and the circular economy. The use of sustainable biocomposites and green composites could contribute to climate change mitigation plans by offering a viable, eco-friendly alternative to conventional petroleum-based and nonrenewable materials. Moreover, sustainable biocomposites and green composites materials are also characterized by lightweight properties, high acoustic properties, health and safety features (Raj et al. 2023; Hassan et al. 2020). Hence, these materials offer industrial applicability, reduce long-term waste management burdens, and demonstrate a pathway toward sustainable technologies.

14.4.2 Key Challenges in Sustainable Biocomposites and Green Composites

Bio-based polymers, which are generally used during the production of biocomposites, and green composites are often 15%–30% more costly than conventional petroleum-based and nonrenewable materials raw alternatives. On the other hand, scaling up the production of biocomposites and green composites for commercial use remains challenging (Sayam 2026; Hubbe 2023).

This characteristic could lead to poor filler–polymer matrix adhesion, poor long-term durability, and the reduction of mechanical performance over time. The inconsistency of the natural fibers' properties as a result of the source or harvesting process also affects the performance and quality of the end products.

14.5 FUTURE OUTLOOK

14.5.1 Future of Sustainable Biocomposites and Green Composites

The new environmental regulations necessitated the global research community to conduct research and development for new sustainable biocomposites and green composite materials, as an alternative to nonbiodegradable materials. Particularly, conventional plastics from fossil-fuel origin create serious environmental pollution and greenhouse gas emissions. In the future, the world will face a shift in population, which means more conventional plastics will be disposed of or burnt. As a result, industries must now shift toward more sustainable materials such as biocomposites and green composite material development and application; this strategy will successfully reduce conventional plastic-based composites waste pollution. The increased awareness in preserving the natural environment and recycling has resulted in efforts being made to develop access to biodegradable polymer composites and green composites.

According to recent market analyses, the global green composites market is expected to reach US$ 86.43 billion by 2030 with a CAGR of 15.30% (Oh et al. 2024). On the other hand, according to recent market analyses, the global biocomposites market is expected to reach approximately US$ 195 billion by 2034, expanding to a CAGR of about 18%, driven by increasing demand for sustainable and bio-based materials across automotive, construction, and packaging sectors (Kushwaha 2026). Some experts estimated the biocomposite market share of about US$ 31.76 billion in 2025, with a projection to reach US$ 79.35 billion by 2033, if the CAGR is 12.1%. Sustainable biocomposites and green composites penetrate almost all the market segments due to their lightweight (30% lighter than glass-fiber alternatives), inherent environmental friendliness, and circular economy capabilities. This potential benefit and approach are appropriate for the industry to meet sustainability goals and comply with global environmental regulations and consciousness. Secondly, conventional plastics are substituted with biocomposites and green composite materials to protect and combat microplastic pollution, and lower carbon emissions.

14.5.2 The Key Market Growth Contributors and Prominent Industries in This Transition

14.5.2.1 Clothing and Textile Industry

Biocomposites and green composites are naturally designed to be biodegradable, hence suitable to reduce the high environmental impact of clothing production and products. On the other hand, the use of waste-to-fiber technology allows for the recycling of textile waste materials.

14.5.2.2 Agriculture Sector

Polybutylene adipate-co-terephthalate (PBAT) and thermoplastic starch (TPS) blends have the potential to replace traditional polyethylene (PE) mulch films, due to their natural degradability properties, thus impacting the global plastic waste pollution (Gao et al. 2021). PBAT/TPS mulch films can reduce labor costs as they don't require disposal or removal after harvest season, making them a sustainable alternative to PE. In addition, biocomposites and blends can be used in plant pots and seed coating applications.

14.5.2.3 Packaging Industry

In recent years, the packaging industry has gained momentum to replace nonbiodegradable single-use plastics with more sustainable materials, such as biocompostable and green composite materials, in response to environmental pollution (Tewary et al. 2025; Makgwane et al. 2025). Packaging products that are actively being investigated and produced from biocomposites and green composites include biomedical test kits, household food containers, fast-food trays, fresh produce packaging, and beverage bottles. Most of the current and future bioproducts include, but are not limited to, PLA, polybutylene adipate terephthalate, and starch-based plastics.

14.5.2.4 Consumer Goods and Electronics

Electronics companies are adopting biocomposites and green composites to improve sustainability in portable devices such as laptops and smartphones (Pickering et al. 2016; Karavasili et al. 2026). In brief, the electronic industry has gained recyclability momentum and is increasingly focusing on the use of biodegradable materials for the production of portable devices, aiming to reduce the environmental impact of electronic waste.

14.5.2.5 Construction and Building Materials

Biocomposites technology has potential in the construction industry. For example, materials such as bamboo, waste wool, and other natural fibers can be used in construction due to their sustainability, high tensile strength, flexibility, and low carbon footprints, as alternatives to steel and concrete (Sharma and van der Vegte 2020; Patnaik et al. 2015).

14.5.2.6 Automotive Industry

Driven by the need to eliminate traditional, petroleum-based, and nonbiodegradable materials, this sector is shifting toward biocomposites and green composite materials. This material is nonabrasive and nontoxic (Balda et al. 2021). Automotive parts that are actively produced from biocomposites include interior door panels, dashboards, seat backs, trunk liners, and headliners. These production adoption materials could help automotive manufacturers to achieve lightweight automobiles with reduced fuel consumption characteristics.

14.6 CONCLUSION

This chapter offers a comprehensive understanding of biocomposites and green composites, and their consideration as the novel generation of eco-friendly materials with critical lightweight characteristics, high performance, and circular economic capabilities. This chapter presents an overview of biocomposites and green composite materials for their potential applications in clothing and textile, agriculture, packaging, consumer goods and electronics, construction and building materials, and the automotive industry. However, the cost issues associated with the biodegradable materials used in the development of biocomposites and green composites limit their widespread and large-scale use. This chapter also highlights the market trends and benefits of biocomposites and green composites for different advanced applications.

REFERENCES

Ashby, M. F. 2012. *Materials and the Environment: Eco-Informed Material Choice*. Elsevier, Amsterdam, Netherlands.

Ashori, A. 2008. Wood–plastic composites as promising green-composites for automotive industries! *Bioresource Technology* 99(11): 4661–4667.

Asrafali, S. P., Periyasamy, T. and Lee, J. 2026. Biopolymer-based active and intelligent food packaging: recent advances in materials, technologies, and applications. *Polymers* 18(2): 196.

Auras, R., Harte, B. and Selke, S. 2004. An overview of polylactides as packaging materials. *Macromolecular Bioscience* 4(9): 835–864.

Balda, S., Sharma, A., Capalash, N. and Sharma, P. 2021. Banana fiber: a natural and sustainable bioresource for eco-friendly applications. *Clean Technologies and Environmental Policy* 23(5): 1389–1401.

Balla, E. and Bikiaris, D. N. 2025. Editorial outline: overview of recent developments in bio-based and biodegradable polymers. *Polymers* 17(24): 3264.

Bledzki, A. K. and Gassan, J. 1999. Composites reinforced with cellulose based fibers. *Progress in Polymer Science* 24(2): 221–274.

Dufresne, A. 2017. *Nanocellulose: From Nature to High Performance Tailored Materials*. Walter de Gruyter GmbH & Co KG, Leiden.

Ekundayo, G. and Adejuyigbe, S. 2019. Reviewing the development of natural fiber polymer composite: a case study of sisal and jute. *American Journal of Mechanical and Materials Engineering* 3(1): 1–10.

Fambri, L. and Migliaresi, C. 2010. Crystallization and thermal properties. In *Poly (Lactic Acid): Synthesis, Structures, Properties, Processing, and Applications;* (Eds. Auras, R., Lim, L.-T., Selke, S. E. M. and Tsuji, H.), Wiley, Hoboken, NJ, 113–124.

Faruk, O., Bledzki, A. K., Fink, H. and Sain, M. 2012. Biocomposites reinforced with natural fibers: 2000–2010. *Progress in Polymer Science* 37(11): 1552–1596.

Gandini, A. 2011. The eruption of polymers from renewable resources on the scene of macromolecular science and technology. *Green Chemistry* 13(5): 1061–1083.

Gao, X., Xie, D. and Yang, C. 2021. Effects of a PLA/PBAT biodegradable film mulch as a replacement of polyethylene film and their residues on crop and soil environment *Agricultural Water Management* 255: 107053.

Gaur, V. K., Sirohi, R., Bhat, M. I., Gautam, K., Sharma, P., Srivastava, J. K. and Pandey, A. 2023. A review on the effect of micro-and nano-plastics pollution on the emergence of antimicrobial resistance. *Chemosphere* 311: 136877.

Gautam, R., Bassi, A. S. and Yanful, E. K. 2007. A review of biodegradation of synthetic plastic and foams. *Applied Biochemistry and Biotechnology* 141(1): 85–108.

Geyer, R., Jambeck, J. R. and Law, K. L. 2017. Production, use, and fate of all plastics ever made. *Science Advances* 3(7): e1700782.

Ghasemlou, M., Barrow, C. J. and Adhikari, B. 2024. The future of bioplastics in food packaging: an industrial perspective. *Food Packaging and Shelf Life* 43: 101279.

Gregory, M. R. 2009. Environmental implications of plastic debris in marine settings: entanglement, ingestion, smothering, hangers-on, hitch-hiking and alien invasions. *Philosophical Transactions of the Royal Society B: Biological Sciences* 364(1526): 2013–2025.

Hassan, T., Jamshaid, H., Mishra, R., Khan, M. Q., Petru, M., Novak, J., Choteborsky, R. and Hromasova, M. 2020. Acoustic, mechanical and thermal properties of green composites reinforced with natural fibers waste. *Polymers* 12(3): 654.

Hopewell, J., Dvorak, R. and Kosior, E. 2009. Plastics recycling: challenges and opportunities. *Philosophical Transactions of the Royal Society B: Biological Sciences* 364(1526): 2115–2126.

Hubbe, M. A. 2023. Sustainable composites: a review with critical questions to guide future initiatives. *Sustainability* 15(14): 11088.

Huerta Lwanga, E., Gertsen, H., Gooren, H., Peters, P., Salánki, T., Van Der Ploeg, M., Besseling, E., Koelmans, A. A. and Geissen, V. 2016. Microplastics in the terrestrial ecosystem: implications for *Lumbricus terrestris* (Oligochaeta, Lumbricidae). *Environmental Science & Technology* 50(5): 2685–2691.

Islam, T., Li, Y., Rob, M. M. and Cheng, H. 2022. Microplastic pollution in Bangladesh: research and management needs. *Environmental Pollution* 308: 119697.

Jeyakanthan, D. M., Schuster, J. and Shaik, Y. P. 2023. Development of ecological absorbent core sanitary pads in combination of kenaf and chitosan fibers. *Materials Sciences and Applications* 14(07): 363–381.

Karavasili, D., Lazaridou, K., Ntrivala, M. A., Pitsavas, A. C., Baziakou, Z., Papadimitriou, M., Bikiaris, N. D., Balla, E. and Terzopoulou, Z. 2026. Biobased polymers in printed electronics: from renewable resources to functional devices. *Polymers* 18(2): 301.

Keya, K. N., Kona, N. A., Koly, F. A., Maraz, K. M., Islam, M. N. and Khan, R. A. 2019. Natural fiber reinforced polymer composites: history, types, advantages and applications. *Materials Engineering Research* 1(2): 69–85.

Kushwaha, A. D. 2026. Market and technological trends in composites for solar energy applications. https://doi.org/10.2139/ssrn.6334009

La Mantia, F. P. and Morreale, M. 2011. Green composites: a brief review. *Composites Part A: Applied Science and Manufacturing* 42(6): 579–588.

Lee, C. H., Lee, S. H., Padzil, F., Ainun, Z., Norrrahim, M. and Chin, K. L. 2021. Biocomposites and nanocomposites. In *Composite Materials: Properties, Characterization, and Applications* (Eds. Sachdeva, A., Singh, P. K. and Rhee, H. W.), CRC Press, Boca Raton, FL, 29–60.

Lu, P. and Hsieh, Y. 2012. Preparation and characterization of cellulose nanocrystals from rice straw. *Carbohydrate Polymers* 87(1): 564–573.

Makgwane, P., Muniyasamy, S., Hlekelele, L., Swanepoel, A., Sypu, V., Mdlalose, L., Naidoo, S., Cele, Z., Maity, A., Balogun, M. and Botlhoko, O. 2025. Biobased biodegradable polybutylene succinate polymers and composites: synthesis, structure properties and applications: a review. *Journal of Renewable Materials* 13(3): 449.

Makwakwa, D., Motloung, M. P., Ojijo, V., Bandyopadhyay, J. and Ray, S. S. 2022. Influencing the shape recovery and thermomechanical properties of 3DP PLA using smart textile and boehmite alumina and thermochromic dye modifiers. *Macromol* 2(3): 485–499.

Mendez, J., Annamalai, P. K., Eichhorn, S. J., Rusli, R., Rowan, S. J., Foster, E. J. and Weder, C. 2011. Bioinspired mechanically adaptive polymer nanocomposites with water-activated shape-memory effect. *Macromolecules* 44(17): 6827–6835.

Mohanty, A. K., Misra, M. and Drzal, L.T. 2002. Sustainable bio-composites from renewable resources: opportunities and challenges in the green materials world. *Journal of Polymers and the Environment* 10(1): 19–26.

Moraru, D., Cortés, A., Martinez-Diaz, D., Prolongo, S. G., Jiménez-Suárez, A. and Sangermano, M. 2024. Sustainable electrically conductive bio-based composites via radical-induced cationic frontal photopolymerization. *Polymers* 16(15): 2159.

Ngo, T. D., Kashani, A., Imbalzano, G., Nguyen, K. T. and Hui, D. 2018. Additive manufacturing (3D printing): a review of materials, methods, applications and challenges. *Composites Part B: Engineering* 143: 172–196.

Oh, E., Godoy Zúniga, M. M., Nguyen, T. B., Kim, B., Trung Tien, T. and Suhr, J. 2024. Sustainable green composite materials in the next-generation mobility industry: review and prospective. *Advanced Composite Materials* 33(6): 1368–1419.

Osman, A. I., Fang, B., Zhang, Y., Liu, Y., Yu, J., Farghali, M., Rashwan, A. K., Chen, Z., Chen, L. and Ihara, I. 2024. Life cycle assessment and techno-economic analysis of sustainable bioenergy production: a review. *Environmental Chemistry Letters* 22(3): 1115–1154.

Patnaik, A., Mvubu, M., Muniyasamy, S., Botha, A. and Anandjiwala, R. D. 2015. Thermal and sound insulation materials from waste wool and recycled polyester fibers and their biodegradation studies. *Energy and Buildings* 92: 161–169.

Pickering, K. L., Efendy, M. A. and Le, T. M. 2016. A review of recent developments in natural fiber composites and their mechanical performance. *Composites Part A: Applied Science and Manufacturing* 83: 98–112.

Pullawan, T., Wilkinson, A. N. and Eichhorn, S. J. 2010. Discrimination of matrix–fiber interactions in all-cellulose nanocomposites. *Composites Science and Technology* 70(16): 2325–2330.

Qing, Y., Sabo, R., Zhu, J. Y., Agarwal, U., Cai, Z. and Wu, Y. 2013. A comparative study of cellulose nanofibrils disintegrated via multiple processing approaches. *Carbohydrate Polymers* 97(1): 226–234.

Raj, H., Tripathi, S., Bauri, S., Choudhury, A. M., Mandal, S. S. and Maiti, P. 2023. Green composites using naturally occurring fibers: a comprehensive review. *Sustainable Polymer & Energy* 1(2): 10010.

Rasal, R. M., Janorkar, A. V. and Hirt, D. E. 2010. Poly (lactic acid) modifications. *Progress in Polymer Science* 35(3): 338–356.

Saba, N., Jawaid, M., Alothman, O. Y. and Paridah, M. T. 2016. A review on dynamic mechanical properties of natural fiber reinforced polymer composites. *Construction and Building Materials* 106: 149–159.

Satyamurthy, P., Jain, P., Balasubramanya, R. H. and Vigneshwaran, N. 2011. Preparation and characterization of cellulose nanowhiskers from cotton fibers by controlled microbial hydrolysis. *Carbohydrate Polymers* 83(1): 122–129.

Satyanarayana, K. G., Arizaga, G. G. and Wypych, F. 2009. Biodegradable composites based on lignocellulosic fibers: an overview. *Progress in Polymer Science* 34(9): 982–1021.

Sayam, S. 2026. Natural fibers in sustainable materials: extraction technologies, fiber modification, and performance–sustainability relationships. *RSC Advances* 16(12): 10495–10537.

Sharma, B. and van der Vegte, A. 2020. Engineered bamboo for structural applications. In *Nonconventional and Vernacular Construction Materials, Characterization, Properties and Applications*; (Eds. Harries, K. A. and Sharma, B.), Woodhead Publishing, Cambridge, UK, 597–623.

Siracusa, V., Rocculi, P., Romani, S. and Dalla Rosa, M. 2008. Biodegradable polymers for food packaging: a review. *Trends in Food Science & Technology* 19(12): 634–643.

Taj, A., Bharath, K. N., Hiremath, A., Madhu, P., Sanjay, M. R. and Siengchin, S. 2025. Mechanical enhancement of sustainable natural fiber composites through filler additives: a comprehensive review. *Journal of Umm Al-Qura University for Engineering and Architecture*: 1–19. https://doi.org/10.1007/s43995-025-00240-y

Tănase, M. and Veres, C. 2025. Green composites in additive manufacturing: a combined review and bibliometric exploration. *Journal of Manufacturing and Materials Processing* 9(9): 301.

Tewary, A., Mishra, A. and Sharma, I. 2025. Green packaging: exploring biodegradable and eco-friendly composite materials. In *Polymers and Composite Materials for Packaging: Smart Food Packaging and Solutions;* (Eds. Verma, A., Gupta, H. S. and Sethi, S. K.), Springer Singapore, Singapore, 247–272.

Vert, M., Doi, Y., Hellwich, K., Hess, M., Hodge, P., Kubisa, P., Rinaudo, M. and Schué, F. 2012. Terminology for biorelated polymers and applications (IUPAC Recommendations 2012). *Pure and Applied Chemistry* 84(2): 377–410.

Wang, Z., Chen, Y., Ma, Y. and Wang, J. 2024. Bioinspired stimuli-responsive materials for soft actuators. *Biomimetics* 9(3): 128.

Wilczyński, K., Buziak, K. and Wilczyński, A. 2025. Wood–plastic composites: manufacturing, rheology and processing and process modeling. *Materials* 18(17): 4042.

Xie, Y., Hill, C. A., Xiao, Z., Militz, H. and Mai, C. 2010. Silane coupling agents used for natural fiber/polymer composites: a review. *Composites Part A: Applied Science and Manufacturing* 41(7): 806–819.

Zhang, O., Takaffoli, M., Ertz, M. and Addar, W. 2023. Environmental impact assessment of plastic waste management scenarios in the Canadian context. *Environments* 10(12): 213.

Zhu, B., Zhong, Y., Wang, D. and Deng, Y. 2023. Active and intelligent biodegradable packaging based on anthocyanins for preserving and monitoring protein-rich foods. *Foods* 12(24): 4491.

15 Sustainable Nanotextiles

Rasiah Ladchumananandasivam,
Iris Oliveira da Silva, and
Francisco Claudivan Da Silva

15.1 INTRODUCTION

15.1.1 Background and Definition

Nanotextiles represent a convergence of nanotechnology and textile science that has transformed the performance and functionality of fibers and fabrics. By manipulating materials at the nanoscale (1–100 nm), textile engineers can impart unique surface and structural features such as enhanced mechanical strength, water repellency, ultraviolet (UV) protection, antibacterial activity, and self-cleaning properties (Hasan et al. 2025; Tamulevičius 2025). The advent of nanofinishes based on metallic or metal-oxide nanoparticles (NPs; Ag, ZnO, TiO_2) and fluorinated compounds enabled revolutionary performance improvements, but it also raised growing concerns regarding toxicity, environmental persistence, and micro- and nanoplastic release during wear and laundering (Boruah and Jose 2025).

In the 21st century, sustainability has emerged as a defining criterion for innovation in advanced materials. Conventional nanofinishes are increasingly scrutinized for their potential ecotoxicological impact, prompting a paradigm shift toward eco-friendly nanotextiles that rely on renewable, biodegradable, and non-toxic constituents (Saleem and Zaidi 2020; Catarino et al. 2025). Sustainable nanotextiles therefore seek to couple nanotechnology's functional advantages with green chemistry principles, circular economy design, and responsible end-of-life management.

15.1.2 Sustainability Context

The textile industry is among the world's largest generators of waste and pollution, and nanotechnology, while promising, can aggravate these impacts if poorly managed (EEA 2022). Sustainable nanotextiles aim to mitigate such effects through the selection of biopolymeric nanofibers (NFs) (cellulose, chitosan, polylactic acid [PLA], polyhydroxyalkanoates [PHAs]), bio-based nanocomposites (NCs) (lignin NPs, starch nanocrystals, nano-clays), and green manufacturing processes (electrospinning with benign solvents, sol–gel synthesis from plant-derived precursors, plasma surface treatment without effluents) (Patnaik 2024).

DOI: 10.1201/9781003689225-15

This sustainability framework addresses the three pillars of modern textile engineering:

1. **Environmental**: minimizing solvent use, energy demand, and hazardous emissions;
2. **Economic**: promoting resource efficiency and circular reuse of fibers;
3. **Social**: ensuring safety for workers and consumers while maintaining product quality.

15.1.3 Evolution of Nanotextiles toward Sustainability

Figure 15.1 illustrates the transition from conventional nanofinishing routes—dominated by metallic NPs and perfluorinated agents—to modern green approaches employing natural polymers, hybrid bio-NCs, and energy-efficient processing. The new generation of sustainable nanotextiles not only performs competitively with conventional systems but also aligns with global sustainability goals such as the UN Sustainable Development Goals (SDG 12: Responsible Consumption and Production and SDG 13: Climate Action).

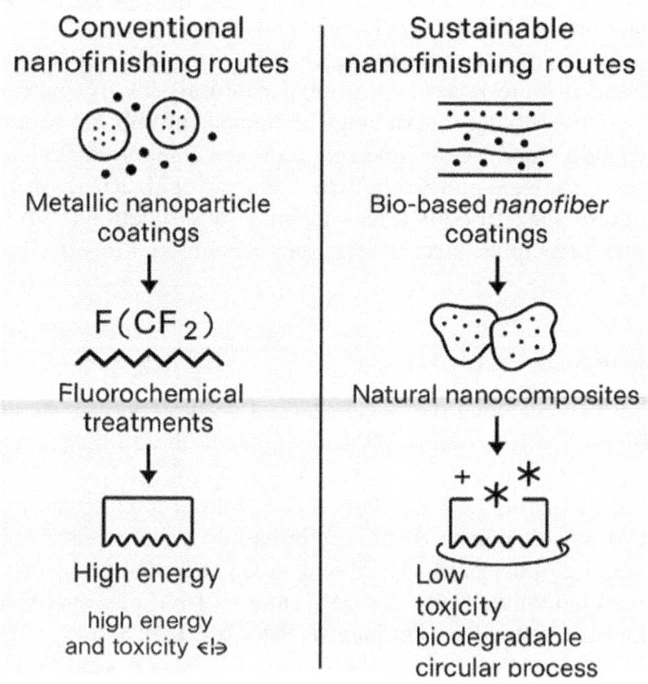

FIGURE 15.1 Conventional vs. sustainable nanofinishing routes.

TABLE 15.1
Classification of Nanotextiles by Function and Sustainability Level

Category	Typical Nanomaterial	Functionality	Sustainability Profile
Antimicrobial	Ag NPs/chitosan NPs	Bactericidal surface	High if biopolymer-based; low if metallic
UV-protective	TiO_2, ZnO, lignin NPs	UV-shielding	Moderate-high depending on source
Water-repellent	Fluoropolymers/wax NPs	Hydrophobicity	Low (fluoro)/high (wax-based)
Self-cleaning	TiO_2 NPs/graphene oxide	Photocatalytic degradation	Moderate, depending on recyclability
Conductive/smart	Carbon nanotubes (CNT)/ Cellulose–CNT hybrids	Electrical conductivity, sensing	High if natural matrix; low if pure CNT
Barrier/filtration	Carbon NFs/clay NCs	Particle and pathogen filtration	High biodegradability potential

15.1.4 Classification of Nanotextiles

Nanotextiles can be classified according to functional purpose, material origin, and environmental performance (Table 15.1). Sustainable systems favor bio-derived matrices and recyclable nanofillers, whereas traditional counterparts rely heavily on nonrenewable synthetics.

15.1.5 Relevance and Chapter Scope

This introductory chapter sets the conceptual foundation for understanding sustainable nanotextiles—from material design to life-cycle considerations. The subsequent sections will

- examine recent developments in sustainable NFs and NCs,
- present global case studies illustrating eco-innovations,
- evaluate the opportunities and challenges facing industry adoption, and
- discuss future research and policy directions.

By combining nanotechnology with sustainability principles, this field is redefining how advanced textiles are conceived, produced, and reused in a circular economy.

15.2 RECENT DEVELOPMENTS IN SUSTAINABLE NANOFIBERS AND NANOCOMPOSITES

15.2.1 Overview

Recent advances in nanotechnology have enabled the fabrication of NFs and NCs from renewable and biodegradable resources, shifting the focus of textile science from performance alone toward environmental responsibility. Sustainable NFs combine nanoscale architecture with biopolymeric chemistry, while NCs integrate natural fillers into green matrices to enhance strength, durability, and functionality (Kolya and Kang 2024). These materials offer viable alternatives to petroleum-based synthetics and have already entered applications in filtration, wound dressings, wearable electronics, and eco-packaging.

15.2.2 Biopolymer-Based Nanofibers

15.2.2.1 Cellulose Nanofibers (CNFs) and Nanocrystals (CNCs)

Derived from wood pulp, bacterial cellulose, or agricultural residues, CNF and CNC exhibit high crystallinity and remarkable tensile strength (Zhang et al. 2021). Their hydroxyl-rich surfaces promote chemical functionalization for hydrophobic, antimicrobial, or conductive coatings. CNF-reinforced cotton and polyester blends have demonstrated improved moisture management and biodegradability, aligning with circular economy goals.

15.2.2.2 Chitosan NFs

Chitosan, obtained from chitin in crustacean shells, possesses intrinsic antibacterial and hemostatic properties. Electrospun chitosan NFs are widely used in medical and hygiene textiles. Recent work employing water-based electrospinning and natural cross-linkers (citric acid, genipin) has reduced environmental burdens (Li et al. 2026).

15.2.2.3 PLA and PHA NFs

PLA and PHA are fully biodegradable aliphatic polyester derived from plant sugars or microbial fermentation. PLA-based electrospun mats have found use in biodegradable packaging and nonwoven filters. By blending PLA with PHA or cellulose acetate, researchers achieved improved flexibility and compostability (Naser et al. 2021). Table 15.2 depicts some of the applications of the NFs along with their source and processing routes.

A typical green electrospinning process flow is the transition from polymer solution preparation using water or ethanol to electrospinning under low voltage, fiber deposition,

TABLE 15.2
Representative Sustainable NFs and Their Key Attributes

NF Type	Source Material	Processing Route	Key Properties	Typical Applications
Cellulose (CNF/CNC)	Wood pulp, cotton linters	Aqueous fibrillation	High strength, renewable, biodegradable	Reinforced fabrics, filtration
Chitosan	Shellfish waste	Water-based electrospinning	Antibacterial, biocompatible	Medical masks, wound dressings
PLA	Corn starch, sugarcane	Melt or solution electrospinning	Compostable, transparent	Packaging, hygiene textiles
PHA	Microbial fermentation	Solvent-free melt spinning	Flexible, marine, biodegradable	Smart wearables, bio-filters

and post-treatment (plasma or UV) without toxic solvents. A grouped bar chart comparing the carbon-footprint reduction (%), energy-consumption reduction (%), and biodegradation rate (%) for four representative NF systems, CNF, Chitosan, PLA, and PHA, is shown in Figure 15.2 and Table 15.3 (EEA 2022; Naser et al. 2021; Zhang et al. 2021).

Figure 15.2 illustrates average life-cycle reductions in carbon emissions and energy demand relative to conventional polyester fibers, together with measured biodegradation rates under industrial-composting conditions (EEA 2022; Naser et al. 2021; Zhang et al. 2021).

Table 15.4 shows the replacement of hazardous solvents with recoverable alternatives in green electrospinning. The use of low-toxicity, bio-based materials improve worker safety, facilitates solvent recovery, and reduces life-cycle impacts.

15.2.3 Nanocomposites in Textiles

NCs combine polymer matrices with nanoscale fillers to improve mechanical, thermal, or barrier properties. The sustainable approach replaces synthetic matrices with biopolymers and inorganic fillers with naturally sourced NPs.

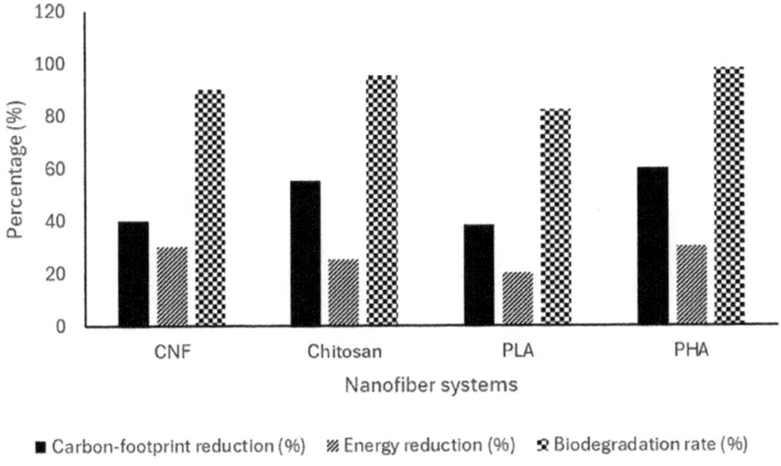

FIGURE 15.2 Comparative environmental performance of sustainable NF systems.

TABLE 15.3
LCA Data for Different NFs

NF System	Carbon-Footprint Reduction (%)	Energy Reduction (%)	Biodegradation Rate (%)
CNF	40 ± 5	30 ± 4	85–90
Chitosan	55 ± 6	25 ± 3	95+
PLA	38 ± 4	20 ± 2	80–85
PHA	60 ± 7	28 ± 3	98+

TABLE 15.4
Solvent Replacement Matrix in Green Electrospinning

Conventional Solvent	Common Polymer Used	Toxicity/Environmental Hazard	Green Alternative	Recovery Efficiency (%)	Remarks
N,N-Dimethylformamide	Polyacrylonitrile, polyurethane	High (carcinogenic, persistent)	Ethanol/water	85–90	Used in aqueous electrospinning for CNF and chitosan
Tetrahydrofuran	PLA, polycaprolactone	High (volatile, flammable)	Ethanol/acetic acid	75–85	Enables safer PLA/PHA spinning
Chloroform	Cellulose acetate	Very high (toxic, ozone-depleting)	Acetone/lactic acid	70–80	Applied in bio-solvent systems
Dimethyl sulfoxide	Polyvinyl alcohol, gelatin	Moderate	Water	90+	Suitable for protein-based nanofibers

- **Natural nanofillers**: Nano-clays, lignin NPs, and starch nanocrystals enhance strength and gas barrier properties while remaining biodegradable (Dong et al. 2024).
- **Hybrid NCs**: Blending bio-nanofillers with small fractions of metal oxides (ZnO, TiO_2) achieves a balance between antibacterial efficacy and environmental safety.
- **Bio-based membranes**: Cellulose–clay or chitosan–starch systems have demonstrated outstanding filtration efficiency (>99%) and reusability in water-purification textiles.

A grouped bar chart (Figure 15.3) compares the carbon-footprint reduction (%), biodegradation rate (%), and water-saving efficiency (%) across the five international case studies presented in Section 15.3—Japan (2019), India (2020), European Union (EU, 2021), Brazil (2022), and South Africa (2023):

- Brazil (2022) achieves the highest water-saving potential (≈60%) using hybrid cellulose–clay membranes.
- EU (2021) displays a balanced environmental profile, with significant carbon reduction (≈38%) from PLA-based nonwovens (Antony et al. 2025; Shen et al. 2023; Rajendran et al. 2024; Silva et al. 2022; Ferreira et al. 2025; EEA 2022).

Table 15.5 shows a comparative summary of environmental performance metrics (carbon footprint, water use, biodegradation) for representative sustainable nanotextile case studies (2019–2023). Data compiled from Antony et al. (2025), Shen et al. (2023), Rajendran et al. (2024), Silva et al. (2022), Ferreira et al. (2025), and EEA (2022).

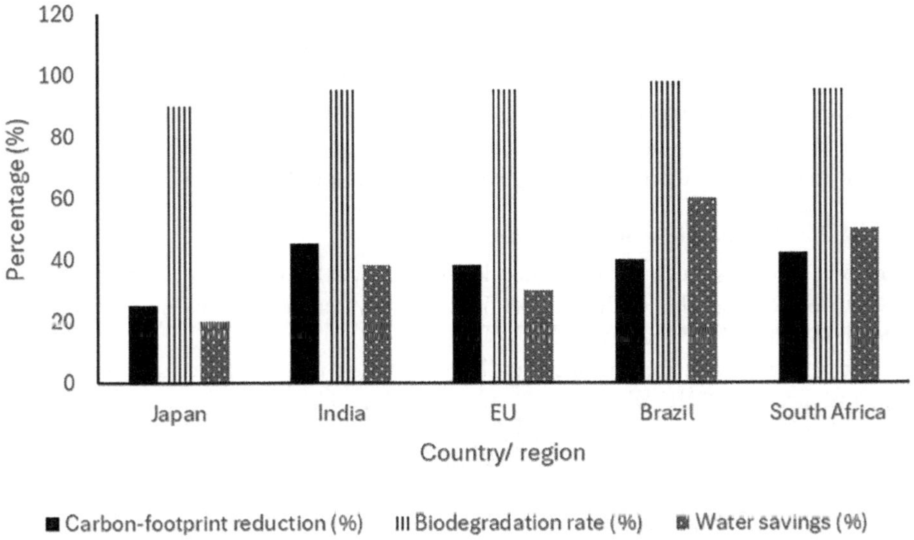

FIGURE 15.3 Comparative life-cycle performance of sustainable nanotextile case studies.

TABLE 15.5
Life-Cycle Performance Summary of Sustainable Nanotextile Case Studies

Country/ Region	Key Material	Carbon-Footprint Reduction (%)	Water Savings (%)	Biodegradation Rate (%)	Sustainability Highlights
Japan (2019)	Cellulose NFs	25	20	90	Low-temperature curing; biodegradable cotton composite
EU (2021)	PLA NFs nonwoven	38	30	92	Horizon 2020 BIO4PACK—38% lower CO_2 footprint
Brazil (2022)	CNF + nano-clay	40	60	88	Waste-cellulose valorization; 95% bacterial removal

15.2.4 Green Manufacturing Techniques

15.2.4.1 Electrospinning with Benign (Green) Solvents

Conventional electrospinning often employs N,N-dimethylformamide or tetrahydrofuran. Replacing them with ethanol, acetic acid, or water significantly reduces toxicity (Berdimurodov et al. 2023). Advances in needleless electrospinning and scaled rotary systems have improved energy efficiency.

15.2.4.2 Sol–Gel Coatings Using Plant-Derived Precursors

Green sol–gel synthesis utilizes silica or titania derived from rice husk ash or sugarcane bagasse. Such precursors yield uniform nanolayers for UV protection and antibacterial effects while promoting waste valorization (Souza et al. 2023).

15.2.4.3 Plasma Functionalization

Low-pressure plasma can activate textile surfaces for dye fixation or hydrophobicity without wet chemicals. Studies report 30%–40% water savings and no effluent generation (Serrano-Martínez et al. 2024).

The key life-cycle improvement indicators of four representative sustainable NF systems, namely, CNF, chitosan, PLA, and PHA, relative to a conventional synthetic-fiber baseline highlights performance across four sustainability metrics:

- energy savings (%),
- water reduction (%),
- CO_2 emission reduction (%), and
- toxicity index reduction (%).

Interpretation:

- PHA demonstrates the strongest performance in CO_2 reduction (≈60%) and toxicity reduction (≈50%), reflecting its marine biodegradability and low-emission processing.
- Chitosan excels in overall biodegradability and toxicity reduction due to its natural origin.
- CNF shows balanced performance across all indicators, consistent with its renewable cellulose feedstock.
- PLA performs well in carbon and toxicity reduction but offers moderate water and energy savings.

Table 15.6 matrix summarizes how leading policy and standardization frameworks map to SDGs relevant to sustainable nanotextiles. It highlights priority themes—circularity, safety, and transparency—and their implications for industry adoption (OECD 2023; EC 2020; EC 2023).

15.2.5 Environmental Performance and Life-Cycle Considerations

Life-cycle assessments reveal that sustainable NF systems can reduce carbon footprints by 20%–60% compared with conventional synthetic counterparts (EEA 2022). Key impact reductions stem from renewable feedstocks, solvent recovery, and lower end-of-life persistence. Nevertheless, scalability and the energy intensity of electrospinning remain ongoing challenges.

The technology readiness level (TRL) progression from 2025 to 2035 indicates the maturation of bio-based NF systems, green manufacturing, global standardization, and circular smart-textile networks aligned with UN SDGs 12 and 13 (EC 2023; Stramarkou et al. 2024; Silva et al. 2022; Râpă et al. 2020). It aligns with the milestones described in Table 15.7.

Interpretation:

- **By 2028**: Pilot-scale production and harmonized sustainability standards reach TRL 6–7.
- **By 2030**: Full integration of digital product passport and life cycle assessment (LCA) traceability systems.
- **By 2035**: Sustainable nanotextiles achieve TRL 9–10, marking global industrial maturity and carbon-neutral manufacturing.

The expanded Table 15.8 integrates institutional and regional leadership across major innovation themes, reflecting international alignment with SDG 12 (Responsible Consumption and Production) and SDG 13 (Climate Action). Data synthesized from EC (2023), Stramarkou et al. (2024), Silva et al. (2022), and Râpă et al. (2020). This table provides a forward-looking synthesis of innovation themes and institutional leadership shaping sustainable nanotextiles. It includes a new "Lead Institutions/Regions" column that connects technological progress to geographic and institutional actors responsible for global advancement.

TABLE 15.6
Policy–SDG Alignment Matrix for Sustainable Nanotextiles

Policy/Framework	Core Instruments/Documents	Focus for Nanotextiles	SDGs	Key Requirements/Targets (to 2030)	Implications for Industry
EU Green Deal and EU Strategy for Sustainable and Circular Textiles	European Green Deal (2020); Circular Economy Action Plan; EU Strategy for Sustainable and Circular Textiles (2023)	Climate neutrality, circularity, durability/repairability, digital product passports (DPPs), restricted hazardous substances	SDG 12, SDG 13, SDG 9	All textile products durable, repairable, and recyclable by 2030; mandatory DPP; substantiated green claims; eco-design requirements	Traceability systems; redesign for recyclability; substitution of hazardous nanofinishes; LCA disclosure
OECD (2023)—Working party on manufactured nanomaterials (WPMN)	Guidelines for the testing and assessment of manufactured nanomaterials (2023)	Harmonized testing, risk assessment, data sharing, life-cycle safety	SDG 12, SDG 3	Standardized ecotoxicity and human-safety protocols; cross-border data comparability	Adopt harmonized test methods; build tox-data dossiers; enable safe-by-design nanoformulations
ISO/TC 229—nanotechnologies	Terminology, measurement, and environmental safety standards for nanomaterials and products	Measurement protocols, terminology alignment, environmental and occupational safety	SDG 12, SDG 9, SDG 8	Growing suite of ISO standards for nano-enabled textiles and LCA reporting	Implement ISO-aligned QA/QC; enable conformity assessment and certification readiness
UNEP—Global Partnership on Sustainable Textiles	Multi-stakeholder initiatives for sustainable production and consumption in textiles	Capacity building, eco-innovation, circularity in developing regions	SDG 12, SDG 17	Best-practice dissemination; financing and technical assistance for circular projects	Access to funding/consortia; North–South technology transfer; scaling pilots in emerging markets
Global eco-labeling network (GEN)	ISO 14024 Type I eco-label schemes; mutual recognition arrangements	Credible eco-claims, product criteria for chemicals, durability, recyclability	SDG 12	Alignment of eco-label criteria with circular design and restricted substances	Pathway to market differentiation; consumer trust via verified eco-labels

TABLE 15.7
Sustainable Nanotextile Innovation Roadmap (Technology Readiness Level)

Innovation Track	2025	2028	2030	2033	2035
Bio-based nanofiber commercialization	4	6	8	9	10
Green manufacturing scale-up	3	5	7	8	9
Standardization and DPPs	2	4	7	9	10
Circular smart-textile networks	1	3	5	8	9

TABLE 15.8
Emerging Trends and Expected Milestones in Sustainable Nanotextiles (2025–2035)—Expanded Version

Theme	Trend (2025–2035)	Key Innovation	Expected Impact	Lead Institutions/ Regions
Biodegradable NFs	Expansion of PLA, PHA, and protein-based fibers	Biorefinery integration and enzymatic recycling	Reduction in microplastic release; enhanced biodegradability	EU (BIO4PACK), Japan (NanoCell Project), Brazil (UFRN, USP)
Smart eco-textiles	Self-powered, biosensing, and energy-harvesting garments	Renewable conductive nanofillers; wearable Internet of Things platforms	Real-time health and environmental monitoring	South Korea (KAIST), Germany (Fraunhofer IZM), India (IIT Delhi)
Circular production	Closed-loop electrospinning and solvent recovery systems	AI-driven energy optimization and green solvents	Zero-waste manufacturing and solvent reuse efficiency >90%	Denmark (DTU), China (Tsinghua University), Brazil (UFRN)
Policy standardization	EU DPP and ISO eco-certification	Blockchain-based life-cycle tracking and certification	Market transparency; enhanced consumer trust	EU (European Commission, ISO/TC 229), OECD (WPMN)
Global collaboration	North–South research partnerships and open data-sharing platforms	Nanomaterial toxicity databases; harmonized LCA protocols	Knowledge transfer; equitable capacity building	UNEP, BRICS Networks, South Africa (UKZN), India (CSIR)

Highlights:

- EU and Japan are leaders in biodegradable NFs through initiatives like BIO4PACK and NanoCell project.
- South Korea, Germany, and India lead smart eco-textiles integrating the Internet of Things (IoT) and biosensing technologies.
- Denmark, China, and Brazil show innovation in circular production and green electrospinning systems.
- ISO/TC 229 and the OECD Working Party on Manufactured Nanomaterials (WPMN) drive policy harmonization and standardization.
- UNEP and BRICS networks promote equitable technology diffusion and collaborative research in the Global South.

15.3 CASE STUDIES

15.3.1 Overview

Practical demonstrations of sustainable nanotextile technologies have accelerated since 2018, driven by stricter environmental regulations, advances in green NF production, and public awareness of microplastic pollution. The following international case studies illustrate diverse routes toward greener NF systems and their translation from laboratory to industry (Table 15.9).

15.3.2 Japan (2019): Cellulose Nanofiber–Reinforced Textiles

In 2019, Japanese researchers and the company Nippon Paper Industries commercialized CNF-reinforced cotton fabrics (Katagiri et al. 2019). The CNF, obtained from wood pulp by mechanical fibrillation, was dispersed in water and applied by padding, yielding a weight gain of ≈3%. Mechanical testing showed a 25% increase in tear strength and a 20% reduction in moisture absorption, while maintaining full biodegradability. Industrial partners reported energy savings because CNF coatings cured at 120°C instead of 170°C typical for synthetic binders.

TABLE 15.9
Summary of Sustainable Nanotextile Case Studies

Year	Country	Key Material	Fabrication Route	Performance Outcome	Sustainability Features
2019	Japan	Cellulose NFs	Aqueous coating on cotton	+25% strength; biodegradable	Low-temperature processing
2021	EU	PLA NFs	Electrospinning nonwoven	High flexibility	38% carbon reduction
2022	Brazil	CNF + nano-clay	Hybrid nanocomposite	>95% bacterial removal	Waste-cellulose utilization

15.3.3 India (2020): Self-Disinfecting Nanofiber Masks

During the COVID-19 pandemic, a team of Indian scientists in collaboration with an industry partner have developed a self-disinfecting copper-based NP-coated antiviral Face Mask to fight against the COVID-19 pandemic (Rao et al. 2020).

15.3.4 European Union (2021): PLA Nanofiber Nonwovens

Within the Horizon 2020 project BIO4PACK, researchers produced biodegradable PLA electrospun nonwovens for single-use packaging and hygiene applications (Bio4Pack 2021). The mats, 40–80 g/m^2, exhibited elongation at break >60% and disintegration >90% after 60 days under industrial composting. An LCA revealed a 38% carbon-footprint reduction compared with polyethylene nonwovens. Several European retailers adopted these materials in eco-labeled hygiene products by 2022.

15.3.5 Brazil (2022): Hybrid Cellulose–Clay Water-Filtration Fabrics

At the Federal University of Rio Grande do Norte, cellulose NFs were combined with natural bentonite nano-clays to fabricate low-cost filtration membranes (Silva et al. 2022). The CNF–clay network formed a tortuous micro-nano structure capable of removing >95% of *Escherichia coli* and heavy-metal ions from contaminated water.

The process consumed 60% less solvent than polymeric membranes and used locally available waste cellulose from the paper industry—demonstrating regional sustainability.

15.3.6 South Africa (2023): Lignin-Based Nanofibers for Packaging Applications

In 2022, researchers developed lignin–CNF-filled thermoplastic starch composite film for potential packing applications (AbdulRasheed-Adeleke et al. 2022). The addition of lignin and CNF to starch composite was found to have reduced the moisture absorption tendency and increased the mechanical properties of the films due to the good filler/matrix interfacial adhesion.

Figure 15.4 shows global distribution of sustainable nanotextile initiatives.

The figure shows a world map indicating case study locations: Japan (CNF), India (mask), EU (PLA nonwoven), Brazil (cellulose–clay membrane), and South Africa (lignin NF).

15.3.7 Discussion

These examples reveal a consistent movement toward bio-derived NFs, solvent-free manufacturing, and regional material circularity. Cross-institutional projects, such as EU Horizon, Japan's NanoCell, and Latin America's Sustainable Fibers Network, show that collaboration accelerates both technological maturity and environmental accountability. Despite challenges in cost and scale, commercial pathways are emerging in biomedical, filtration, and packaging textiles.

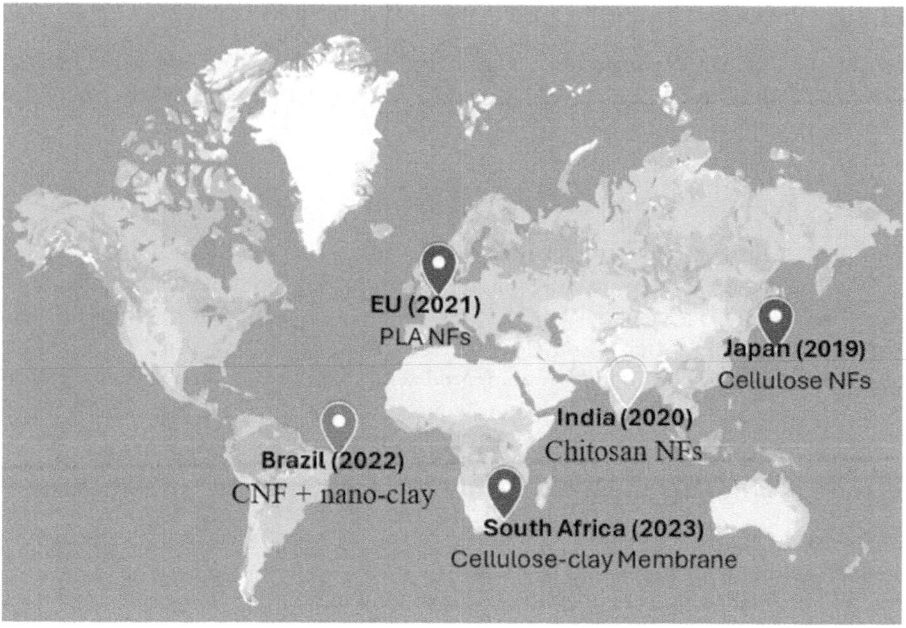

FIGURE 15.4 Global distribution of sustainable nanotextile initiatives.

15.4 OPPORTUNITIES AND CHALLENGES

15.4.1 Introduction

The transformation toward sustainable nanotextiles represents both a technological revolution and a systemic reconfiguration of the textile value chain. With nanotechnology now entering mainstream manufacturing, sustainability becomes the guiding framework for innovation, trade, and regulation. Key opportunities lie in biodegradable NFs, green surface engineering, and digital monitoring of life cycles, while challenges persist in economic feasibility, environmental safety, and consumer awareness (OECD 2023; EC 2020).

The textile sector's alignment with circular economy principles, design for durability, reuse, recycling, and resource efficiency, mirrors policy frameworks such as the European Green Deal (EGD), the OECD Green Growth Strategy, and the ISO/TC 229 Nanotechnologies sustainability guidelines. These initiatives collectively define sustainability metrics for nanomaterials and promote life-cycle transparency.

15.4.2 Opportunities

15.4.2.1 Circular Economy Integration

Sustainable nanotextiles naturally complement circular economy strategies. By employing biodegradable NFs (PLA, PHA, cellulose, chitosan) and recyclable composites, manufacturers can reduce end-of-life waste and microplastic pollution. For example, closed-loop electrospinning systems now recover up to 90% of solvents (Berdimurodov et al. 2023).

15.4.2.2 Green Product Innovation

Nanotechnology enhances textile performance without compromising environmental objectives. Biodegradable NC coatings based on lignin or starch nanocrystals achieve high UV protection and self-cleaning, supporting eco-label certification (ISO 14024).

Smart, responsive textiles integrating renewable nanomaterials enable real-time environmental sensing—a frontier of sustainable wearables.

15.4.2.3 Regulatory and Market Drivers

The EU Strategy for Sustainable and Circular Textiles (2023) aims to make all textile products durable, repairable, and recyclable by 2030. Similar directives are emerging in Asia and Latin America, supported by OECD-led innovation clusters. Consumer demand for eco-branded and biodegradable apparel is projected to grow by 12%–15% annually through 2035 (OECD 2023).

15.4.2.4 Collaborative Research Networks

Global initiatives, such as the Sustainable Nanofiber Consortium (2024), encourage open innovation and shared databases of nanomaterial toxicity data. These collaborations enhance standardization and harmonization under ISO/TC 229 and ASTM E56, bridging academia and industry.

A loop diagram (Figure 15.5) shows interconnected stages of the sustainable nanotextile life-cycle integrating phases: The loop connects the following stages:

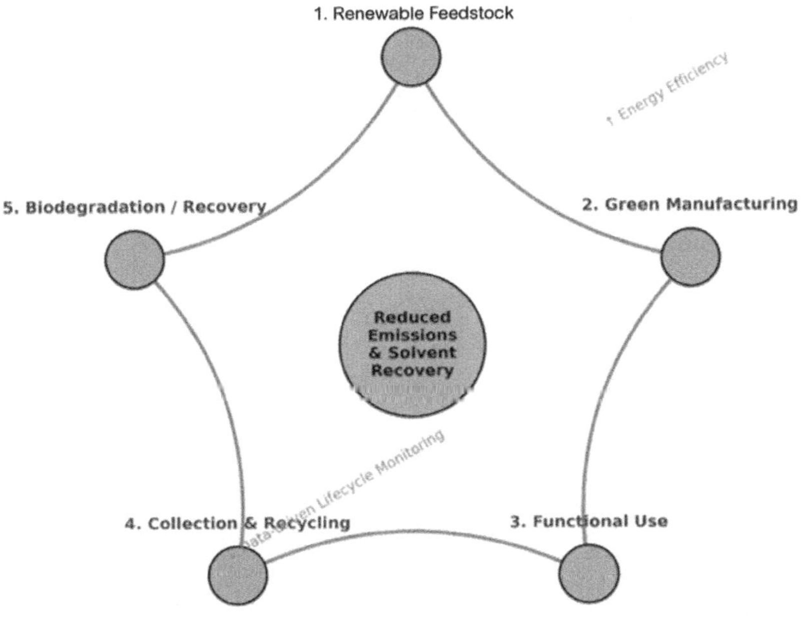

FIGURE 15.5 Circular economy model for sustainable nanotextiles.

1. **Renewable feedstock**: Bio-based raw materials such as cellulose, chitosan, and PLA derived from plant or microbial sources.
2. **Green manufacturing**: Low-toxicity electrospinning, sol–gel coating, and plasma finishing using solvent recovery systems.
3. **Functional use**: High-performance fabrics with added value (UV protection, filtration, sensing).
4. **Collection and recycling**: Material recovery, reuse, and mechanical/chemical reprocessing.
5. **Biodegradation or recovery**: Composting or enzymatic degradation of biopolymer NFs returning nutrients or monomers to the production cycle.

At the core, icons represent reduced emissions, solvent recovery, and closed-loop NF systems, symbolizing energy efficiency and digital life-cycle monitoring.

15.4.3 Challenges

Despite significant progress, multiple structural barriers constrain widespread adoption of sustainable nanotextiles (Table 15.10).

15.4.3.1 Scalability and Cost

Green electrospinning, sol–gel coating, and plasma modification remain capital-intensive and less efficient at industrial scale. Pilot production lines often require 30%–50% higher energy/kg of NF output compared with traditional wet processing (Serrano-Martínez et al. 2024).

15.4.3.2 Toxicity Assessment and Regulation

While bio-based NFs are considered safer, standardized ecotoxicity testing protocols are lacking. The OECD's Guidelines on the Testing of Manufactured Nanomaterials (2023) emphasize harmonization of data reporting, but implementation is uneven across jurisdictions (OECD 2023).

TABLE 15.10
SWOT Analysis of Sustainable Nanotextile Development

Strengths	Weaknesses	Opportunities	Threats
Renewable and biodegradable nanomaterials	High production cost and limited scalability	Regulatory incentives (EU Green Deal, ISO eco-labels)	Market volatility and raw-material shortages
Reduced microplastic pollution	Lack of global toxicity standards	Circular economy integration	Technological lock-in from established synthetic industries
High-performance, multifunctional fabrics	Complex recycling for hybrid composites	Growing demand for eco-certified textiles	Potential nanomaterial misuse and greenwashing
Compatibility with smart textiles	Energy-intensive electrospinning	Collaboration under ISO/TC 229	Regulatory delays in developing nations

15.4.3.3 Consumer Acceptance

Higher production costs and limited awareness slow consumer transition. Surveys indicate that 60% of buyers prioritize comfort and price over sustainability in textile purchases (EEA 2022). Clear eco-labeling and traceability tools such as QR-coded life-cycle disclosures are essential.

15.4.3.4 End-of-Life Management

NCs containing mixed organic–inorganic phases pose recycling difficulties. While biodegradability has improved, full circular recovery remains elusive, especially when fibers are blended with synthetic backings or metallic sensors.

15.4.4 Policy Alignment and Standardization

Policy frameworks form the backbone of sustainable transition:

- **EU Green Deal (2020–2030)**: Promotes climate neutrality and resource efficiency through the Circular Economy Action Plan, mandating sustainability declarations for textile products.
- **OECD WPMN**: Coordinates testing, safe use, and cross-border data on nanomaterial life cycles.
- **ISO/TC 229 nanotechnologies**: Establishes terminology, measurement, and environmental safety standards for NFs, NCs, and related textile applications.
- **Global Eco-labeling Network (GEN)**: Encourages harmonized certification for green nanofinishes and biodegradable fibers.

Together, these frameworks provide regulatory legitimacy and global compatibility, paving the way for sustainable nanotextile markets.

15.4.5 Discussion

The opportunities offered by sustainable nanotextiles are transformative: greener production, enhanced recyclability, and long-term economic value. Yet real progress depends on interdisciplinary collaboration, transparent certification, and policy-backed investment in pilot-scale infrastructure. With coordinated action among research institutions, governments, and industry, sustainable nanotextiles can become a keystone of the 2030 circular economy vision.

15.5 FUTURE OUTLOOK

15.5.1 Introduction

Sustainable nanotextiles are poised to redefine the global textile landscape by 2035.

Their evolution will depend not only on advances in material science but also on harmonized policies, digital traceability, and public engagement.

Nanotechnology, once perceived as resource-intensive, is now transitioning toward a bio-circular, data-driven ecosystem, integrating renewable chemistry, smart sensing, and responsible end-of-life strategies (Stramarkou et al. 2024).

15.5.2 Biodegradable and Renewable Nanotextiles

The next generation of nanotextiles will prioritize biodegradability, compostability, and circular reuse.

Emerging work focuses on the following:

- Protein-based NFs (e.g., silk fibroin, keratin, soy protein) that combine mechanical resilience with rapid degradation in soil or seawater (Râpă et al. 2020).
- Algae- and bacterial-cellulose systems, cultivated using CO_2-capturing bioreactors, producing carbon-negative NF matrices.
- Enzymatic degradation routes, allowing selective recovery of nanofillers from composite waste.

Biorefineries that co-process agricultural residues into nanocellulose, lignin NPs, and natural dyes are expected to close the carbon loop and localize raw-material supply.

15.5.3 Smart and Functional Eco-Nanotextiles

The convergence of sustainability and functionality will dominate the next decade.

By embedding renewable conductive nanofillers such as carbonized cellulose, graphene from biomass, or ZNO_2 NPs from plant precursors, researchers are creating textiles that monitor health, harvest energy, and interact with the environment (Serrano-Martínez et al. 2024). Energy-autonomous garments powered by flexible piezoelectric NFs are already under prototyping. Digitalization and IoT platforms will enable life-cycle tracking, ensuring full transparency from feedstock origin to postconsumer recovery.

This "digital passport" concept is expected to be standardized within the EU by 2030 (EC 2023). A horizontal timeline displaying milestones is shown in Figure 15.6 highlights policy integration, technological maturity, and adoption curves aligned with UN SDGs 12 and 13.

15.5.4 Recycling and End-of-Life Management

A major frontier involves closed-loop recovery of nanomaterials.

Current strategies include the following:

1. Mechanical recycling with selective solvent re-dissolution to recover nanofillers.
2. Chemical depolymerization of PLA/PHA matrices to monomers for re-polymerization.
3. Biological upcycling uses enzymes or microbial consortia that degrade NC films into benign by-products.

Pilot plants in Europe and Japan have demonstrated 70%–80% recovery efficiencies from postconsumer NF mats (EEA 2022). The next step is to integrate such systems into textile industrial parks to achieve "zero-nanowaste" manufacturing.

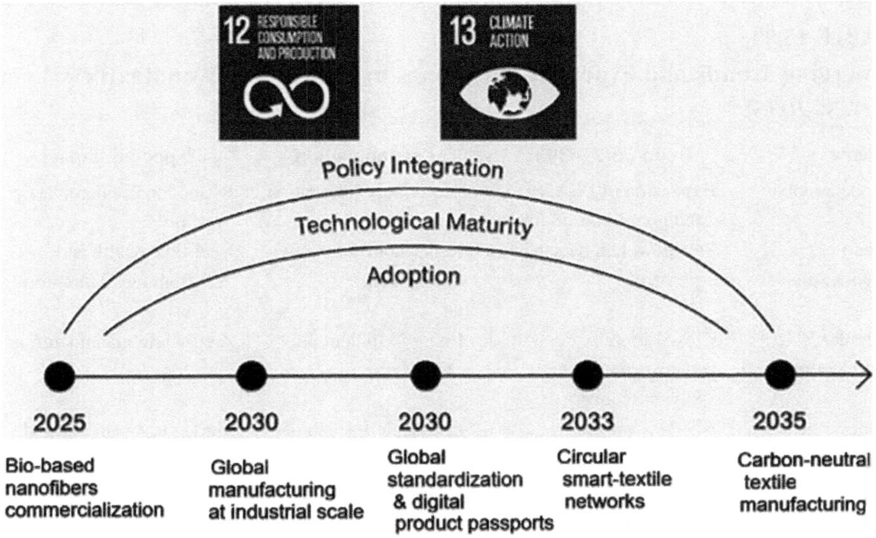

FIGURE 15.6 A horizontal timeline displaying five milestones.

15.5.5 Global Collaboration and Policy Roadmap

Future success will depend on international governance and harmonized sustainability criteria. ISO/TC 229, the OECD WPMN, and UNEP's Global Partnership on Sustainable Textiles are expected to establish cross-referenced indicators for (Table 15.11)

- carbon and water footprints per kilogram of NFs,
- toxicity thresholds for bio-NCs, and
- circular-design certification schemes.

Developing nations, especially in Latin America, Africa, and South Asia, will play a decisive role by leveraging agricultural residues and low-carbon energy in green NF production (Silva et al. 2022). Public–private partnerships and BRICS-level research networks can ensure equitable technology diffusion.

15.5.6 Concluding Remarks

The transition from laboratory research to industrial practice is already underway.

By 2035, sustainable nanotextiles are projected to represent 20%–25% of global technical-textile output, anchored in renewable chemistry, green nanomanufacturing, and intelligent life-cycle management. Their success will not only mitigate pollution but also exemplify how advanced materials can evolve within planetary boundaries, a decisive step toward a circular, climate-neutral textile economy.

TABLE 15.11
Emerging Trends and Expected Milestones in Sustainable Nanotextiles (2025–2035)

Theme	Trend (2025–2035)	Key Innovation	Expected Impact
Biodegradable NFs	Expansion of PLA, PHA, and protein-based fibers	Biorefinery integration	Reduction in microplastic release
Smart eco-textiles	Self-powered, biosensing garments	Renewable conductive nanofillers	Real-time health and environmental monitoring
Circular production	Closed-loop electrospinning and solvent recovery	Energy-efficient pilot lines	Zero-waste manufacturing
Policy standardization	EU DPP and ISO eco-certification	Life-cycle traceability	Market transparency and consumer trust
Global collaboration	North–South research partnerships	Open-data toxicity databases	Knowledge transfer and capacity building

REFERENCES

AbdulRasheed-Adeleke, T., Egwim, E. C., Sadiku, E. R. and Ochigbo, S. S. 2022. Optimization of lignin–cellulose nanofiber-filled thermoplastic starch composite film production for potential application in food packaging. *Molecules* 27(22): 7708.

Antony, J. S., Cowan, N., Davidson, M., Godina, G., Smith, I., Xin, J. and Menezes, P. L. 2025. A comprehensive review on cellulose nanofibers, nanomaterials, and composites: manufacturing, properties, and applications. *Nanomaterials* 15: 356.

Berdimurodov, E., Dagdag, O., Berdimuradov, K., Wan Nik, W. M. N., Eliboev, I., Ashirov, M., Niyozkulov, S., Demir, M., Yodgorov, C. and Aliev, N. 2023. Green electrospun nanofibers for biomedicine and biotechnology. *Technologies* 11: 150.

Bio4Pack. 2021. Biodegradable bio-based bioactive packaging technology for supporting food safety. https://cordis.europa.eu/project/id/101223372 [Accessed 05 September 2025].

Boruah, S. and Jose, S. 2025. Nanotechnology in textiles: Environmental safety and sustainable practices. *Environmental Nanotechnology, Monitoring & Management* 23: 101062.

Catarino, M. L., Sampaio, F. and Gonçalves, A. L. 2025. Sustainable wet processing technologies for the textile industry: A comprehensive review. *Sustainability* 17: 3041.

Dong, M., Mastroianni, G., Bilotti, E., Zhang, H. and Papageorgiou, D. G. 2024. Biodegradable starch-based nanocomposite films with exceptional water and oxygen barrier properties. *ACS Sustainable Chemistry & Engineering* 12(29): 11056–11066.

EC. 2020. *A European Green Deal: Striving to be the First Climate-Neutral Continent*. European Commission (EC), Brussels: Publications Office of the European Union, https://ec.europa.eu/newsroom/know4pol/items/664852 [Accessed 10 September 2025].

EC. 2023. *EU Strategy for Sustainable and Circular Textiles*. European Commission (EC), Brussels: Publications Office of the European Union. https://www.europarl.europa.eu/legislative-train/theme-a-new-plan-for-europe-s-sustainable-prosperity-and-competitiveness/file-eu-textiles-strategy [Accessed 15 September 2025].

EEA. 2022. Microplastics from textiles: Towards a circular economy for textiles in Europe. European Environment Agency (EEA) Briefing.https://www.eea.europa.eu/en/analysis/publications/microplastics-from-textiles-towards-a-circular-economy-for-textiles-in-europe [Accessed 10 September 2025].

Ferreira, S., Pais, V., Bessa, J., Cunha, F., Hsia, L. D. A., Mai, E. F., Sborchia, G. and Fangueiro, R. 2025. Lignin-based coatings: A sustainable approach to produce antibacterial textiles. *International Journal of Molecular Sciences* 26(3): 1217.

Hasan, T., Hossen, Md. R., Rimon, Md. I. H. and Mobarak, Md. H. 2025. Advances of nanotechnology in fabric and clothing. *Nano Trends* 11: 100140.

Katagiri, K., Honda, S., Minami, S., Yamaguchi, S., Kawakita, S., Sonomura, H., Ozaki, T., Uchida, S., Nedu, M., Yoshioka, Y. and Sasaki, K. 2019. Enhancement of the mechanical properties of the CFRP by cellulose nanofiber sheets using the electro-activated deposition resin molding method. *Composites Part A: Applied Science and Manufacturing* 123: 320–326.

Kolya, H. and Kang, C.-W. 2024. Eco-friendly polymer nanocomposite coatings for next-generation fire retardants for building materials. *Polymers* 16(14): 2045.

Li, S., Li, Y., Zhang, X., Jiang, D. and Kong, L. 2026. Electrospinning of chitosan-based nanofibers: Innovations in fabrication and applications. *Journal of Agriculture and Food Research* 25:102599

Naser, A. Z., Deiab, I. and Darras, B. M. 2021. Poly(lactic acid) (PLA) and polyhydroxyalkanoates (PHAs), green alternatives to petroleum-based plastics: A review. *RSC Advances* 11(28): 17151–17196.

OECD. 2023. *Guidelines for the Testing and Assessment of Manufactured Nanomaterials* (2nd ed.). Organization for Economic Co-operation and Development (OECD), Paris. https://www.oecd.org/chemicalsafety/nanosafety [Accessed 18 September 2025].

Patnaik, A. 2024. Recent developments in application of nanofibers. *Processes* 12(9): 1894.

Rajendran, D. S., Venkataraman, S., Jha, S. K., Chakrabarty, D. and Kumar, V. V. 2024. A review on bio-based polymer polylactic acid potential on sustainable food packaging. *Food Science and Biotechnology* 33(8): 1759–1788.

Rao, T. N., Hembram, K. and Sarada, B. V. 2020. Scientists develop self-disinfecting, biodegradable face masks to combat COVID-19. https://dst.gov.in/scientists-develop-self-disinfecting-biodegradable-face-masks-combat-covid-19#:~:text=combat%20COVID%2D19-,Scientists%20develop%20self%2Ddisinfecting%2C%20biodegradable%20face%20masks%20to%20combat%20COVID,virus%20load%20is%20very%20high [Accessed 18 September 2025].

Râpă, M., Gaidău, C., Stefan, L. M., Matei, E., Niculescu, M., Berechet, M. D., Stanca, M., Tablet, C., Tudorache, M., Gavrilă, R., Predescu, C. and Vidu, R. 2020. New nanofibers based on protein by-products with bioactive potential for tissue engineering. *Materials* 13(14): 3149.

Saleem, H. and Zaidi, S. J. 2020. Sustainable use of nanomaterials in textiles and their environmental impact. *Materials* 13: 5134.

Serrano-Martínez, V. M., Ruzafa-Silvestre, C., Hernández-Fernández, C., Bañón-Gil, E., Arán-Ais, F. and Orgilés-Calpena, E. 2024. improving cotton fabric dyeability by oxygen plasma surface activation. *Surfaces* 7(4): 1079–1095

Shen, R., Guo, Y., Wang, S., Tuerxun, A., He, J. and Bian, Y. 2023. Biodegradable electrospun nanofiber membranes as promising candidates for the development of face masks. *International Journal of Environmental Research and Public Health* 20: 1306.

Silva, I. O., Ladchumananandasivam, R. and Souza, F. A. 2022. Hybrid cellulose-clay nanocomposite fabrics for low-cost water filtration. *Materials Today: Sustainability* 18: 100237.

Souza, F. A., Oliveira da Silva, I. and Ladchumananandasivam, R. 2023. Sol–gel nanocoatings from agricultural waste precursors for sustainable textiles. *Materials Today: Sustainability* 21: 100257.

Stramarkou, M., Tzegiannakis, I., Christoforidi, E. and Krokida, M. 2024. Use of electrospinning for sustainable production of nanofibers: A comparative assessment of smart textiles-related applications. *Polymers* 16: 514.

Tamulevičius, T. 2025. Nanotechnologies in textiles. *Materials* 15: 1466.

Zhang, L., Guo, L. and Wei, G. 2021. Recent advances in the fabrication and environmental science applications of cellulose nanofibril-based functional materials. *Materials* 14(18): 5390.

16 Clothing and Textile Industries

Microfiber Sources, Impacts, Threats and Innovative Sustainable Case Studies and Practices

Conrad Sparks

16.1 INTRODUCTION

The clothing and textile industries are among the most resource-intensive and polluting sectors globally. With fast fashion accelerating consumption and waste, the environmental footprint of garments—from production to disposal—has become a pressing concern. This chapter conceptualizes the issues related to the clothing and textile industries and introduces the concept of sustainability in the textile and clothing industries. The chapter will emphasize the critical role producers, industries and consumers play in shaping practices and impacts. It sets the stage for a deeper exploration of microplastic (fiber) pollution, case studies of sustainable innovation and the future of consumer-driven change.

16.1.1 Terms and Definitions

Natural (cotton, wool, hemp, etc.) and synthetic textiles such as polyester, nylon and acrylic shed microfibers during routine washing. These microscopic fibers enter wastewater systems, bypass many filtration processes and ultimately settle in marine and terrestrial ecosystems. The term microplastics generally refers to synthetic particles composed of polymers within the context of environmental pollution (GESAMP 2019). The term microfibers will refer to natural and synthetic fibers, whereas microplastic fibers will refer to synthetic fibers. Based on size, "micro" refers to particles smaller than 5 mm, but larger than 0.001 mm (GESAMP 2019).

DOI: 10.1201/9781003689225-16

16.2 MICROFIBERS: SOURCES, RELEASE ROUTES AND SETTLEMENT

Textiles are woven materials from various types of materials. The origin of textile industry dates back to centuries and has since evolved in many areas and is influenced by various cultures and environmental factors. It has evolved based on societal developments and often changed based on migration of communities due to existing conditions. The textiles industry started by harvesting natural products to create raw materials for weaving garments and textiles.

The clothing people wear is also cultural and based on available resources. The textile industry today is focused on clothing, bedding, flooring, furniture and its decorations. These sectors have developed into multibillion-dollar industries and the number of textiles being produced is linked to the global increase in the population, people becoming more affluent to buy more goods, and societies setting trends in what people use in their homes and surroundings. The global textile market is hence an important economic sector for all countries, and the market is affected by various factors such as the costs of textiles, demand for textile products, the versatility of textile products and the affordability of textiles (Cao et al. 2008). The economy of societies is based on the premise that an increase in output results in lower production costs. The mass market of textile products is not only producing more textile products, but also creating environmental issues related to the impacts of textiles on the environment, the economy and ultimately the health of humans (Brander et al. 2024). The fundamental environmental and human health issue related to the textile industry is waste, in the form of microplastic fibers.

16.3 MANUFACTURING PROCESS

Microfiber generation starts during the various manufacturing processes of textile industries, during fiber production and textile processing. The basic fiber shape, surface roughness and substructural defects that affect future shedding of synthetic fibers are determined by the polymer extrusion, spinning and texturizing processes (Vassilenko et al. 2021). Filaments with surface microvoids and fibrillation points produced by high-throughput spinning and drawing become initiating sites for fiber breakage when they are subsequently subjected to mechanical stress (Mehdikhani et al. 2019). Mechanical tensions introduced by yarn formation processes (spinning, twisting, plying) and fabric building (weaving, knitting, nonwovens) result in loose fiber ends and short-length pieces (fly) that are easily removed during downstream handling.

Mechanical abrasion is the release of fibers for the textile industry and can be in various forms, depending on the textile product being produced. It includes spinning, weaving and knitting. The mechanical abrasion depends on the scale of the production and where it is taking place. Industrial-scale operations are generally characterized by large centralized facilities, often located in urban areas, to which raw materials are transported. Conversely, small-scale operations may be decentralized and located in rural regions adjacent to the source of raw material production or harvesting.

Brushing, sanding, sueding, elevating and enzymatic surface modification are examples of mechanical finishing techniques that modify fiber-bundle cohesion to produce desired tactile and aesthetic qualities, but frequently at the expense of decreased fiber retention (Rashid et al. 2024). The way fibers interact with one another under stress is impacted by wet operations (dyeing, washing, chemical finishing) that alter surface chemistry and lubrication. Manufacturing produces airborne and aqueous losses of short fibers even before clothing is delivered to customers. If these losses are not caught by process controls or effluent treatment systems, they may deposit close to facilities or find their way into wastewater streams. An estimated 35% of microplastics released into the oceans are from synthetic textiles (25% from wastewater) and up to 22 million tons of microplastic fibers can be released into the oceans by 2050 (Boucher and Friot 2017).

Both natural and regenerated cellulosic fibers are not immune: carding, combing and spinning cotton result in significant short fibers and dust (Simpson and Fiori 1974). Chemical treatments of cellulosic fibers (such as crosslinkers, resins and flame retardants) can produce composite particles that endure or affect biodegradation rates (Toldy et al. 2011). Additionally, blended yarns introduce composite particles that contain both synthetic and cellulosic fractions, making analytical detection and environmental destiny more difficult.

It is also most likely that the production of the fibers is taking place in developing countries (where the raw products are sourced from). The countries include China and India, which are major exporters of textile raw products (Chen et al. 2023). China's textile exports were estimated to be US$ 34 billion in 2004 (Chan and Au 2007), increasing to US$ 292 billion in 2024 (OEC 2026). In addition to these exports, it is also likely that the scale of production is linked to the scale of waste produced. Developing countries spend less resources on management of waste, in order to reduce operating costs (Chen et al. 2023). There is hence the likelihood that there is less management of waste stemming from mechanical abrasion in fiber production and textile processing. Fiber and textile waste produced is discarded in wastewater or moved to landfill sites. The annual consumption of textiles is estimated at 13 kg/person with greater than two-thirds of textiles produced ending up in landfills (Shirvanimoghaddam et al. 2020).

Chemical treatment of textiles serves various purposes, depending on the end product. The chemical treatments include dyeing, coating and "finishing" of textiles. When chemicals are added to fibers and textiles, it serves a particular purpose that is limited to a certain time and subject to the usage of the product. We can imagine the "smell" of new clothing, furniture, carpets, etc. Cleaning the product most often degrades the integrity of the chemical that was added, and this results in the weakening of the fibers and textiles. This weakening then results in increased rates of fiber and textile degradation and hence, the increase in the release of microfibers (Akyıldız et al. 2025).

Besides weakening the tensile strength, an effect not measured in many cases is the release of chemicals sorbed onto microfibers to the environment (Hahladakis et al. 2018). Often, we do not know the chemical composition of the coating due to the proprietary ownership of the chemical. The chemicals sorbed onto fibers and textiles are released into the air and environments that are most likely to end up in aquatic systems, where the chemicals have the potential to be ingested by biota and subsequently affect the health of ecosystems.

Wastewater discharge includes the waste from textile production facilities that use water during the manufacturing and production processes. Fiber and textile waste includes microfibers that end up in wastewater from production facilities and processing plants, which include textile mills and factories. Water at textile plants is either extracted from natural water bodies or from municipal reticulation systems. The high volumes of water used are an integral part of production processes, including cleaning, staining and washing systems. Once the water has served its purpose, it is discarded in most cases. In many countries, there are existing laws and regulations regarding water quality and the manner in which the wastewater must be treated before it can be discarded (Xu et al. 2019).

There are often national environmental standards that must be met before wastewater can be discarded into the environment. Standards may include the temperature, pH, turbidity, oxygen levels, biological oxygen demand, bacterial loads, etc. (Willis et al. 2018). Measuring the many parameters, and the frequency of the measurements are expensive and increase running costs of businesses. In addition, should the water quality not meet the set standards, it will need to be treated by businesses before it can be released into the environment. This further adds to the operational costs and increases the price of textile products. To avoid these costs, businesses may either ignore requirements to meet environmental standards or manipulate the regulatory framework by providing inaccurate (or intentionally incorrect) information. Should monitoring and enforcement agencies be absent or nonfunctional, businesses will ignore regulatory and legislative frameworks and laws and discard wastewater containing microplastic fibers.

16.4 CONSUMER USAGE OF CLOTHING AND TEXTILES

The use phase is the most persistent and widely distributed source of microfiber emissions. Mechanical stresses (friction, bending, tensile loading) are applied throughout every interaction of clothing and include wearing, rubbing, laundering, drying and handling. The results are gradual breaks of interfiber connections and the release of microfibers. Globally, an estimated 98 million tons of textile waste is produced each year, costing the industry approximately US$ 500 billion (Tang 2023).

16.4.1 Wear and Tear of Clothing and Textiles

Shedding occurs because of wear and tear of fibrous and textile materials. It occurs at different stages of the life cycle of clothing and textiles, with an estimated 49% shedding during manufacturing, 28% during washing and 23% during wearing (Lim et al. 2022). The wide array of fibrous materials used to produce garments and textiles means that the products will shed fibers at different rates, depending on the fiber type and how the fiber was produced (Akyıldız et al. 2025). Low-cost and cheaply produced clothing and textiles may shed at a higher rate than high-end products. Given the scale of economies, it is likely that more low-cost products will be sold, resulting in higher rates of shedding.

As clothing and textile shed fibers, it also reduces the quality of the product. Often clothing will lose its tensile strength and stretch, making it less attractive, that is, lower quality and less appealing to use. In the end, such clothing and textiles are stored and ultimately discarded as waste. Once discarded, the clothing and textiles end up in landfills that are exposed to direct sunlight where chemical degradation and mechanical abrasion rates increase (Wan et al. 2022). It is at this stage that the tensile strength is lowest and likely the rate of microfiber production is highest. Landfills are often exposed open fields or adjacent water bodies (canals and rivers), where it becomes conduits for microfibers to be translocated into the environment (Wan et al. 2022).

Domestic laundry activities include washing and drying of clothing (and textiles). Laundering has a particularly significant impact on microfiber release since both home and commercial washing procedures expose clothing to shear pressures, turbulent flows and contact with abrasive surfaces and attachments (buttons, zips) that encourage separation (Table 16.1). Shedding rates are influenced by several factors, including temperature, cycle length, detergent chemistry, drum geometry, load size, water volume and washing machine design (front- vs. top-loader) (Lim et al. 2022). Polyester clothing and garments can shed up to 1,900 fibers per single wash, with a 6 kg washing load releasing between 700,000 and 140,000 fibers (Sudheshna et al. 2022).

Given that clothing is washed and dried the most, this section will focus on clothing (but may include other textiles [e.g., carpets]). Once purchased, all clothing is washed after it has been worn. In developing countries and rural communities, the laundry process may involve washing clothes in natural water bodies (rivers, canals, lakes, etc.), collecting water from a natural water body or extracting water from a water reticulation system (de Villiers 2019). Once washed, clothes are often hung on lines (or the ground in some areas) where the sun and wind contribute to drying the clothes.

TABLE 16.1
Laundry Parameters and Their Qualitative Influence on Microfiber Shedding

Parameters	Direction of Effect on Shedding	Rationale/Notes	Practical Mitigation
Wash temperature	↑ at higher temperatures	Heat + chemistry weaken fiber–yarn cohesion	Prefer cold/low-temp cycles
Cycle duration/agitation	↑ with longer, more turbulent cycles	Prolonged mechanical stress	Select “delicate/low-shear” programs
Load size	↑ when under-loading	Higher garment-drum friction	Wash full, balanced loads
Detergent chemistry	Mixed (formulation-dependent)	Surfactant/enzymes alter fiber surfaces	Use mild detergents; avoid overdosing
Tumble drying	↑ line-dry	Heat + abrasion → lint/fiber loss	Line-dry where possible; clean lint filters
Fabric construction	↑ for loose constructions, low twist	Easier fiber pull-out	Choose tighter weaves/knits; higher twist yarns

The distance to a natural water body will influence the amount and frequency of laundry activities. People living adjacent to water bodies are likely to wash less laundry but more frequently, whereas people living further away may amass laundry and wash more, less frequently. The proximity to water bodies, particularly flowing water, has resulted in more people wanting to reside next to water bodies, not only to wash clothes, but also to access drinking water and water for irrigation purposes. Communities residing further away from water bodies may have water trucked in. This water is often stored in containers and used for various purposes (e.g., potable water) as well as for laundry.

Cities and developed residential areas have water reticulation systems for households. In low-income communities, the water from the reticulation system is used to hand-wash clothes and textiles and the wastewater is discarded into municipal wastewater or sewage systems. Generally, clothes are then hung on lines allowing them to dry in the sun and wind (de Villiers 2019).

Higher-income communities and densely populated cities use washing machines for laundry, either in their homes or commercial laundry services (Vassilenko et al. 2021). Wastewater is discarded in municipal waste systems (either sewage or wastewater systems). Due to limited space, it is often in these communities that drying machines (tumble dryers) are used to dry clothes. It is in these communities, where the highest amounts of microfibers are released, due to high volumes of laundry in washing machines and tumble dryers (McIlwraith et al. 2019). Even if lint filters catch some of the released fibers, exhausts and leftover particles continue to be an emission route to the air and settle indoors, with tumble drying producing more mechanical action and higher temperatures.

A key contributing factor is that both washing machines and tumble dryers can be set to various temperatures. In cold and wet regions, higher temperatures are used to wash and dry clothes in the shortest time span. This results in high microfiber shedding rates due to laundry and drying at higher temperatures (Lim et al. 2022). In addition to being directly caused by wear, airborne fiber emissions can also be indirectly caused by drying, abrasions, vacuuming or cleaning indoor areas where fibers build up as dust. Microfibers are frequently concentrated indoors, where they can be resuspended by human activity or heating, ventilation and air conditioning systems and finally vented outside, increasing atmospheric microfiber loads and enabling long-distance transmission (Chen et al. 2020).

16.5 MICROFIBERS IN TEXTILES: A HIDDEN POLLUTANT FOR MECHANISMS OF UPTAKE

Microfibers released into the environment have the potential to be distributed via aerial and aquatic pathways. The aerial movement applies to fewer fibers, but the distance traveled can be extensive (Prata 2018). The distance depends on the location of the release and prevailing weather conditions. Microfibers' compact size and low settling velocities enhance atmospheric transport. Fibers released from urban surfaces, laundry exhausts and textile mills can aerosolize and spread both locally and internationally. Even localized emissions can have far-reaching effects, as seen by deposition onto distant habitats like mountain snow and marine gyres (Batool et al.

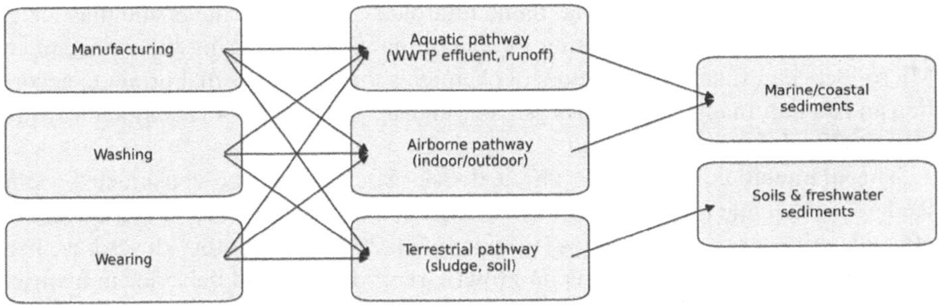

FIGURE 16.1 A conceptual flow of microfibers from source to environmental sinks (not to scale).

2021). An estimated 81.6 microfibers/m^2, per day, can be deposited in urban areas (Napper et al. 2023).

It is also probable that airborne microfibers will adsorb chemicals that are dangerous or hazardous. The greatest risk of airborne microfibers is the potential to be inhaled by animals and humans (Prata 2018). Most microfibers are released into aquatic and terrestrial systems. Point sources of microfiber release include outflows from wastewater treatment plants (WWTPs), stormwater systems, rivers and other water bodies (Simmerman and Wasik 2020). Nonpoint sources include downstream from aquatic systems. Vertical and horizontal dispersion in aquatic environments is determined by hydrodynamics, particle buoyancy, biofouling and aggregation into marine snow; many fibers finally settle into sediments where they can last for decades (Alimi et al. 2018). As much as 500,000 microfibers can be released from WWTPs per day (Napper et al. 2023). Once prevalent in the aquatic system, microfibers have the potential to adsorb chemicals, including metals, organics, pharmaceuticals and biological material and microbes (viruses, bacteria and other microbiota) (Joo et al. 2021). Depending on the time in the environment, the propensity to adsorb chemicals and development of microbes result in a "wicked problem" where the chemical cues from microbes (and even microbiomes) on microfibers increase the potential for it to be misidentified as a potential source of food to organisms (Wagner 2022). The attraction to microbes results in microfibers being ingested. The uptake of microfibers and adsorbed contaminants has the potential to result in environmental and human health hazards and risks that can ultimately affect the health of ecosystems and societies (Figure 16.1). These effects ultimately result in impacts on the economies of countries.

16.6 ENVIRONMENTAL RISKS ASSOCIATED WITH MICROFIBER ACCUMULATION

Microfibers' physical shape, polymeric identity and related chemical additives influence how they affect the environment. During the entire life cycle (manufacturing to waste) of clothing and textiles, the abrasions and continuous fragmentation result in the production of smaller particles (micro- and nanoplastics) that are released into the environment (Brander et al. 2024). Textile and clothing additives such as dyes, ultraviolet (UV) stabilizers, flame retardants, water repellents and plasticizers leach from the fibers, contributing to contaminants also entering the environment. Microfibers can function as vectors of chemicals and microbes that promote adsorption and contaminant transfer to organisms and are bioavailable in ecosystems (Wang et al. 2020; Holmes et al. 2012).

The filamentous form poses physical risks since fibers are consumed by filter feeders and aquatic invertebrates, which can result in obstructions, decreased feeding efficiency or energy expenditures (Brander et al. 2024). Laboratory doses have been shown to have sublethal impacts on growth, reproduction and behavior in a variety of organisms, and studies have documented ingestion across trophic levels (Davies et al. 2021; Brander et al. 2024). The health effects of long-term low-level exposure are still being studied. Human exposure pathways include ingestion through contaminated seafood, inhalation of airborne microfibers and unintentional ingestion through dust (Barboza et al. 2018).

The impacts of microfibers start at the subcellular level and can lead to effects on entire ecosystems. Once ingested, the first potential impact of microfibers is to cause physical harm to the digestive tract. This may be in the form of gut irritation or tearing and physical damage to the gut lining. The organism's biological response will be to repair the damage caused by the physical harm of the microfiber. Another reaction will be to the chemical nature of the microfiber, which will depend on the chemical composition of the microfibers. However, the most damaging impact of microfibers will be the amount and type of microbe and chemicals sorbed onto the microfiber at the time of ingestion (Duan et al. 2020).

Biological responses to chemicals are diverse and depends on the type of contaminant (metals, organics, pharmaceuticals). The body will first defend itself by reacting to foreign chemicals at the subcellular level. The responses will also depend on the fitness and health of the organism and the duration of the exposure to contaminants (acute vs. chronic). Should the organism not be able to deal with the response at the subcellular level, the effect will start to be manifested at a higher level, from cells to organs, systems and organisms. The ultimate revelation of the impact at the organism level will be the death of individuals. The death of individuals will result in declining communities and populations. Once populations die off, this will eventually affect ecosystem structure and function. Depending on the ecosystem, biospheres may change and this may include the loss of living resources that societies depend on for their livelihood (Browne et al. 2013).

16.7 HUMAN HEALTH AND SOCIETAL RISKS OF MICROFIBER

The human health risks are evident in the inhalation, ingestion and skin contact with microfibers (Barboza et al. 2018). There is a growing body of evidence that the ingestion of microfibers is causing human health risks that may be costing countries billions of US dollars per annum.

Regardless of political and administrative boundaries, microfiber pollution has an impact on all ecosystems and human health worldwide. Concerns about environmental injustice are raised by the global socioeconomic inequities that confuse the picture of plastic pollution (Prata 2018). Developing countries and Small Island

Developing States are more impacted than their own domestic microfiber pollution due to microfiber pollution produced by other far-off nations through long-distance movement of microfibers. Ecosystems and communities residing in those locations are prone to environmental and human health risks associated with microfiber exposure and uptake (inhalation and ingestion).

16.8 CASE STUDIES

This section will present examples of sustainable practices in the clothing and textile industries, focusing on consumer engagement and innovation (Table 16.2). Each case will be analyzed for its *approach, impact and replicability, offering practical insights* for readers. Designing clothing and textiles for circularity and sustainable life cycles has become more strategic in the textile sector. One approach is close-loop systems that endeavor to entertain materials in a cyclical system by collection of used materials, repairs, remanufacturing and recycling.

16.8.1 International Case Studies

16.8.1.1 H&M (Global)

H&M is an international fashion and design company with global impact that entails 4,000 stores, 79 markets and 60 online sale markets (H&M 2026). In 2024, the company used 89% recycled or sustainably sourced materials in its commercial products, including 29% recycled materials (H&M 2026). The company has a global in-store "garment-collecting program". Partnering with Looper Textile Co. since 2023, customers can drop off any brand of textile, which is subsequently sorted for resale, repurposing and recycling (Garment Collecting at H&M 2026). The system has collected over 172,000 tons of clothing and textiles since 2013, indicating that the clothing industry has the potential to encourage consumer

TABLE 16.2
International/African Case Studies—Approach, Impact and Replicability

Case	Approach	"Loop" Type	Indicative Impact (Qual.)	Replicability Notes
H&M garment collection	In-store take-back, sorting for reuse/recycling	Postconsumer material loop	Scale for collection; awareness	Needs local sorters/offtakes
Patagonia Worn Wear	Repair/resale of own brand	Lifetime extension loop	Fewer new units; less washing	Brand-anchored; strong comms
TCB (South Africa)	Retail returns → resell/repair/remake	Retail waste loop	Diversion + livelihoods	Needs retailer memorandum of understanding (MOU), training
Hawassa (Ethiopia)	Park-level ZLD for dye/finish	Water loop	Compliance; reuse of water	

participation in a closed-loop system if convenient, visible and incentivized. Although only a small percentage of the system is fiber-to-fiber recycling, there is an increase in feedstock for recycling and capacity for future extended producer responsibility (EPR) obligations.

16.8.1.2 Patagonia (USA)

Patagonia is an international clothing brand that focuses on outdoor apparel (Patagonia 2026), including trail running, climbing and skiing. In 2012, the company introduced Worn Wear®, which started with swapping clothing and evolved into a pop-up to resell Patagonia clothing, as well as repairing existing Patagonia clothing (Patagonia Worn Wear 2026). Guides for do it yourself (DIY) repairs and product care are included in the list of offers. The impact of this initiative is the diversion of garments from landfills and incineration. Reusing and fixing garments result in less virgin material use. The company also engaged more with its consumers by hosting workshops and storytelling sessions. From a fiber perspective, this program provides an example of how increasing product lifetime decreases the possibility of new shedding sources and lowers the need for laundering of clothing because garments are more durable.

16.8.1.3 Taking Care of Business (South Africa)

Previously known as The Clothing Bank, Taking Care of Business (TCB) is a closed-loop social enterprise based in South Africa that repurposes retail returns and excess stock for local markets and livelihoods (Brand 2025). Programs offered include reselling clothing and homeware, training in repair and remaking garments and recycling of clothing waste materials. TCB has diverted about 23 million tons of items from landfills since 2010 and contributed to the growth of small enterprises, job creation and skills development. The impact on the sustainability of fiber is that the resale and repair of clothing have resulted in reduced demand for the production/importing of clothing and lowered laundry demand per wear.

16.8.1.4 Hawassa Industrial Park (Ethiopia)

The Hawassa Industrial Park (HIP) is an eco-industrial park situated in the Hawassa region of Ethiopia. The HIP includes a centralized zero-liquid discharge (ZLD) facility that treats and recycles processed wastewater from dyeing and finishing, creating a near-closed-loop system in the park (IPDC 2026). The ZLD removes solids and returns permeates to production plants. This system reduces the abstraction of freshwater and minimizes pollution in Lake Hawassa. The ZLD provides an enabling environment for tenants of the park to meet increasingly stringent international sustainability requirements, as well as creating opportunities to access international markets. The HIP is an example for governments on how to manage effluent, shared infrastructure finances and effective sludge management. The ZLD includes an advanced filtration system (membrane bioreactor) that captures fibers prior to discharge, providing an immediate benefit for microplastic pollution mitigation.

16.9 INNOVATIONS TO ADDRESS REDUCING MICROPLASTIC RELEASE INTO THE ENVIRONMENT

16.9.1 Opportunities and Challenges

The recognition of the need to transition toward a globally sustainable textile industry presents both opportunities and challenges. Increased public awareness of the impacts of plastics (Beaumont et al. 2019), technology innovations and new science and policy frameworks have enabled opportunities to reduce and repurpose production, consumption and end-of-life systems of microfibers. The opportunities lie with the clothing and textile industries to change with the environmental requirements. However, current industrial, societal and economic dysfunctional systems (including intensive natural-based extraction and production methods, international trade agreements, societal trends and disjunct governance) potentially inhibit progress and change toward sustainable practices. This section of the chapter discusses key opportunities that can enable a sustainable transition in the clothing and textile industries. It discusses potential technical, financial and societal obstacles that may inhibit change, and proposes strategic opportunities for industrial, societal and economic transformation.

16.9.2 Opportunities for Transformation

16.9.2.1 Innovation and Alternative Fibers—Development of Eco-Friendly Materials

Technological and innovative developments have created opportunities for new types and repurposed fibers. Biodegradable plastics are derived from biological or petroleum sources and feedstocks. If exclusively sourced from renewable resources, it is classified as bio-based biodegradable plastics. If sourced from nonrenewable fossils fuels, they are considered to be petroleum-based (Goel et al. 2021). Examples of bio-based, biodegradable plastics are the starch-based polyhydroxyalkanoates and sugar-based polylactic acid. Biodegradable plastics are designed to degrade more quickly than polymeric plastics and are intended to break down into natural particles (such as water, carbon dioxide and biomass).

Bioplastics are plastics derived from renewable organic material, but may not readily biodegrade (Jabeen et al. 2015). They include bio-based polyethylene terephthalate that is manufactured from ethanol sourced from sugarcane or other biomasses. Although sourced from a renewable source, it has the same chemical properties as PE and does not degrade rapidly (Tang and Zhou 2025). Although biodegradation is a challenge, these substitutes can lessen the need for polymers derived from fossil fuels, reduce embodied greenhouse gas emissions and, if they are biodegradable, they can reduce the environmental prevalence and impacts of fibers.

There is a need for innovation related to material production that focuses more on changing synthetic fibers that have low-shedding characteristics, such as closer filament bonding, higher yarn twist and surface treatments that reduce fibrillation. Factors influencing fiber shedding include type of washing (hand or machine), age of fabrics, water temperature, type of detergent, textile construction and chemical finishing (Vassilenko et al. 2021). All these factors are opportunities to (re-)design manufacturing and cleaning of clothing and textiles to reduce microfiber release.

16.9.2.2 Circular Economy and Product Design for Longevity

The highest potential for opportunities is to switch from the linear clothing and textile market that focuses on "take-make-waste" approaches to circular models. An economic model focusing on the circulation of clothing and textiles can entail the inclusion of more durable designs, ease of repair and design-for-disassembly (Jin et al. 2011). Such approaches will result in lowering the per-capita consumption of clothing and textiles and create new business opportunities and models that offer repair, resale, rental and subscription services that steer clothing and textile industries away from increasing production and fiber loss to the environment.

16.9.2.3 Opportunities in Laundering and Wastewater Treatment

Technological innovations in commercial and home washing and laundry equipment offer opportunities to mitigate fiber pollution, since laundry is a major source of microfiber pollution (Vassilenko et al. 2021). Wash fiber release rates can be reduced by drum and filtration systems, low-shear washing cycles, and appliance redesigns that reduce mechanical abrasion (McIlwraith et al. 2019). Such intervention can be incentivized by regulatory conditions, and industry-wide adoption of sustainable laundry retrofit filters or point-of-use capture devices (Browne et al. 2020).

WWTPs are major sources of fiber release into the aquatic environments (Xu et al. 2019). Fiber from WWTP effluent streams must be improved, and this is possible by establishing a minimum standard of fiber release in municipal areas. Further innovations are possible with improving the capacities of membrane bioreactors, tertiary filtering and targeted particle-capture technologies (Long et al. 2021). By developing innovations and technologies that extract fibers from sludge and treat biosolids can reduce the release of fibers into the environment.

16.9.2.4 Digital Platforms Promoting Sustainable Choices

Blockchain, material passports and digital product passports are examples of digital systems that create material life cycle data that enables a system to track various facets of fibers, from chemical inputs, the manufacturing processes, end-of-life pathways and international supply chains (Aklashie et al. 2026). Greater transparency within the clothing and textile industries will create an enabling downstream environment (consumers, recyclers and brands) that will promote well-informed decisions, validate claims and promote circular material flows (Fares et al. 2025).

Although the principles of sustainability include a circular economy (CE), the management and tracking of life cycles provide unique opportunities. An aspect of CE management about fibers is the opportunity to create transparency in supply chains and sustainability reporting. Markets are moving toward promoting ecolabels and sustainability standards that are supported by legislative frameworks (Plakantonaki et al. 2023). Factors of ecolabels include ecological responsibility at all stages of the life cycle of products, which includes the clothing and textile industries. For example, the "Ecolabel Index" is the biggest global ecolabel registry that includes 456 labels in 199 countries and 25 business sectors (Ecolabel Index 2025). Such initiatives create opportunities for uniform benchmarks created by standards.

16.9.2.5 Policy Instruments and Financing Mechanisms

EPR, eco-design rules, product labeling, public procurement systems and targeted subsidies for sustainable manufacturing are examples of policy levers that create opportunities for change (Tumu et al. 2023). Environmental externalities are internalized and financial incentives are created for creating durable clothing and lower-emission products through EPR systems that force companies to subsidize collection, sorting, recycling and safe disposal. The funds needed for plant improvements, chemical recycling facilities and increased recycling infrastructure can be raised through blended finance models (public–private partnerships, green bonds, concessional loans). Companies are encouraged to make longer-term sustainability investments when policy certainly lowers investor risk and indicates market demand for low-impact textiles.

16.9.2.6 Increased Consumer Awareness

Shifting consumer behavior is an opportunity to reduce fiber emissions and overall environmental effects. Household emissions can be considerably decreased by awareness programs that encourage customers to wash less frequently, utilize cold water cycles, wash complete loads, use fiber-capturing laundry accessories and choose long-lasting clothing (Vassilenko et al. 2021). Expanding the second-hand marketplaces, rental platforms and repair services creates opportunities for communities to have access to alternatives to fast-fashion consumption (Jin et al. 2011). Behavioral changes, when combined with inexpensive low-shedding product options, create opportunities for new markets in an enabling economy that include repair networks and resale platforms. Such economies foster awareness and collaboration among consumers and various stakeholders such as governments, retailers and brands. The impact of this may be increased reach and normalized sustainable consumption habits.

16.9.2.7 Cross-Sectoral Synergies and Co-Benefits

Beyond the reduction of microfibers, the sustainable textile transition offers additional benefits such as decreased water use and pollution during the dyeing process. It further enables lower greenhouse gas emissions through material substitution and circularity, better worker health due to a decrease in the use of hazardous chemicals and new local economic opportunities from repair and recycling businesses. The business case is strengthened, and stakeholder buy-in is expanded when initiatives are framed to capture numerous environmental and social advantages.

16.10 CHALLENGES

Despite the aforementioned prospects, a number of interconnected issues in the technological, institutional, economic and behavioral realms stand in the way of a sustainable textile sector.

16.10.1 Fragmented Global Supply Chains and Uneven Capacities

The clothing and textile industries are characterized by geographically dispersed and highly fragmented supply networks: fiber production, spinning, dyeing, assembly and finishing, sometimes in different regions, each with unique industrial capabilities, environmental impacts and regulatory frameworks (Cao et al. 2008). In addition to these sectors, the clothing and textile supply chain includes housing, transportation and sustenance (Mishra et al. 2025). Hence, this fragmentation negatively impacts traceability, standards enforcement and technology transfers. Given that a significant proportion of the clothing and textile industry is located in low- and middle-income countries, which often face limited investment, constrained technical and research capacities, and weakened enforcement capabilities, international agreements that transfer greater responsibilities to these countries may give rise to equity challenges.

Informal sectors of low- and middle-income countries are often significant contributors to the local economy. Despite this, these sectors sometimes lack official recognition, financial resources and safe working conditions. Co-design policies and inclusive economic strategies are necessary to integrate these actors into formal circular systems without uprooting livelihoods.

16.10.2 Technical Limits in Recycling and Mixed-Fiber Handling

Blended textiles and polymer combinations provide challenges for current recycling processes. Fiber length and quality are deteriorated by mechanical recycling, which frequently results in downcycled materials unfit for premium fabrics (Tumu et al. 2023). Mechanical recycling includes identification, sorting, washing, shredding and reprocessing of plastics. Developments in the mechanical recycling sector include development of automated sorting (equipment, spectroscopic methods and machine learning principles) (Lubongo et al. 2024). Despite these innovative approaches, these developments are mainly in the northern hemisphere and in developed countries. Chemical recycling has the potential to recover monomers or purified feedstocks, but it requires economies of scale to be cost-competitive, is energy- and capital-intensive, and is sensitive to impurities (finishes, dyes) (Hahladakis et al. 2018). Although sorting technologies (such as near-infrared and hyperspectral imaging) show promising potential, they are still not widely used at the necessary scale.

Blended clothing is frequently seen in consumer goods and performance wear (Pranta et al. 2024). However, blended clothing has serious circularity issues because it is still technically and financially challenging to separate polymers at scale without requiring extensive mechanical processing. Mixed-fiber clothing will remain a barrier to a closed-loop textile economy until separability is enhanced by design or new, scalable recycling techniques are launched.

16.10.3 High Cost of Sustainable Production

Many clothing companies, particularly in the fast-fashion industry, use low-margin, high-volume business methods that do not encourage investing in more expensive, sustainable materials or long-lasting designs. Customers often do not have options in selecting products and frequently have to make price-sensitive decisions due to a lack of options. The mass market creates demand not because of options to choose from, but because only low-quality and cheap products are available. Despite the potential for sustainable production options, supply chains focus on cost reductions, scales of production and preferential economic and legislative factors (government levies and subsidies) to maximize profits (or to break even).

Consumer options are further complicated by greenwashing and brand reputational issues (de Freitas Netto et al. 2020). There is a need for more verification systems to verify claims related to sustainability or biodegradability. The lack of established measures promotes weak claims that erode customer confidence and negatively affect markets.

16.10.4 Measurement, Standardization and Scientific Uncertainty

Methods for measuring fiber shedding, environmental concentrations and ecological and human health impacts are continuously changing among scientists (Gong and Xie 2020). Monitoring of microplastics is not conducted in national monitoring programs in most countries, and the collection, extraction and reporting of microplastics data (including fibers) are not standardized (Hung et al. 2021). Policies related to plastic (and fibers) are made more difficult by the absence of standardized assessment procedures, which also compromises study comparability (Nielsen et al. 2023; Farrelly et al. 2025). Regulatory and pollution management issues arise from uncertainties surrounding ecological thresholds and human health implications related to fibers (Barboza et al. 2018). Precautionary measures may be warranted, but industry and policymakers frequently wait for more precise dose–response data before investing significant financial resources in mitigation (Brander et al. 2024). Action may be slowed by this conflict between evidence-based regulation and caution.

16.10.5 Resistance to Behavioral Change

Overcoming industry-driven influences regarding low-cost and single-use products and marketing-driven consumption patterns is necessary to achieve significant changes in the clothing and textile industries. Changes to laundering procedures, expectations for the lifespan of clothing and acceptance of textiles with recycled content are needed but will likely not be actively pursued by the industry, as it may lead to higher costs. Frequent purchases and disposability that are sustained by convenience, fashion innovation, and social signaling and major barriers to behavioral change. Behavioral interventions can be successful, but to prevent placing undue burden on consumers alone, they must be maintained, culturally appropriate and supported by systemic improvements (such as easily available repair services, product labeling and affordable sustainable options) (de Freitas Netto et al. 2020).

16.10.6 Regulatory Gaps

Waste management, water quality, chemical safety, product standards and trade policy are just a few of the areas that entail the governance of plastic fiber and textile sustainability (Tiller et al. 2022; Farrelly et al. 2025). These areas are frequently managed by international Multilateral Environmental Agreements, national regulations and policies (Tiller et al. 2022), that include many organizations and involve various levels (and multiple departments) of government engagement (Farrelly et al. 2025). Inconsistent incentives and regulatory loopholes result from fragmented approaches. Because producing and consuming nations have different economic agendas and capacities, it is challenging to harmonize policy internationally. The UN global plastics treaty initiated in 2022 is aimed at being a legally binding international instrument to end plastic pollution, throughout the entire life cycle of plastics (Farrelly et al. 2025). In addition to international legal instruments, the adoption of localized circular solutions may also be hampered by trade policy if tariff and nontariff obstacles do not support sustainability objectives (Abalansa et al. 2020).

16.10.7 Social Justice and Equity Concerns

Transitions to sustainable practices may have unforeseen social effects. Stricter environmental regulations in producing nations, for example, could raise costs and jeopardize jobs if they are not backed by supportive policies (Willis et al. 2018). On the other hand, informal laborers who presently depend on textile recovery for their livelihoods may be displaced by the development of official recycling infrastructure (Tumu et al. 2023). In order to facilitate workforce changes into higher-value CE occupations, policies that protect and uplift workers, integrate social discussion and invest in skill training are necessary to ensure an equitable transition.

16.11 FUTURE OUTLOOK: INNOVATION AND SUSTAINABILITY

Looking ahead, emerging trends and technologies are set to reshape the clothing, textile and fashion industries. Prudent innovation that focuses on sustainable economies will influence sustainability frameworks and sustainability practices. It will require integrated, multilevel strategies that coordinate technology, politics, finance and behavior to address these issues. A feasible transition can be guided by the strategic initiatives listed below.

16.11.1 Policy Developments and Global Sustainability Frameworks

Coherent market signals are produced by policy packages that incorporate EPR, required product information (material passports), minimum durability requirements and incentives for recycling investments. Supply chains have time to adjust while receiving clear guidance thanks to phased deployment, which starts with transparent reporting and incremental performance requirements. Mutual recognition agreements or internationally coordinated frameworks (such as the plastics treaty) help level the playing field and reduce economic friction.

16.11.2 Innovations in Textile Recycling and Regenerative Design

Consortia of companies, suppliers, utilities and academics, as well as public–private partnerships, can reduce risks through innovation and technology development.

Economies of scale and diffuse costs can be maximized by shared chemical recycling plants, coordinated low-shedding material sourcing (mechanical recycling) and cooperative investment in sorting and recycling logistics. Incorporating members from informal sectors guarantees locally relevant solutions and protects livelihoods.

Downstream processing loads can be decreased by including circularity criteria—material simplicity, mono-material design (when feasible), detachable components and design-for-repair into product development. At the same time, procurement norms and specifications should encourage industry-wide use of low-shedding textile constructions and finishes. Product value and behavior modification are supported by labeling (ecolabeling) that informs consumers about best-care procedures (washing, drying) and shedding performance. The foundation of reliable ecolabels and procurement standards is the development of globally standardized testing procedures for shedding, durability and recyclability. Greenwashing is lessened and customers are empowered to choose verified low-impact products thanks to third-party certification and open reporting. Standards should be adaptable to different geographical and economic production conditions and continuously updated to reflect new scientific knowledge.

Scientific and technological innovations should drive regenerative designs. Loan guarantees, concessional capital and outcome-based financing are examples of blended finance tools that can help manufacturers and recyclers overcome initial cost obstacles in designs. Adoption can be accelerated by targeted grants for recycling infrastructure in producing nations, tax incentives for durable and sustainable product design and subsidies for appliance retrofits. When properly managed, EPR income can assist worker transitions through social programs, innovation awards and collection systems.

16.11.3 The Evolving Role of Consumers as Co-Creators of Sustainable Value

Campaigns to modify behavior should be backed by enabling infrastructure (simple repair services, accessible second-hand platforms), based on scientific evidence and be culturally sensitive. Social marketing that makes use of influencers and local conventions can change attitudes about repair, rental and reuse. Uptake of sustainable value propositions will require partnerships with retailers, marketing agencies and platforms that can include care instructions and environmental effect information at the time of purchase (Table 16.3).

16.11.4 Investments in Measurement, Monitoring and Research

It is important to strive for financing and implementation of standardized laboratory procedures, harmonized monitoring networks and studies on the effects on the environment and human health. Community science methods and open data platforms can increase the scope of monitoring and make information more accessible (Barrows et al. 2018). Scalable recycling methods for blended textiles, low-energy chemical recycling and life cycle assessments that incorporate microfiber emission indicators should be given top priority in research.

TABLE 16.3
Policy and Financing Levers Mapped to Actors and Time Horizons for Microfiber Mitigation and Circularity

Lever	Primary Actors	Time Horizon	Expected Outcome	Notes
Extended producer responsibility	Brands, PROs, regulators	3–5 years (phase-in)	Financed take-back, sorting, repair, safe recycling	Allocate funds for wastewater treatment plant (WWTP) upgrades and reuse/repair networks
Eco-design and durability standards	Standards bodies, brands	0–3 years (pilot → adoption)	Lower shedding; longer lifetimes; recyclability	Include shedding tests and design-for-disassembly
Digital product passports	Brands, tech providers, EU/UNEP	2–5 years	Traceability of materials and chemicals; anti-greenwashing	Link to care guidance and verified recycled content
Blended finance for recycling and WWTPs	Government, private capital	1–5 years	Scale fiber-to-fiber and tertiary capture	Use green bonds, concessional loans, guarantees
Consumer behavior programs	Retailers, NGOs, municipalities	0–2 years	Reduced wash frequency and heat; uptake of filters	Targeted, culturally sensitive campaigns

16.11.5 The Rise of Digital Fashion and Virtual Wardrobes

The clothing and textile industries have an opportunity to separate elements of style, identity and consumption from actual production to digital fashion and virtual wardrobes. In addition to creating new revenue streams, digital fashions (clothing and accessories) can be advertised in virtual environments, social media platforms, augmented reality overlays and in gaming or metaverse spaces. Such platforms also have the potential to minimize impacts on the environment by reducing material demand, physical waste and pressure on global supply chains. By permitting stylistic experimentation without the need to buy new pieces, virtual wardrobe management services further curb overconsumption by assisting users in finding, reusing and recomposing existing clothing digitally. When combined, these advancements produce an ecosystem of new consumer markets, business models and product innovation that can support initiatives to promote circularity and lower microfiber emissions.

Brands and startups are drawn to the various commercial vectors created by digital fashion (Global Fashion Agenda 2025). Designers can work with influencers to create avatar-only collections, release limited-edition digital drops and monetize the production of looks through subscription services, licensing and non-fungible token frameworks (Sereti et al. 2025). Without the lead times, raw material procurement and manufacturing emissions associated with physical releases, digital lines enable established clothing firms to quickly test products, segment markets and expand their brands into gaming and social media platforms. This feature is especially useful for lowering the churn of trend-driven quick fashion: if consumers' needs for novelty and self-expression can be met digitally, there may be less incentive to create fleeting physical clothing.

16.12 CONCLUSION: BALANCING AMBITION WITH PRAGMATIC PATHWAYS

A sustainable clothing and textile sector is both necessary and achievable, but the route to attain it is neither linear nor quick. Microfiber pollution offers a clear indication of the system's wider dysfunctions: materials and design choices that shed fibers, manufacturing and laundering that externalize costs, fragmented governance that struggles to align incentives, and social inequities that make some communities bear disproportionate risks. The cases and analyses in this chapter show that meaningful progress emerges when innovation, technological solutions, policy instruments, market signals and everyday practices are coordinated rather than pursued in isolation.

In practical terms, five levers must be advanced together: (1) materials and design: prioritize durability, mono-material or separable constructions and lower-shedding yarns and finishes; (2) process upgrades: scale wastewater and laundry to cut releases at source; (3) circular business models: normalize repair, resale and rental to lower throughput and extend lifetimes; (4) governance and finance: use EPR, eco-design rules verified product data (e.g., digital passports) and blended finance to de-risk investment; and (5) measurement and transparency: standardize shedding tests and monitoring and publish comparable scientific data to prevent greenwashing and enable accountability. These levers reduce microfiber emissions, conserve water and energy and support healthier labor and community outcomes.

A credible pathway forges actions with structural investments and just-transition safeguards:

- **Now (0–2 years)**: Fit simple capture technologies (appliance filters, tertiary filtration where feasible); publish care guidance that measurably reduces shedding; specify longer-lasting, repairable designs; and establish city-level take-back and sorting where informal and formal actors can collaborate.
- **Next (3–5 years)**: Build sorting and recycling capacity matched to the real waste mix (cotton rich, synthetics, blends); adopt common test methods and labels for durability, shedding and recyclability; phase in EPR with funds earmarked for collection, repair and safe recycling; and link public procurement to verified low-impact textiles.
- **Beyond (to 2030 and after)**: Scale advanced fiber-to-fiber pathways, where net benefits are clear; redesign product portfolios away from fast-churn items; and embed worker protections, upskilling and financing so that circularity creates opportunity rather than displacing livelihoods.

As regulation improves and technology advances, policies must ensure equity, safe jobs, fair pricing and participation for informal sectors and small enterprises, especially in producer countries and coastal communities that are already burdened by environmental risks. Social justice requires inclusive co-design of standards and support programs.

The textile sector has the potential to change because designers adopt sustainable approaches by using new and different materials; manufacturers invest in capture and reuse; retailers focus on longevity over rapid turnover of stock; government agencies set and enforce clear rules; financiers back infrastructure and innovation; educators shape new curricula; and consumers practice care, repair and recirculation. If these roles align, we can reduce microfiber contamination, lighten the industry's ecological footprint and build resilient circular systems that protect livelihoods and environments. The invitation is simple and urgent: wear the change—and help design the system that makes it last.

REFERENCES

Abalansa, S., Mahrad, B. E., Vondolia, G. K., Icely, J. and Newton, A. 2020. The marine plastic litter issue: A social-economic analysis. *Sustainability* 12(20): 20.

Aklashie, S., Agyekum, K., Tchao, E. T., Adinyira, E. and Nani, G. 2026. Blockchain-digital twin integration in construction material passports: Preparing for digital product passport implementation. *Built Environment Project and Asset Management* 21: 1–18.

Akyıldız, S. H., Sezgin, H., Eniş, İ. Y., Balestra, V., Marini, P. and Bellopede, R. 2025. From production to pollution: A review of microfiber release mechanisms and mitigation strategies in the textile industry. *The Journal of The Textile Institute,* 1–21. https://doi.org/10.1080/00405000.2025.2547125.

Alimi, O. S., Budarz, J. F., Hernandez, L. M. and Tufenkji, N. 2018. Microplastics and nanoplastics in aquatic environments: Aggregation, deposition, and enhanced contaminant transport. *Environmental Science and Technology* 52(4): 1704–1724.

Barboza, L. G. A., Vethaak, A. D., Lavorante, B. R. B. O., Lundebye, A. -K. and Guilhermino, L. 2018. Marine microplastic debris: An emerging issue for food security, food safety and human health. *Marine Pollution Bulletin* 133: 336–348.

Barrows, A. P. W., Christiansen, K. S., Bode, E. T. and Hoellein, T. J. 2018. A watershed-scale, citizen science approach to quantifying microplastic concentration in a mixed land-use river. *Water Research* 147: 382–392.

Batool, I., Qadir, A., Levermore, J. M. and Kelly, F. J. 2021. Dynamics of airborne microplastics, appraisal and distributional behaviour in atmosphere: A review. *Science of the Total Environment* 806(4): 150745.

Beaumont, N. J., Aanesen, M., Austen, M. C., Börger, T., Clark, J. R., Cole, M., Hooper, T, Lindeque, P. K., Pascoe, C. and Wyles, K. J. 2019. Global ecological, social and economic impacts of marine plastic. *Marine Pollution Bulletin* 142: 189–95.

Boucher, J., and Friot, D. 2017. *Primary Microplastics in the Oceans: A Global Evaluation of Sources.* International Union for Conservation of Nature. https://doi.org/10.2305/IUCN.CH.2017.01.en.

Brand, H. 2025. Reuse, repair, empower: TCB's circular economy impact. *TCB|Taking Care of Business.* https://www.tcb.org.za/taking-care-of-the-circular-economy-empowering-people-reducing-waste/ [Accessed 10 January 2026].

Brander, S., De Falco, F., Kelly, M., Weis, J., Almroth, B. C., Baztan, J., Sparks, C., Courtene-Jones, W. and Thompson, R. C. 2024. Microfibers from textiles - A key source of microplastics to the environment: fate, effects, and mitigation strategies: Key sources of microplastics to the environment: Working papers developed by members of the scientists' coalition. September. https://ikhapp.org/wp-content/uploads/2024/09/Brander-et-al.-2024.-Microfibres-from-textiles-a-key-source-of-microplastics-to-the-environment-fate-effects-and-mitigation-strategies.pdf [Accessed 10 January 2026].

Browne, M. A., Niven, S. J., Galloway, T. S., Rowland, S. J. and Thompson, R. C. 2013. Microplastic moves pollutants and additives to worms, reducing functions linked to health and biodiversity. *Current Biology* 23(23): 2388–2392.

Browne, M. A., Ros, M. and Johnston, E. L. 2020. Pore-size and polymer affect the ability of filters for washing-machines to reduce domestic emissions of fibers to sewage. *PLOS ONE* 15(6): e0234248.

Cao, N., Zhang, Z., To, K. M. and Ng, K. Po. 2008. How are supply chains coordinated?: An empirical observation in textile-apparel businesses. *Journal of Fashion Marketing and Management* 12(3): 384–397.

Chan, E. M. H., and Au, K. F. 2007. Determinants of China's textile exports: An analysis by gravity model. *The Journal of the Textile Institute* 98(5): 463–469.

Chen, F., Ahmad, S., Jiang, G. and Chen, J. 2023. Factors affecting textiles products exports of major producers: A gravity model approach. *Sage Open* 13(4). https://doi.org/10.1177/21582440231213688

Chen, G., Feng, Q. and Wang, J. 2020. Mini-review of microplastics in the atmosphere and their risks to humans. *Science of the Total Environment* 703(10): 135504.

Davies, H. L., Robb, H., Cox, K. D., Covernton, G. A., Eastham, T. M., Alexander, H. J. and Juanes, F. 2021. A preliminary analysis of ingestion and egestion of microplastic fibers in the acorn barnacle *Balanus glandula*. *Journal of Experimental Marine Biology and Ecology* 542–543: 151589.

Duan, Y., Xiong, D., Wang, Y., Zhang, Z., Li, H., Dong, H. and Zhang, J. 2020. Toxicological effects of microplastics in *Litopenaeus vannamei* as indicated by an integrated microbiome, proteomic and metabolomic approach. *Science of the Total Environment* 761: 143311.

Ecolabel Index. 2025. All ecolabels on textiles, Ecolabel Index. https://www.ecolabelindex.com/ecolabels/?st=category,textiles [Accessed 15 January 2026].

Fares, N., Islam, M. S., Jauhar, S. K. and Kucukaltan, B. 2025. Towards an international digital product passport: The new paradigm of a worldwide circular economy. *Circular Economy and Sustainability* 5(6): 5475–5495.

Farrelly, T., Brander, S., Thompson, R. and Almroth, B. C. 2025. Independent science key to breaking stalemates in global plastics treaty negotiations. *Cambridge Prisms: Plastics* 3: e6.

Freitas Netto, V. de, S., Sobral, M. F. F., Ribeiro, A. R. B. and da Luz Soares, G. R. 2020. Concepts and forms of greenwashing: A systematic review. *Environmental Sciences Europe* 32(1): 19.

Garment Collecting at H& M. 2026. https://www2.hm.com/en_ph/sustainability-hm/services/garment-collecting.html [Accessed 15 January 2026].

GESAMP. 2019. Guidelines for the monitoring and assessment of plastic litter in the ocean. *GESAMP Reports & Studies* 99: 130.

Global Fashion Agenda. 2025. Fashion CEO Agenda 2025. https://globalfashionagenda.org/resource/fashion-ceo-agenda-2025/ [Accessed 16 January 2026].

Goel, V., Luthra, P., Kapur, G. S. and Ramakumar, S. S. V. 2021. Biodegradable/bio-plastics: Myths and realities. *Journal of Polymers and the Environment* 29(10): 3079–104.

Gong, J. and Xie, P. 2020. Research progress in sources, analytical methods, eco-environmental effects, and control measures of microplastics. *Chemosphere* 254: 126790.

H&M. 2026. About us, H&M Group. https://hmgroup.com/about-us/ [Accessed 16 January 2026].

Hahladakis, J. N., Velis, C. A., Weber, R., Iacovidou, E. and Purnell, P. 2018. An overview of chemical additives present in plastics: Migration, release, fate and environmental impact during their use, disposal and recycling. *Journal of Hazardous Materials* 344: 179–199.

Holmes, L. A., Turner, A. and Thompson, R. C. 2012. Adsorption of trace metals to plastic resin pellets in the marine environment. *Environmental Pollution* 160(1): 42–48.

Hung, C., Klasios, N., Zhu, X., Sedlak, M., Sutton, R. and Rochman, C. M. 2021. Methods matter: Methods for sampling microplastic and other anthropogenic particles and their implications for monitoring and ecological risk assessment. *Integrated Environmental Assessment and Management* 17(1): 282–291.

IPDC. 2026. Industrial Parks Development Corporation. https://www.ipdc.gov.et/ [Accessed 18 January 2026].

Jabeen, N., Majid, I. and Nayik, G. A. 2015. Bioplastics and food packaging: A review. *Cogent Food & Agriculture* 1(1): 1117749.

Jin, G. H., Cao, H., Bennett, J., Helmkamp, C. and Farr, C. 2011. Application of design for disassembly in men's jacket: A study on sustainable apparel design. *International Journal of Clothing Science and Technology* 23(2–3): 83–94.

Joo, S. H., Liang, Y., Kim, M., Byun, J. and Choi, H. 2021. Microplastics with adsorbed contaminants: Mechanisms and treatment. *Environmental Challenges* 3: 100042.

Shirvanimoghaddam, K., Motamed, B., Ramakrishna, S. and Naebe, M. 2020. Death by waste: Fashion and textile circular economy case. *Science of The Total Environment* 718: 137317.

Lim, J., Choi, J., Won, A., Kim, M., Kim, S. and Yun, C. 2022. Cause of microfibers found in the domestic washing process of clothing; Focusing on the manufacturing, wearing, and washing processes. *Fashion and Textiles* 9(1): 24.

Long, Y., Zhou, Z., Yin, L., Wen, X., Xiao, R., Du, L., Zhu, L., Liu, R., Xu, Q., Li, H., Nan, R. and Yan, S. 2021. Microplastics removal and characteristics of constructed wetlands WWTPs in rural area of Changsha, China: A different situation from urban WWTPs. *Science of the Total Environment* 811: 152352.

Lubongo, C., Daej, M. A. A. B. and Alexandridis, P. 2024. Recent developments in technology for sorting plastic for recycling: The emergence of artificial intelligence and the rise of the robots. *Recycling* 9(4): 59.

McIlwraith, H. K., Lin, J., Erdle, L. M., Mallos, N., Diamond, M. L. and Rochman, C. M. 2019. Capturing microfibers: Marketed technologies reduce microfiber emissions from washing machines. *Marine Pollution Bulletin* 139: 40–45.

Mehdikhani, M., Gorbatikh, L., Verpoest, I. and Lomov, S. V. 2019. Voids in fiber-reinforced polymer composites: A review on their formation, characteristics, and effects on mechanical performance. *Journal of Composite Materials* 53(12): 1579–669.

Mishra, A., Soni, G., Ramtiyal, B., Dhaundiyal, M., Kumar, A. and Sarma, P. R. S. 2025. Building risk mitigation strategies for circularity adoption in Indian textile supply chains. *Annals of Operations Research* 355(2): 1649–1684.

Napper, I. E., Parker-Jurd, F. N. F., Wright, S. L., and Thompson, R. C. 2023. Examining the release of synthetic microfibers to the environment via two major pathways: Atmospheric deposition and treated wastewater effluent. *Science of the Total Environment* 857: 159317.

Nielsen, M. B., Clausen, L. P. W., Cronin, R., Hansen, S. F., Oturai, N. G. and Syberg, K. 2023. Unfolding the science behind policy initiatives targeting plastic pollution. *Microplastics and Nanoplastics* 3(1): 3.

OEC. 2026. Textiles in China trade. The observatory of economic complexity. https://oec.world/en/profile/bilateral-product/textiles/reporter/chn [Accessed 19 January 2026].

Patagonia. 2026. Patagonia outdoor clothing and gear. https://www.patagonia.com/home/ [Accessed 19 January 2026].

Patagonia Worn Wear. 2026. Used Patagonia® clothing and gear. https://wornwear.patagonia.com/ [Accessed 18 January 2026].

Plakantonaki, S., Kiskira, K., Zacharopoulos, N., Chronis, I., Coelho, F., Togiani, A., Kalkanis, K. and Priniotakis, G. 2023. A review of sustainability standards and ecolabeling in the textile industry. *Sustainability* 15(15): 11589.

Pranta, A. D., Rahaman, Md. T., Repon, Md. R. and Shikder, A. A. R. 2024. Environmentally sustainable apparel merchandising of recycled cotton-polyester blended garments: Analysis of consumer preferences and purchasing behaviors. *Journal of Open Innovation: Technology, Market, and Complexity* 10(3): 100357.

Prata, J. C. 2018. Airborne microplastics: Consequences to human health? *Environmental Pollution* 234: 115–126.

Rashid, Md. E., Islam, Md. A., Kanon, T. A., Khan, Md. R., Uddin, Md. B., Haque, R. U., Tonmoy, Md. W., Hasan, Md. R. and Haque, Md. M. 2024. Valorization of dyed brush fiber waste through production of upcycled mélange yarn: A sustainable approach. *Journal of Cleaner Production* 447: 141464.

Sereti, Z., Mavrikos, E., Cholevas, C. and Tsekouras, G. E. 2025. Green NFTs: Technologies related to energy-efficient non-fungible tokens. *Information* 16 (4): 305.

Simmerman, C. B. and Wasik, J. K. C. 2020. The effect of urban point source contamination on microplastic levels in water and organisms in a cold-water stream. *Limnology and Oceanography Letters* 5: 137–146.

Simpson, J. and Fiori, L. A. 1974. An Analysis of carding efficiency and processing performance of cotton /polyester blends. *Textile Research Journal* 44(10): 813–822.

Sudheshna, A. A., Srivastava, M. and Prakash, C. 2022. Characterization of microfibers emission from textile washing from a domestic environment. *Science of the Total Environment* 852: 158511.

Tang, K. H. D. 2023. State of the art in textile waste management: A review. *Textiles* 3(4): 454–467.

Tang, K. H. D. and Zhou, J. 2025. Ecotoxicity of biodegradable microplastics and bio-based microplastics: A review of in vitro and in vivo studies. *Environmental Management* 75(3): 663–679.

Tiller, R., Booth, A. M. and Cowan, E. 2022. Risk perception and risk realities in forming legally binding agreements: The governance of plastics. *Environmental Science & Policy* 134: 67–74.

Toldy, A., Szolnoki, B. and Marosi, G. 2011. Flame retardancy of fiber-reinforced epoxy resin composites for aerospace applications. *Polymer Degradation and Stability* 96(3): 371–376.

Tumu, K., Vorst, K. and Curtzwiler, G. 2023. Global plastic waste recycling and extended producer responsibility laws. *Journal of Environmental Management* 348: 119242.

Vassilenko, E., Watkins, M., Chastain, S., Mertens, J., Posacka, A. M., Patankar, S. and Ross, P. S. 2021. Domestic laundry and microfiber pollution: Exploring fiber shedding from consumer apparel textiles. *PLoS One* 16(7): e0250346.

de Villiers, S. 2019. Microfiber pollution hotspots in river sediments adjacent to South Africa's coastline. *Water SA* 45(1): 97.

Wagner, M. 2022. Solutions to plastic pollution: A conceptual framework to tackle a wicked problem. In *Microplastic in the Environment: Pattern and Process* (Ed. Bank, M. S.), Springer International Publishing. 333–352.

Wan, Y., Chen, X., Liu, Q., Hu, H., Wu, C. and Xue, Q. 2022. Informal landfill contributes to the pollution of microplastics in the surrounding environment. *Environmental Pollution* 293: 118586.

Wang, Z., Dong, H., Wang, Y., Ren, R., Qin, X. and Wang, S. 2020. Effects of microplastics and their adsorption of cadmium as vectors on the cladoceran *Moina monogolica* Daday: Implications for plastic-ingesting organisms. *Journal of Hazardous Materials* 400: 123239.

Willis, K., Maureaud, C., Wilcox, C. and Hardesty, B. D. 2018. How successful are waste abatement campaigns and government policies at reducing plastic waste into the marine environment? *Marine Policy* 96: 243–249.

Xu, X., Jian, Y., Xue, Y., Hou, Q. and Wang, L. P. 2019. Microplastics in the wastewater treatment plants (WWTPs): Occurrence and removal. *Chemosphere* 235: 1089–1096.

17 Microplastics Release and Its Effect during Tumble Drying

Rasiah Ladchumananandasivam,
Iris Oliveira da Silva, and
Clovis Medeiros de Bezerra

17.1 INTRODUCTION

Microplastics (MPs) are generally defined as polymer particles ranging between 1 μm and 5 mm in size, released from primary source (manufactured at small size) or secondary source (fragmented from larger plastics). They are found across environmental compartments, water, air, and soil, and pose risks to ecosystems and human health (Sheikhi et al. 2025; Hazlehurst et al. 2023; Periyasamy and Tehrani-Bagha 2022).

Textiles (especially synthetic fabrics such as polyester, polyamide, and acrylic) are major contributors to MP pollution. The process of wearing, washing, and disposal of textiles is well recognized, but less attention has been paid to tumble drying as a release pathway. Though many studies focus on washing (laundering), recent research indicates that tumble drying may emit large numbers of microfibers (MFs), a form of MPs into the air or exhaust to outdoors (EEA 2022; Cummins et al. 2023; Kärkkäinen and Sillanpää 2021; Dris et al. 2016; WHO 2019; OECD 2022).

The rationale for this chapter is that tumble drying is widespread in many parts of the world, particularly in countries with cold or humid climates, or where indoor space or climate conditions make outdoor drying difficult or impractical. Understanding how much MPs are released during drying, the mechanisms by which release occurs, what environmental and health pathways result, and how to mitigate these emissions are important for improving sustainability and public health.

This chapter seeks to

- define the mechanisms by which MPs (especially MFs) are released during tumble drying;
- quantify what is known so far about rates of release (number, mass) under different conditions;
- examine how dryer type, fabric type, load, lint filter, and lint disposal affect emissions;
- discuss environmental and human exposure pathways from drying emissions; and
- review mitigation strategies, policy implications, and research gaps.

DOI: 10.1201/9781003689225-17

17.2 MECHANISM OF MPS RELEASE IN TUMBLE DRYING

Tumble drying causes MP release via several overlapping physical, thermal, and mechanical processes. In the following, we detail the main contributing mechanisms (Table 17.1).

Tumble drying mobilizes fibers through combined mechanical and thermal/moisture effects. Mechanical abrasion arises from garment–garment and garment–drum contact, especially for fabrics with raised or brushed surfaces. Loop pull-out, fiber-end breakage, and surface fibrillation are common detachment modes. Thermal exposure and moisture gradients influence polymer softening and repeated swelling/shrinkage cycles, weakening interfiber bonds. Dryer design determines airflow and capture: vented dryers exhaust fiber-laden air outdoors; condenser and heat-pump dryers internalize moisture removal, shifting fibers toward lint screens and condensate streams. Cycle parameters—temperature, duration, load size, residual moisture—scale the total stress experienced by textiles. Fabric parameters such as fiber type (filament vs. staple), yarn structure, fabric construction (knit vs. woven), finishing chemistry, and garment age modulate both the shedding rate and size distribution of released MFs (Hossain et al. 2025; EEA 2022; Tao et al. 2012; De Falco et al. 2020).

17.2.1 Physical Abrasion and Fiber Fragmentation

- **Friction between fibers and drum or between garments**: The tumbling motion causes garments to rub against themselves, against the dryer drum walls, and the lint trap. This friction abrades fibers, causing tiny fragments or complete fibers to shear off. Synthetics especially are prone, because their fibers are less elastic in some conditions and might have weaker bonds when exposed to heat (Akyildiz et al. 2025; Sheikhi et al. 2025; Cummins et al. 2023).

TABLE 17.1
Mechanisms of Microplastic Release in Tumble Drying and Influencing Factors

Mechanism Category	Primary Factors	Expected Effect on MFs Release	Relative Importance
Mechanical abrasion	Garment–garment; garment–drum; surface pile/loops	Direct fiber detachment; coarse and fine fibers	High
Thermal and moisture	Temperature; residual water; swelling/shrinkage cycles	Weakening; increased brittleness under stress	Medium–high
Dryer design and airflow	Vented vs. condenser/heat-pump; lint-screen pore size; velocity	Governs capture vs. escape to vent	High
Fabric properties	Fiber type; filament/staple; structure; finishes; garment age	Different shedding rates and size distributions	High
Cycle parameters	Duration; load size; spin-dry efficiency; frequency of use	Cumulative exposure to stress	Medium

Source: Compiled by the authors from published literature.

- **Repeated mechanical stress over many cycles**: As garments are washed and dried multiple times, fiber strengths degrade, finishing chemicals wear off, and the microstructural integrity of the yarn (twist, staple vs. filament) is compromised. Thus, older garments may release more fibers during drying (Cummins et al. 2023; Hazlehurst et al. 2023; Kärkkäinen and Sillanpää 2021).

17.2.2 Thermal and Moisture Effects

- **High temperature exposure**: Drying often involves elevated temperatures (e.g., 60°C–80°C or more), which can soften or degrade synthetic polymers, reduce elasticity, and weaken fiber–fiber or fiber–matrix bonds. This makes fibers more susceptible to fragmentation under mechanical stress (Cummins et al. 2023; Hazlehurst et al. 2023).
- **Moisture content and residual dampness**: Fabrics exiting washing cycles into the dryer may still retain moisture. The combination of moisture plus heat can cause fiber swelling and contraction, which when combined with motion (tumbling) can lead to microcracking or fiber breakage. Natural fibers (or blends) may behave differently, sometimes shedding less or with different size distributions (Hazlehurst et al. 2023; Kärkkäinen and Sillanpää 2021).

17.2.3 Dryer Design, Lint Filtration, and Airflow

- **Lint-filter efficiency and maintenance**: Many dryers have built-in lint traps or filters intended to catch larger lint. The design (pore size, surface area, ease of cleaning) strongly influences how many MFs are captured vs. how many pass into the exhaust or vent. Studies show that filters are often not fine enough to capture the smallest fibers (Hazlehurst et al. 2023; Kärkkäinen and Sillanpää 2021).
- **Vented vs. condenser dryers**: Vented dryers exhaust air (and any entrained particles) to the outdoor environment; condenser or heat-pump dryers may recirculate air or condense vapor, sometimes removing moisture more gradually. The exhaust path and whether there is a vent outdoors affect how much MP escapes vs. being trapped (Hazlehurst et al. 2023).
- **Airflow and tumble motion parameters**: Drum rotation speed, direction changes, load fill fraction (how full the drum is), drum material, and internal agitation all affect how garments move, the intensity of rubbing, and likelihood of abrasion (Kapp and Miller 2020).

17.2.4 Fabric Properties and Textile Treatments

1. **Fiber type (synthetic, natural, blend)**: Polyester, polyamide, and acrylic tend to release more persistent synthetic fibers than natural fibers like cotton. Cotton fibers are biodegradable over time, though sometimes treated with chemical finishes which may delay or reduce decomposition (Tao et al. 2022; EEA 2022).
2. **Fabric structure and construction**: Knit vs. woven, fleece or raised surfaces (e.g., pile, brushed surfaces) vs. smooth surfaces, staple vs. filament fibers,

yarn twist, density, raised or brushed/loose surface fabrics tend to shed more fibers (Hossain et al. 2025; Napper and Thompson 2016; Pirc et al. 2016).

3. **Chemical finishes and pre-treatments**: Dyes, softeners, coatings, and finishing treatments (e.g., antiwrinkle, water repellent) may change how fibers interact, reduce or increase abrasion, change fiber surface roughness, or modify fiber strength. Also, pre-washing or wear may remove loose surface fibers, leading to lower shedding in later cycles (Hossain et al. 2025; Akyildiz et al. 2024).

17.2.5 Cycle Parameters: Duration, Temperature, Load, and Residual Water

1. **Drying time and temperature**: Longer drying cycles and higher temperatures can increase the total mechanical and thermal stress on fibers. However, some studies suggest that beyond a certain temperature or time, incremental increases produce diminishing returns in additional fiber release (Cummins et al. 2023; Hazlehurst et al. 2023).
2. **Load size and drum fill**: A dryer that is underloaded (few garments) may allow garments to tumble more freely and with more impact, leading to increased abrasion per garment; overloading may reduce free motion but increase compression and friction, also possibly increasing fiber breakage. Optimal load balancing is important (Cummins et al. 2023).
3. **Residual moisture (spin efficiency, pre-drying spin cycle)**: Wet garments with more residual water are heavier and may impact more within the drum; also, moisture–heat cycles cause fiber swelling/shrinkage, which can lead to microcracks (De Falco et al. 2020).

Figure 17.1 shows the flow of MPs from laundry to food chain. Conceptual flow of MPs released from textiles during laundering and drying, illustrating transport through environmental compartments and entry into the food chain.

17.2.5.1 MPs Release Mechanisms in a Tumble Dryer

The schematic shows garments inside a dryer drum show MPs release (arrows indicating heat + moisture + tumbling → fiber shedding), with paths for lint filter, vented exhaust, airborne release, and fiber fragments collecting in lint traps or through vent. Dryer drum → Vent exhaust (airborne) → local deposition/atmospheric transport; Lint filter → Solid waste (lint) → landfill/open burning; or Rinsed filter (waterborne) → sewer → waste water treatment plan → effluent/sludge.

17.3 QUANTIFICATION AND CASE STUDIES

17.3.1 Mechanism of Fiber Shedding

MFs are released from synthetic textiles through mechanical abrasion, detergent action, and water agitation in washing machines. Factors influencing fiber shedding include the following:

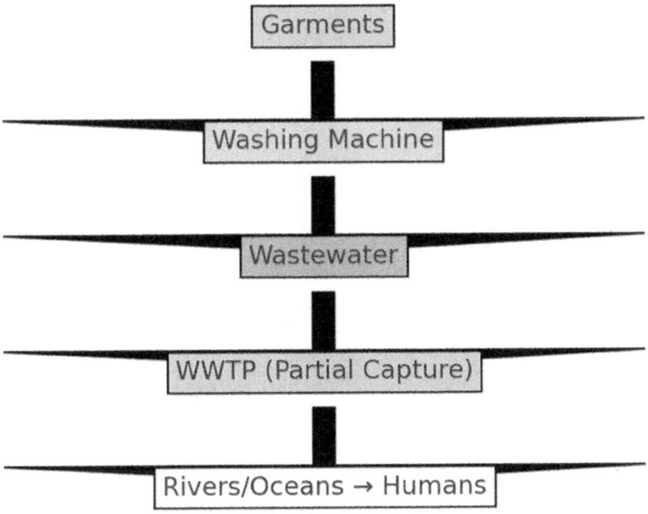

FIGURE 17.1 Laundry-to-plate microplastics pathway. (Created by the authors based on the synthesis of published literature.)

1. **Fabric type**: Loose-knit and low-quality fabrics shed more fibers.
2. **Washing parameters**: Higher water temperature, longer washing cycles, and stronger detergents increase fiber release.
3. **Garment age**: Older garments shed more fibers due to fiber weakening from repeated washes.
4. **Mechanical drying**: Tumble dryers can release MFs into the air, contributing to atmospheric MPs.

17.3.2 Shedding Rates by Fabric Types

Studies have quantified the number of MFs released per kilogram of fabric per wash. Table 17.2 summarizes typical shedding rates.

Research on tumble-drying emissions has accelerated in recent years, complementing the earlier focus on washing. Collectively, results show that vented dryers can release substantial quantities of airborne MFs to the outdoor environment,

TABLE 17.2
Fiber Shedding Rates from Different Fabrics

Fabric Types	Average Fibers Released (per kg per wash)	Sources
Polyester fleece	736,000	Napper and Thompson (2016)
Polyester (woven)	496,000	De Falco et al. (2019)
Acrylic	728,000	Browne et al. (2011)
Nylon	307,000	Hartline et al. (2016)
Blended fabrics	250,000–600,000	Gavigan et al. (2020)

TABLE 17.3
Evidence-Based Mitigation Options for MFs from Tumble Drying

Source	Dryer Type/Load	Metric	Result (Typical)	Notes
Tao et al. (2022)	Vented; polyester and cotton	Count	4.33×10^5–5.62×10^5 fibers/15 minutes	Airborne to ambient; polyester scaled with load mass.
Kapp and Miller (2020)	Vented; field sampling	Mass	35 ± 16 and 70 ± 77 mg across three cycles (two sites)	Deposition in snow near vents; local atmospheric fallout.
Cummins et al. (2023)	Vented vs. condenser; real loads	Fraction and pathway	~86% retained on lint filter; airborne (vented) and waterborne if filter rinsed under tap	User behavior governs pathway.
Kärkkäinen and Sillanpää (2021)	Sequential wash/dry; synthetics	Mass	10–1,700 mg/kg (first drying), decreasing thereafter	Gravimetric; point-of-use traps assessed.
Zambrano et al. (2019)	Household laundering and drying	Count (comparative)	Dry mechanical action produced substantial MF counts	Demonstrates significance of dry processes.

Methodological Note: Units differ (counts vs. mass) due to analytical choices (e.g., microscopy vs. gravimetry). Cross-study comparisons should normalize by load mass, cycle duration, and fiber identification (polymer type).

while condenser/heat-pump dryers primarily accumulate fibers on the lint filter and within internal condensate streams; in both cases, user practices (e.g., how the lint filter is cleaned) critically affect environmental release (De Falco et al. 2020; Kapp and Miller 2020; Kärkkäinen and Sillanpää 2021; Tao et al. 2022). Some of the findings are summarized in Table 17.3.

17.3.3 Measurement Approaches

Studies quantify release as (1) counts (number of fibers/cycle or per kilogram), (2) mass (mg/cycle or mg/kg), or (3) concentration in air or water captured from vents/condensers. Typical capture pathways include vent exhaust sampling, lint-trap mass/particle analysis, and controlled drying of known fabric loads (polyester, cotton, blends) in standardized cycles.

17.3.4 Key Studies and Magnitudes

1. Airborne counts from a vented dryer. Tao et al. (2022) measured 433,128–561,810 fibers in 15 minutes (polyester and cotton test loads) released to ambient air; extrapolated annual totals for a single household dryer reached ~120 million MFs (site-specific). Fiber abundance scaled with polyester load mass.

2. Mass and deposition near outdoor vents. Kapp and Miller (2020) sampled snow near dryer vents and measured 35 ± 16 mg (Site 1) and 70 ± 77 mg (Site 2) lint mass across three consecutive dry cycles, evidencing local deposition from vented exhaust.
3. Vented vs. condenser dryers; role of lint-filter cleaning. Cummins et al. (2023) showed that most fibers generated during drying accumulate on the lint filter (~86% ± 5.5% in real-world loads), but if consumers rinse the lint filter under tap water (as some manuals suggest), dryers become a significant waterborne source; vented units also emit airborne fibers through exhaust ducts.
4. Sequential washing and drying (gravimetric results). Kärkkäinen and Sillanpää (2021) reported 10–1,700 mg/kg (first drying) with declining shedding over subsequent cycles (polyester, polyamide, acrylic) and evaluated point-of-use traps.
5. Dry vs. wet mechanical processes. Zambrano et al. (2019) compared MFs from wet and dry actions during household laundering and drying, highlighting that dry mechanical agitation can rival or exceed some washing scenarios in fiber counts.
6. Reviews collate determinants of shedding (fiber type, fabric structure, finishing, age of garment) and emphasize the growing recognition of tumble drying as a distinct emission pathway requiring dedicated standards (Allen et al. 2019; Periyasamy and Tehrani-Bagha 2022; Akyildiz et al. 2025).
7. Tao et al. (2022) measured that for 1 kg of polyester fabrics, about 93,635 ± 17,026 MFs could be released in a 15-minute drying process. For 1 kg of cotton, it was 72,188 ± 11,813 MFs.
8. Cummins et al. (2023) found that vented tumble dryers are an important source of airborne MF pollution; built-in lint filters were often inefficient in capturing all MFs, so a portion of the fibers get vented outdoors.
9. Sheikhi et al. (2025) compared MF emissions from heat-pump-based tumble dryers vs. conventional ones, showing that dryer type impacts both release to air and on subsequent wash cycles.
10. According to a study by Boucher and Friot (2017), 35% of primary MPs in oceans originate from synthetic textiles, making domestic laundry a major contributor. Annually, it is estimated that 500,000–900,000 tons of synthetic fibers are released globally from laundry alone.
11. Research shows that integrated filters in washing machines can reduce MF emissions by up to 80%. In France, regulations introduced in 2025 mandate that all new washing machines must have MF filters, a policy expected to significantly reduce textile-derived pollution (Okyere et al. 2025).

17.4 ENVIRONMENTAL PATHWAYS

17.4.1 Airborne Emissions (Outdoor and Indoor)

Vented dryers propel warm, particle-laden air outdoors through ducts, releasing fibers to the near-house atmosphere where they can deposit on surfaces (e.g., snow/soil) or remain suspended and be transported regionally. This has been demonstrated by field sampling near vents and by controlling capture of vent exhaust; estimated annual emissions for a single household dryer have reached ~120 million fibers depending on usage and loads (Gavigan et al. 2020).

Indoor exposure can also occur if venting is faulty or absent; however, peer-reviewed quantifications remain limited compared to outdoor vent measurements. Continued work is needed to characterize fiber size distributions and polymer identities in indoor air linked to dryers, and to connect these data with inhalation risk frameworks.

17.4.2 Waterborne Releases via User Practices

Dryers do not discharge water, but waterborne release emerges when users rinse lint filters in sinks or flush condensate (from condenser/heat-pump units) without adequate filtration. Cummins et al. (2023) showed that because most shed fibers are retained on the lint filter, rinsing the filter can export a large pulse of MFs to sewers, bypassing the advantages of dry capture. This makes consumer guidance and appliance manuals a direct intervention point (City University of Hong Kong 2022).

17.4.3 Solid-Waste Pathways (Lint)

Lint removed from filters is typically disposed of as household solid waste. Where waste management relies on open dumping or open burning, lint becomes a secondary source (windblown or combusted). Even in engineered landfills, mechanical disturbance can entrain light fibers. Practical mitigation includes bagging the lint and discouraging open burning (Cummins et al. 2023).

17.5 MITIGATION STRATEGIES

We organize mitigation measures across consumer practice, appliance design, textile design, and policy/standards. Traditional practice like line-drying and regular lint-filter cleaning (dry disposal, not rinsing) remain reliable first steps.

17.5.1 Consumer Practices (Low-Cost, Immediate)

a. Prefer line-drying (weather/space permitting). Eliminates tumble-induced abrasion and airborne venting.
b. **Clean lint filters dry after every cycle**: Avoid rinsing under water to prevent transferring retained fibers into sewers. Evidence shows ~86% of fibers generated during drying are captured on the lint screen; rinsing exports this load to wastewater (Hazlehurst et al. 2023).
c. Optimal loads (neither under- nor overfilled) and shorter/lower temperature cycles, to reduce mechanical/thermal stress (supports lower shedding; see mechanistic basis and bench data across studies).
d. **Use of fabric conditioners/dryer sheets**: Single-variable testing indicates certain conditioning products, especially paired with finer-mesh lint screens, reduced airborne emissions in controlled experiments; real-world benefits depend on product, dosage, and machine design (Lant et al. 2020)

17.5.2 Appliance Design and Add-Ons

Finer lint-screen mesh: Reducing lint-screen pore size from 0.20 to 0.04 mm^2 reduced airborne release by ~35% in controlled testing; some commercial screens are ~1 mm^2, suggesting substantial room for improvement via original equipment manufacturer (OEM) redesign or retrofits (Lant et al. 2020).

Vented vs. condenser/heat-pump dryers: Vented units release fibers into the external environment; condenser/heat-pump designs retain more fibers internally (screen/condensate), shifting the pathway toward solid waste (lint) and household wastewater (if rinsed). Labels/manuals should discourage rinsing of screens and specify proper lint disposal (Kärkkäinen and Sillanpää 2021).

Secondary filtration on vents: Research groups and appliance developers have proposed auxiliary vent filters; preliminary work and press communications recommend smaller mesh and additional stages to curb airborne emissions (City University of Hong Kong 2022).

17.5.3 Textile/Material Design (Source-Directed)

Low-shedding constructions (tighter knits/weaves, continuous filament yarns, minimized raised/pile surfaces) and durable finishing reduce fiber loss during both washing and drying. Systematic reviews identify yarn/fabric parameters that minimize shedding (Hazlehurst et al. 2023; Napper and Thompson 2016; Zambrano et al. 2019).

Polymer selection and recycling trade-offs show that while recycled polyester advances circularity, recent studies caution that long-term shedding/aging behaviors still demand scrutiny and robust standards (Sheikhi et al. 2025; Akyildiz et al. 2024).

17.5.4 Policy, Standards, and Guidance

Standards development: The ISO 4484 series address material loss during washing, a foundation for harmonizing textile shedding tests. Comparable dryer-specific test methods would enable performance targets for lint capture and vent emissions (SGS 2023).

Regulatory signals: France's law mandating MF filters in new washing machines (from 2025) exemplifies a design-for-environment approach; analogous requirements could be explored for dryers, factory-fitted finer-mesh lint screens; vent-end MF traps; clear no-rinse guidance (EEA 2022; Okyere et al. 2025). Table 17.4 shows mitigation options for MFs from tumble drying.

Public communication: Clear consumer messaging—bag lint, do not rinse, check and replace worn seals/ducts—offers immediate reductions at scale (Cummins et al. 2023; Hazlehurst et al. 2023).

17.6 EFFECTS ON HEALTH AND ENVIRONMENT

17.6.1 Human Exposure via Inhalation (Near-Source and Indoor/Outdoor)

Vented tumble dryers release warm, particle-laden air outdoors; field and laboratory studies show substantial MF loads in vent exhaust and local atmospheric deposition near vents. In controlled trials, a single vented dryer emitted ~ 430,000–560,000 fibers

TABLE 17.4
Evidence-Based Mitigation Options for MFs from Tumble Drying

Level	Measure	Expected Effect	Notes
Consumer	Line-drying/sun-drying	↓↓↓ airborne and abrasion-driven shedding	Eliminates tumbling; climate/space-dependent
Consumer	Clean lint screen, dry after each cycle; do not rinse	↓ air and waterborne pathways	Most fibers retained on screen
Consumer	Optimize load size; lower temperature/time	↓ mechanical/thermal stress	Consistent with mechanism and tests
Appliance	Finer lint-screen mesh	↓ through-pass to vents (~35% in tests)	Pore size reductions improve capture
Appliance	Condenser/heat-pump designs	Shift to lint/condensate pathways	Requires clear disposal guidance
Appliance	Auxiliary vent filters (vented units)	↓ vent emissions	Pilot concepts; smaller mesh
Textile	Tight weaves; filament yarns; durable finishes	↓ shedding at source	Systematic review consensus
Policy/ Standards	EPR; dryer-specific tests; labels	System-level reduction	Builds on ISO 4484 and EU policy

in 15 minutes, with annual household totals estimated on the order of 10^8 fibers depending on usage and load mass. These emissions included both polyester and cotton fibers, with polyester emissions scaling with load mass. Near vents, mass deposition of lint has been measured in tens of milligrams across a few cycles, confirming localized fallout (Cummins et al. 2023; De Falco et al. 2020).

Indoors, faulty venting or recirculating systems can elevate airborne concentrations. Although comprehensive exposure assessments linked specifically to dryers are still limited, broader indoor studies using a breathing thermal manikin indicate that people readily inhale airborne MPs in typical homes, with measured indoor concentrations from ~1.7 to 16.2 particles/m^3 (≥11 µm measured via micro-fourier transform infrared spectroscopy-µFTIR imaging). These data, coupled with dryer-vent findings, point to a plausible near-source inhalation pathway (Boucher and Friot 2017; Allen, et al. 2019).

17.6.2 Evidence of MPs in Human Tissues (Weight-of-Evidence)

Beyond exposure, multiple biomonitoring studies have reported MPs in human lung tissue (µFTIR-identified polymers including polypropylene and polyester), placentas, and blood, demonstrating bioavailability and tissue residence of plastic particles, though exposure sources are not uniquely attributable to dryers. Toxicological reviews on airborne MPs suggest potential mechanisms including particle overload, oxidative stress, and translocation across epithelia, with risks modulated by particle size, shape (fibers), and chemistry (Periyasamy and Tehrani-Bagha 2022; Akyildiz et al. 2024). At present, quantitative risk estimates specific to tumble-dryer exposure remain uncertain; nevertheless, minimizing avoidable emissions aligns with precautionary environmental health principles (Akyildiz et al. 2024).

17.6.3 Environmental Fate and Ecosystem Effects

Airborne fibers from dryer vents can deposit locally (e.g., on snow/soil) or remain suspended and transport regionally, joining broader atmospheric MPs fluxes. For waterborne pathways, condenser/heat-pump dryers and user actions (notably rinsing lint filters under the tap) can transfer captured fibers to sewers, where a fraction may pass through wastewater treatment or be redistributed via sludge application (Cummins et al. 2023). Given evidence that tumble drying can emit magnitudes comparable to laundering under some conditions, both air and water vectors merit attention in mitigation design (Gavigan et al. 2020; Kapp and Miller 2020).

17.7 OPPORTUNITIES AND CHALLENGES

17.7.1 Opportunities

17.7.1.1 Traditional Practices That Work

1. **Line-drying/sun-drying where climate and space allow**: Eliminates tumble-induced abrasion and vent exhaust altogether; fits naturally with low-energy, low-emission household routines.
2. **Diligent lint care**: Clean lint screens dry after every cycle; do not rinse under the tap—most fibers shed in drying are already on the screen, so rinsing moves them into wastewater. Evidence from real-load tests shows ~86% of generated fibers are retained on lint filters (Lant et al. 2020; Kapp and Miller 2020).

17.7.1.2 Appliance Design and OEM Upgrades

1. Finer-mesh lint screens and better airflow management to cut through-pass to vents.
2. Condenser/heat-pump architectures that internalize moisture removal and reduce direct venting can shift releases from airborne to solid-waste lint, which households can bag and bin responsibly (Lant et al. 2020; Okyere et al. 2025).

17.7.1.3 Standards as Foundations

The ISO 4484 series provide reference methods for material loss during washing (Part 1), qualitative/quantitative analysis of MPs in various matrices (including air and wastewater) (Part 2), and mass released from textile end-products via the washing outlet (Part 3). These give laboratories a common language and data structure to build upon (ISO 2023; SGS 2023).

17.7.1.4 Policy Momentum

1. France has mandated MF filters on all new washing machines from 2025—a clear precedent that could inform dryer-side lint capture policies and consumer-facing guidance (Okyere et al. 2025).

2. The EU Strategy for Sustainable and Circular Textiles (2022) and the revision of the Waste Framework Directive introduce mandatory, harmonized Extended Producer Responsibility (EPR) schemes for textiles, shifting costs of collection/sorting/recycling to producers and encouraging design for durability/low shedding. Parliament's approvals and briefings through 2024–2025 outline implementation (EEA 2022).

17.7.1.5 Research-to-Practice Pathways

Vent-end filters, finer lint screens, and advisory labels/manuals (e.g., "bag lint; do not rinse") are near-term, low-friction interventions aligned with household traditions of care and maintenance—easy to scale if coupled with standards and EPR incentives (EEA 2022; Okyere et al. 2025).

17.7.2 Challenges

1. **Measurement heterogeneity and comparability**: Current tumble-dryer studies report counts vs. mass, with differing load masses, fabrics, and cycle parameters; unlike washing, there is no widely adopted international method specifically for dryer-vent MF emissions. ISO 4484 formalizes washing and analysis, but dryer-vent testing remains a methodological gap (ISO 2023).
2. **Trade-offs and rebound**: Heat-pump/condenser units can reduce airborne emission but shift fibers to lint/condensate; without proper disposal guidance, users may still rinse screens, exporting fibers to sewers (Gavigan et al. 2020).
3. **Costs, adoption, and equity**: Finer screens/retrofits introduce costs; in humid/cold regions, line-drying is not always practical. Policies must respect household realities and provide clear instructions rather than assuming optimal behavior.
4. **Policy timing and global alignment**: While the EU is advancing EPR for textiles, a global plastics treaty remains under negotiation—with timetable slippage and unresolved issues (production caps, additives). Dryer-specific provisions are unlikely without stronger evidence and harmonized tests (UNEP 2025; EEA 2022).

17.8 FUTURE OUTLOOK

17.8.1 Standardization Roadmap (Short–Medium Term)

1. Develop a dryer-specific emission method (vent-stack capture, isokinetic sampling, particle identification/count/mass normalized by kg, and cycle time), building on ISO 4484-2's analytical toolkit and ISO 6330's defined drying procedures. This would allow performance targets for lint capture and vent emissions in product design and regulation (ISO 2023).
2. Reference loads and fabrics (polyester filament vs. staple; knits vs. wovens; brushed fleece) represent high- and low-shedding scenarios, enabling comparable benchmarking across labs (Sillanpää and Sainio 2017).

17.8.2 Appliance Guidance and Labeling

Design features may include no-rinse labels on lint screens, lint bagging icons to promote proper lint disposal, and optional vent-filter readiness indicators for vented units. These "quiet interventions" respect traditional maintenance routines while improving environmental outcomes.

17.8.3 Textile Design and Procurement

Under EPR, buyers and public procurement can privilege low-shedding constructions (tight weaves, filament yarns, durable finishes) validated by ISO-based tests. This sends a market signal upstream toward conservative, long-lasting textiles consistent with a repair-and-reuse culture (EEA 2022).

17.8.4 Policy Alignment

Incorporate tumble-dryer pathways in national MPs action plans and consumer advice (e.g., energy agencies, municipal waste authorities), harmonized with EU rules and national precedents such as France's filter mandate (for washing) as a stepping stone to dryer-side measures (Okyere et al. 2025). Table 17.5 shows standardization gaps with practical actions.

TABLE 17.5
Research and Standardization Gaps with Practical Actions

Gaps	Why It Matters	Practical Action (12–36 Months)	Enablers
Lack of dryer-vent emission test	Cannot set performance targets	Draft a CEN/ISO technical report adapting ISO 4484-2 analytics to vent sampling; round-robin validation	ISO/TC 38; CEN/TC 248; OEM labs Iso
Unit normalization (per kg, per minutes) not consistent	Cross-study comparability	Standardize reporting: counts and mass per kg and cycle; fiber ID (μFTIR/Raman)	Journal/editorial policies; standards bodies
User behavior dominates outcomes	Rinsing exports fibers to sewer	Mandatory no-rinse manual language, icons on lint screens; public campaigns	Consumer-protection agencies; OEMs
Vent retrofits not common	Easy win for vented units	Approve auxiliary vent filters; encourage OEM-ready ports	Municipal rebates; EPR eco-modulation
Policy linkage to product design	Incentives unclear	Integrate lint capture into EPR fee modulation; recognize low-shedding products	EU water framework directive (WFD) revision/EPR; national eco-labels

17.9 CONCLUSION

Tumble drying, long treated as a mere afterthought to laundering, is a meaningful source of MF release. The good news is that traditional practice line-drying, careful lint maintenance, and conservative cycle settings already deliver practical reductions. From an engineering standpoint, finer lint screens, optional vent-end filtration, and clear no-rinse guidance can substantially curb emissions with minimal disruption to household routines.

Policy and standards are catching up. The ISO 4484 series provide the analytical and washing-loss foundations: the EU's EPR framework for textiles and France's filter mandate (washing) supply real-world regulatory levers. The next step is explicit dryer-vent test methods so that manufacturers, regulators, and consumers can rely on comparable, well-anchored data. With these elements aligned—and by valuing established, low-tech habits—we can materially reduce MF pollution from tumble drying while preserving the durability, cleanliness, and comfort that textiles have offered for generations.

REFERENCES

Akyildiz, S. H., Fiore, S., Bruno, M., Sezgin, H., Yalcin-Enis, I., Yalcin, B. and Bellopede, R. 2024. Release of microplastic fibers from synthetic textiles during laundering: State of knowledge and influencing factors. *Environmental Pollution* 341: 123132.

Allen, S., Allen, D., Phoenix, V. R., Le Roux, G., Jiménez, P. D., Simonneau, A., Binet, S. and Galop, D. 2019. Atmospheric transport and deposition of microplastics in a remote mountain catchment. *Nature Geoscience* 12: 339–344.

Boucher, J. and Friot, D. 2017. *Primary Microplastics in the Oceans. A Global Evaluation of Sources*. International Union for Conservation of Nature and Natural Resources (IUCN), Gland.

Browne, M. A., Crump, P., Niven, S. J., Teuten, E., Tonkin, A., Galloway, T. and Thompson, R. 2011. Accumulation of microplastic on shorelines worldwide: Sources and sinks. *Environmental Science & Technology* 45(21): 9175–9179.

City University of Hong Kong. 2022. CityU's research: Clothes dryers are an overlooked source of airborne microfibers. https://www.cityu.edu.hk/en/research/stories/2022/04/08/cityus-research-found-clothes-dryers-are-overlooked-source-airborne-microfibers [Accessed 15 September 2025].

Cummins, A. M., Malekpour, A. K., Smith, A. J., Lonsdale, S., Dean, J. R. and Lant, N. J. 2023. Impact of vented and condenser tumble dryers on waterborne and airborne microfiber pollution. *PLoS One* 18: e0285548.

De Falco, F., Cocca, M., Avella, M. and Thompson, R. C. 2020. Microfiber release to water, via laundering, and to air, via everyday use: A comparison between polyester clothing with differing textile parameters. *Environmental Science & Technology* 54(6): 3288–3296.

De Falco, E., Di Pace, E., Cocca, M. C. and Avella, M. 2019. The contribution of washing processes of synthetic clothes to microplastic pollution. *Scientific Reports* 9: 6633.

Dris, R., Gasperi, J., Saad, M., Mirande, C., and Tassin, B. 2016. Synthetic fibers in atmospheric fallout: A source of microplastics in the environment? *Marine Pollution Bulletin* 104(1–2): 290–293.

EEA. 2022. Microplastics from textiles: Towards a circular economy for textiles in Europe. European Environment Agency (EEA) report. https://www.eea.europa.eu/en/analysis/publications/microplastics-from-textiles-towards-a-circular-economy-for-textiles-in-europe [Accessed 20 September 2025].

Gavigan, J., Kefela, T., Macadam-Somer, I., Suh, S. and Geyer, R. 2020. Synthetic microfiber emissions to land rival those to waterbodies and are growing. *PLoS One* 15(9): e0237839.

Hartline, N. L., Bruce, N. J., Karba, S. N., Ruff, E. O., Sonar, S. U. and Holden, P. A. 2016. Microfiber masses recovered from conventional machine washing of new or aged garments. *Environmental Science and Technology* 50(21): 11532–11538.

Hazlehurst, A., Tiffin, L., Sumner, M. and Taylor, M. 2023. Quantification of microfiber release from textiles during domestic laundering: Influence of fabric parameters and test methods. *Scientific Reports* 13: 5332.

Hossain, M. I., Zhang, Y., Haque, A. N. M. A. and Naebe, M. 2025. Fibrous microplastics release from textile production phases: A brief review of current challenges and applied research directions. *Materials* 18: 2513.

ISO. 2023. *ISO 4484-1:2023 Textiles and Textile Products — Microplastics from Textile Sources — Part 1: Determination of Material Loss from Fabrics during Washing*. International Organization for Standardization, Geneva.

Kapp, K. J. and Miller, R. Z. 2020. Electric clothes dryers: An underestimated source of microfiber pollution. *PLoS One* 15(10): e0239165.

Kärkkäinen, N. and Sillanpää, M. 2021. Quantification of different microplastic fibers discharged from synthetic cloth washing and tumble drying. *Environmental Science and Pollution Research* 28(13): 16253–16263.

Lant, N. J., Hayward, A. S., Peththawadu, M. M. D., Sheridan, K. J. and Dean, J. R. 2020. Microfiber release from real soiled consumer laundry and the impact of fabric care products and washing conditions. *PLoS One* 15(5): e0233332.

Napper, I. E. and Thompson, R. C. 2016. Release of synthetic microplastic plastic fibers from domestic washing machines: Effects of fabric type and washing conditions. *Marine Pollution Bulletin* 112(1–2): 39–45.

Okyere, E. Y., Danquah, J. A., Miyittah, M. K. and Amoah-Nuamah, J. 2025. Tackling microfiber pollution: A review of policy responses and circular economy innovations in textile sustainability. *Waste Management Bulletin* 3(4):100253.

Organization for Economic Co-operation and Development (OECD). 2022. *Global Plastics Outlook: Economic Drivers, Environmental Impacts and Policy Options*. OECD Publishing, Paris.

Periyasamy, A. P. and Tehrani-Bagha, A. 2022. A review on microplastic emission from textile materials and its reduction techniques. *Polymer Degradation and Stability* 199: 109901.

Pirc, U., Vidmar, M., Mozer, A. and Kržan, A. 2016. Emissions of microplastic fibers from microfiber fleece during domestic washing. *Environmental Science and Pollution Research* 23(21): 22206–22211.

SGS. 2023. ISO launches new test method to determine material loss from fabrics during washing (ISO 4484-1). https://www.sgs.com/en-us/news/2023/02/safeguards-3123-iso-launches-new-test-method-to-determine-material-loss-from-fabrics-during-washing [Accessed 10 September 2025].

Sheikhi, M., Bianco, C., Tiraferri, A. and Sethi, R. 2025. Evaluating microfiber emissions and point-of-use filtration for laundry and drying. *Journal of Hazardous Materials* 489 137646.

Sillanpää, M. and Sainio, P. 2017. Release of polyester and cotton fibers from textiles in machine washings. *Environmental Science and Pollution Research* 24(23): 19313–19321.

Tao, D., Zhang, K., Xu, S., Lin, H., Liu, Y., Kang, J., Yim, T., Giesy, J. P. and Leung, K. M. Y. 2022. Microfibers released into the air from a household tumble dryer to ambient air. *Environmental Science and Technology Letters* 9(2): 120–126.

UNEP. 2025. Everything you should know about microplastics. United Nation Environment Program (UNEP). https://www.unep.org/news-and-stories/story/everything-you-should-know-about-microplastics [Accessed 20 October 2026].

World Health Organization (WHO). 2019. *Microplastics in Drinking-Water*. WHO, Geneva. https://www.who.int/publications/i/item/9789241516198 [Accessed 10 September 2025].

Zambrano, M. C., Pawlak, J. J., Daystar, J., Ankeny, M. A., Cheng, J. J. and Venditti, R. A. 2019. Microfibers generated from the laundering of cotton, rayon and polyester based fabrics and their aquatic biodegradation. *Marine Pollution Bulletin* 142: 394–407.

18 Textile Waste Management and Recycling in South Africa

Competing Approaches in the Pursuit of a Circular Textile Economy

Guera Massyn Romo and
Nadine Cynthia Sonnenberg

18.1 INTRODUCTION

The textile industry plays a pivotal role in the global economy, driving industrialization, economic development and social value through supply chains that connect manufacturers, retailers and consumers worldwide (UNEP 2023). Yet, this industry is also known for its devastating environmental and social impacts (McKinsey and Company 2020), which necessitate an urgent need for transformation in how business models, products, workers, production processes and natural resources are valued in pursuing a more sustainable and circular future (UNEP 2023). This transformation requires collective engagement among all stakeholders in textile value chain, which, according to the UNEP (2023), should focus on three priorities: shifting consumption patterns, improving practices and investing in infrastructure. To achieve these priorities, UNEP (2023) proposes several building blocks, which, among others, aim to avoid shifting the responsibility for waste management and disposal (typically embedded in the trade of second-hand clothing) to locations that are ill-equipped to process textile waste adequately.

Textile waste has become a contentious issue in the African context, with vast volumes of second-hand clothing diverted to African shores on a relentless basis (Nkatha 2023). While some African countries, such as South Africa, have restricted the import of second-hand clothing, other factors, such as cheap imports (Whitfield and Mkhabela 2022) and rising consumption levels (Cavusgil et al. 2018), have nonetheless contributed to mounting levels of textile waste in the local context (Sonnenberg et al. 2024). Although limited data are available on the exact scale of textile waste disposal in South Africa, a report by the Department of Environmental Affairs (2020). Estimated that approximately 100,000 tons of textile waste end up in

DOI: 10.1201/9781003689225-18

landfills annually, concluding that textiles constitute a substantial share of municipal solid waste. Consequently, various stakeholders perceive the management and recycling of textile waste as key drivers in achieving a more sustainable and circular textile value chain in South Africa (GreenCape 2023).

Despite a consensus on the importance of textile waste management and recycling initiatives, debates about how to build a circular textile economy in South Africa increasingly revolve around recurring approaches with distinct implementation logics or strategic models that organize how policymakers, funders, NGOs, industrial actors and social enterprises interpret the problem of textile waste and devise solutions. These approaches are not simply rhetorical "narratives" but rather function as action-guiding frameworks that influence investment decisions, Program design, and the selection of technologies, beneficiaries and organizational partners. This chapter provides an overview of five conventional approaches that have characterized efforts focused on adopting circular principles in the local textile waste management context. The "technology-centric" approach, infused by views such as those of Reike et al. (2022) and Eppinger (2022), emphasizes industrial-scale plant installation as a primary catalyst for circularity, while the "human-capital-led" approach foregrounds training and entrepreneurial skills development as the main levers of change (Jayarathna et al. 2025). A third, the "pilot-driven" approach, relies on grant-funded demonstration projects. This approach has, for example, facilitated transitions in the energy, water and waste sectors (Andersen and Gulbrandsen 2020; Burnstine and Camargo 2025; Grönwall and Jonsson 2017). The fourth approach, labeled the "informal-sector-led" approach assumes that South Africa's extensive informal recycling economy can absorb textile waste streams (Biyada and Urbonavičius 2025; Schröder 2020). A fifth approach, the "production-led livelihood," presumes that once recycled textile products are produced, markets and income will naturally follow. This approach is often exemplified in international examples of successful eco-brands and products with recycled content (Swilling and Annecke 2012; Reike et al. 2022). These conventional approaches are widely practiced and institutionally embedded but have repeatedly underdelivered on their intended outcomes in the local context. The chapter concludes with an alternative systems-based approach: Ecosystem-building models are emerging as a view applied to vexing societal problems, serving as a method for constructing interdependent relationships, institutions and infrastructures (Papachristos 2014; Fischer and Herrmann 2011).

18.2 CONVENTIONAL MAINSTREAM APPROACHES TO ADOPTING CIRCULARITY IN THE SOUTH AFRICAN CLOTHING AND TEXTILE INDUSTRY

Following transition studies and innovation systems literature (Geels 2002; Geels and Schot 2007; Turnheim and Sovacool 2020), an approach is understood as a coherent set of assumptions, institutional priorities and operational mechanisms that shape how actors attempt to solve a systemic problem. Such approaches may also apply to textile waste management and recycling (Abrishami et al. 2024; Juanga-Labayen et al. 2022; European Commission 2022). The following section will examine the conventional mainstream approaches to accomplishing circular principles, specifically relating to textile waste management and recycling, using the following criteria: defining the scope and justification for each approach; its underlying assumptions about markets, technology, governance and agency; and the challenges and opportunities for reframing each approach by applying a systems-based logic.

This analytical structure enables a rigorous comparison while preserving the detailed, context-specific insights necessary for interpreting circular-economy transitions in South Africa. The chapter conceptualizes these approaches, which, while often reproduced through discourse, function in practice as distinct implementation logics guiding investment decisions, program design and policy interpretation. To emphasize scientific rigor, the criteria used to evaluate the approaches include internal logic and assumptions; historical and institutional origins, operational mechanisms, economic and governance viability, evidence of outcomes (success or underperformance), system-level dependencies (innovation system theory), and inclusivity and justice implications, thus employing a systematic comparative methodology.

18.2.1 The Technology-Centric Approach

The technology-centric approach frames circular textile development primarily as a technical and engineering challenge. Within this logic, the perceived central barrier to circularity is the lack of industrial-scale plant capacity capable of efficiently processing large volumes of textile waste (Reike et al. 2022; Eppinger 2022).

18.2.2 Definition, Scope and Justification

Policy actors, Producer Responsibility Organizations (PROs), development institutions and private investors prioritize the acquisition and development of shredding lines, automated sorting technologies, fiber-regeneration systems, chemical recycling processes and other industrial plant configurations. Circularity is thereby conceptualized as a function of throughput, measured through tonnage processed, plant uptime and the emergence of large formal manufacturing facilities. This interpretation strongly aligns with a long-standing belief that technological upgrading is synonymous with industrial progress and that circularity can be "engineered into existence" through the installation of advanced machinery (Grin et al. 2010; Papachristos 2014; DTIC 2021; European Commission 2022). The model's institutional appeal lies in its visibility and measurability. The creation of large plants is viewed as tangible evidence of circular industrialization and green manufacturing commitments.

The technology-centric approach, while globally practiced, is rooted in South Africa's industrial policy tradition, which has consistently prioritized capital-intensive technological upgrading as a pathway to competitiveness (DTIC 2019, 2021). Sectors such as automotive, chemicals and metals illustrate a policy logic in which industrial revitalization is tied to plant expansion, machinery acquisition and industrial clustering. The *Clothing, Textiles, Footwear and Leather Growth Program (CTFLGP)* (DTIC 2021) specifically offers large investments for factory expansion based on key criteria such as productivity gains and job creation. The *Retail Clothing, Textiles, Footwear and Leather (R-CTFL) Masterplan* (2030) (DTIC 2019) is tied to the incentives offered under the CTFLGP. These policies represent an effort by the South African government to stabilize and grow a historically strained, labor-intensive sector. By encouraging investment in modern equipment, promoting skill development and securing local market access through retailer commitments, these programs aim to improve the economic success of the sector, create sustainable jobs and reduce reliance on imports. This approach of industrialized factory development is further supported by local product content requirements under the *Preferential Procurement Policy Framework Act (PPPFA)* (South Africa 2000).

This orientation is further reinforced by global developments. High-profile fiber-regeneration plants in Europe are often heavily subsidized through enabling legislation such as the *EU Strategy for Sustainable and Circular Textiles* (European Commission 2022; Cernansky 2024). This strategy aims to ensure that textile products available on the European market are durable, repairable or recyclable. The strategy thus provides the basis for funding toward technology development to enable the textile performance and management objectives contained in its scope. Thus, technological change is embedded within existing industrial ecosystems and supported by strong regulatory frameworks, which serve as exemplary circular textile transition (Eppinger 2022). These technology development efforts circulate as aspirational templates at global trade expos and circular-economy forums, shaping expectations among South African policymakers and funders.

Yet the socio-technical evidence suggests that these global technologies emerged within systems characterized by high institutional coherence, dense market linkages and strong governance capacity, conditions that differ significantly from the fragmented, undercapitalized and spatially uneven South African textile waste context (Salmenperä 2023; Snoek 2017; Biyada and Urbonavičius 2025).

18.2.3 Underlying Assumptions of the Technology-Centric Approach

The technology-centric model is supported by a series of strong assumptions concerning feedstock, markets, policy instruments, technology transfer and finance. It assumes that extended producer responsibility (EPR) levies, landfill bans or municipal diversion targets will reliably generate steady volumes of usable feedstock. It assumes that major retailers, brands and industrial buyers are both willing and able to supply consistent quantities of textile waste, and that downstream buyers will switch to recycled materials once they become available. The approach also assumes that sophisticated machinery designed for stable grid conditions, predictable waste streams and robust maintenance ecosystems can be transplanted into the South African context with minimal adaptation. Moreover, it presupposes that economies of scale will eventually make recycled fibers cost-competitive with virgin imports and that financial institutions will finance large capital investments even in the face of uncertain order books or untested markets. At its core, this approach rests on the belief that scale itself will resolve structural constraints, upstream and downstream. These assumptions align with what Turnheim and Sovacool (2020) identify as *cognitive and selection biases* within transition practice, particularly the notion that technological innovation, by itself, can trigger systemic transformation in regimes that are structurally locked-in and resistant to change.

18.2.4 Challenges and Opportunities for Reframing the Technology-Centric Approach

The technology-centric approach often underperforms because it misinterprets the causal logic of socio-technical change. It assumes that installing industrial-scale machinery will catalyze circularity, yet transition scholarship demonstrates that technologies do not create new systems on their own. These technologies only become viable when embedded within aligned configurations of role players, institutions, infrastructures, cultural norms and markets (Geels 2002; Miörner et al. 2021). Miörner et al.'s analysis reinforces this point, showing that sectors with weak institutional coherence, fragmented actor networks and low alignment possess limited "transformative potential," meaning that even advanced technologies struggle to influence entrenched industrial practices (Miörner et al. 2021). South Africa's textile system is marked by precisely such fragmentation. Heterogeneous waste streams, under-capacitated municipal systems, uncoordinated EPR actors and unstable market linkages are all conditions that severely limit the ability of technology to function as a lead driver of system transformation (GreenCape 2019, 2023; Godfrey et al. 2013; Prevent Waste Alliance 2022).

Compounding this structural weakness are deep-seated path dependencies and lock-in mechanisms that reinforce incumbent linear textile systems. Eitan and Hekkert (2023) conceptualize lock-ins as constraints created by past choices that restrict flexibility and hinder the adoption of alternatives, thus reducing a system's capacity to deviate from existing trajectories. South Africa's textile sector exhibits multiple forms of technological, institutional, economic and social lock-in (DTIC 2019, 2021). There is heavy reliance on virgin fiber imports, procurement systems optimized for low-cost global supply chains, diminished domestic manufacturing capacity and weak public waste infrastructure. These entrenched structures actively resist the emergence of new recycling technologies, regardless of their technical sophistication. Klitkou et al. (2015) further demonstrate that learning effects, economies of scale, network externalities and institutional complementarities stabilize incumbent regimes and inhibit the upscaling of new innovations unless broader system shifts are initiated.

The enabling ecosystem required for industrial textile recycling is largely absent. High-capacity recycling technologies depend on stable, high-quality feedstock produced through financed collection, sorting and pre-processing systems. Yet, South Africa's textile waste streams are geographically dispersed, contaminated and unclassified (Greencape 2023), resulting in unreliable volumes and inconsistent quality. Without funded sorting, densification and logistics systems, the material flows required to sustain industrial machinery simply do not exist. Downstream markets are equally underdeveloped. Salmenperä (2023) shows that even in Finland, where public–private coordination is stronger and circularity enjoys broad support, transitions stall when regime actors fail to shift procurement norms or when expectations remain fragmented across institutional levels. South African retailers and manufacturers exhibit similar reluctance. Without quality assurance, price stability and predictable supply volumes, they will not adopt recycled fibers on scale. The result is that expensive machinery operates well below capacity because the upstream and downstream systems needed to support it remain weak.

The operational environment further undermines performance. Industrial recycling technologies require stable electricity, specialized maintenance expertise, ready access to spare parts and reliable operating conditions. Instead, plants in South Africa face high maintenance costs, dependence on imported technical components and limited local expertise. As Eitan and Hekkert (2023) caution, such conditions can transform technology investments into lock-ins themselves. Once capital is invested in large plants that underperform, system flexibility decreases and opportunities for more context-appropriate alternatives diminish. Existing South African nonwoven and shoddy manufacturers exemplify this dynamic, as several operate far below capacity or rely on imported textile waste despite the abundance of local materials, indicating systemic rather than technical failure.

Evidence from global transitions underscores the limits of technology-led change. The shift to Zero Liquid Discharge technology in Tirupur, India, did not emerge organically from technological promise. Rather, it was driven by intense regulatory pressure, litigation, farmer mobilization and court-mandated enforcement (Grönwall and Jonsson 2017). This case demonstrates that industrial technologies succeed when crisis, governance pressure or collective coordination create the institutional alignment necessary for systemic change. These stars will not align when technology is expected to generate such alignment on its own. South Africa does not currently exhibit the regulatory coherence or enforcement capacity seen in Tirupur, further limiting the likelihood that industrial-scale textile recycling technologies can succeed through voluntary uptake.

Taking together, these dynamics suggest that the technology-centric approach may fail not because the technologies themselves are ineffective, but because they are introduced into a socio-technical system that lacks coherence, institutions, market incentives, governance capacity and infrastructural foundations required for them to function effectively (Eitan 2023; Turnheim and Sovacool 2020). The result is predictable. Machinery is procured, but feedstock systems remain weak; market demand remains uncoordinated, and operational challenges mount, resulting in plants running well below their intended capacity. Circularity does not emerge from isolated technological installations but from constructing interdependent systems that allow technology to perform its role (Reike et al. 2022). Socio-technical transitions research emphasizes that complex systems do not transform simply through the insertion of new technologies but depend on the alignment of cultural norms, market structures, regulatory environments, infrastructures and user practices (Geels 2002; Turnheim and Sovacool 2020; Cernansky 2024). The technology-centric approach thus selects one system element, technology, as the primary change lever while underestimating the systemic dependencies required to sustain circular textile practices, particularly in an emerging market context such as South Africa.

A more grounded reframing of the technology-centric approach begins with the recognition that technology must follow ecosystem formation rather than precede it. Instead of assuming that large plants will catalyze new value chains, policymakers and funders should prioritize the development of the foundational elements that enable technology to operate effectively. These include financed collection systems, paid sorting structures, community-level pre-processing hubs, standardized quality protocols and verified demand signals from industrial and retail buyers. Technology choices should emerge from the realities of feedstock composition, market requirements and operational resilience rather than from aspirations for scale. Modular, context-adapted technologies that can be expanded incrementally offer a more realistic starting point than large, inflexible industrial plants (Geels and Schot 2007; Juanga-Labayen et al. 2022; Korhonen et al. 2018). In this reframed model, industrial technology remains important, but it is positioned as one component of a broader, coordinated system rather than as the central driver of circularity.

18.3 THE HUMAN-CAPITAL-LED APPROACH

The human-capital-led approach interprets circular textile development primarily as a challenge of insufficient skills. Within this logic, unemployment, poverty and limited entrepreneurial activity are understood as consequences of a skills deficit rather than structural conditions of the economy (Parliamentary Research Unit 2025; CSIR 2021; Jayarathna et al. 2025).

18.3.1 Definition, Scope and Justification

Training Programs are therefore designed to equip individuals, particularly unemployed youth and women, with sewing, craft, design, upcycling and basic business skills. The implicit expectation is that these newly skilled individuals will either enter existing value chains or establish microenterprises capable of producing products from recycled textiles. Measurement of success is typically expressed in quantifiable outputs such as the number of trainees, certificates issued, workshops delivered and beneficiaries reached. Training organizations, NGOs and corporate social investment (CSI) programs thus become key intermediaries in the circular economy, promoting the idea that skill acquisition is the primary barrier to economic participation.

The origins of this approach lie in South Africa's deep institutional commitment to human capital development as a vehicle for addressing inequality and unemployment. Policy frameworks such as the B-BBEE codes mandate investment in training, creating a steady flow of funding for skills-development initiatives (Republic of South Africa 2003). Historically, South Africa has also hosted a large ecosystem of craft and microenterprise training organizations, many of which produced isolated examples of entrepreneurial success that are frequently celebrated as evidence of the model's effectiveness. In addition, donors and funders often favor training-based interventions because they generate clear, countable metrics and align with normative values associated with empowerment and upliftment. The broader circular-economy discourse itself places strong emphasis on awareness-raising, skills development and behavioral change, which easily translate into training-heavy programs even in contexts where other systemic constraints such as weak markets or unaffordable equipment are far more significant.

18.3.2 Underlying Assumptions of the Human-Capital-Led Approach

The human-capital-led approach is sustained by several interconnected assumptions about markets, entrepreneurship and individual agency. It assumes that once individuals acquire skills, they will naturally create or access economic opportunities. This approach presumes the existence of strong markets for handmade or recycled textile products and assumes that environmental awareness will translate into consumer purchasing behavior. It posits that trained individuals will secure access to inputs such as equipment, workspace, working capital and reliable sources of textile waste, even though such access is often scarce and costly. This emphasis on individual capacity reflects a significant reliance on individual effort and agency in which structural constraints like market saturation, import competition, infrastructural gaps and unequal access to assets are rendered invisible (Geels and Schot 2007; Miörner et al. 2021).

18.3.3 Challenges and Opportunities for Reframing the Human-Capital-Led Approach

In practice, skills-development interventions often result in the creation of skilled individuals who remain economically marginalized. As an example, graduates may gain competencies in sewing, craft or design, but without access to machinery, dedicated workspace, subsidized inputs or market linkages, these skills cannot easily be translated into reliable income streams. Many training programs operate independently of validated market demand, with curricula focused on hypothetical products rather than buyer-driven requirements. As a result, participants develop products that may not meet industry standards for quality, durability, design consistency or price competitiveness. The local market for crafts and recycled items is oversaturated, leading to intense competition among micro-producers who often end up offering similar items, such as tote bags or cushions, at minimal margins. The assumption that individuals can create livelihoods by training on a limited product range often yields short-term empowerment outcomes without generating long-term economic resilience.

Global textile recycling research shows that the viability of circular textile enterprises depends not only on skills but also on access to high-quality feedstock, stable collection systems, specialized sorting infrastructure and predictable markets (Juanga-Labayen et al. 2022; Abrishami et al. 2024). These systemic conditions are particularly weak in South Africa, where textile waste is contaminated, geographically dispersed and lacks dedicated collection channels. Without institutional support for machines, workspaces, logistics, standards and market commitments, trained artisans often remain unemployed after completing their training. Just transitions scholarship reinforces this diagnosis. Swilling and Annecke (2012) argue that capacity-building alone cannot yield meaningful livelihoods in contexts of structural inequality, resource constraints and institutional fragmentation. In such settings, skills programs risk generating short-term empowerment indicators while failing to establish stable economic pathways.

In addressing these challenges, a more context-appropriate reframing requires repositioning training as one component within a much broader ecosystem-building strategy. Training should be introduced only when accompanied by access to shared infrastructure, long-term mentorship, subsidized working capital and structured procurement pathways linked to real buyers. Curricula must be co-designed with industry stakeholders, including brands, retailers, recyclers and manufacturers, to ensure alignment with technical and market requirements. Training Programs could, for example, incorporate modules on process engineering, machine operation, quality standards, traceability, EPR compliance and collaborative production models. Collective enterprise structures, including cooperatives and social franchises, can provide a more viable platform for pooling risk, consolidating supply and achieving economies of scale (Fischer and Herrmann 2011; Schröder 2020). In this reframing, training is understood not as a primary solution but as a secondary intervention that becomes effective only when participants are embedded in a supportive ecosystem capable of translating skills into livelihoods.

18.4 THE PILOT-DRIVEN APPROACH

The pilot-driven approach frames circular textile development as a problem of uncertainty and experimentation. Within this model, funders, NGOs, development agencies and innovation hubs invest in small-scale demonstration projects designed to test new technologies, business models or community engagement strategies. Pilots are expected to demonstrate technical feasibility, highlight social benefits and generate evidence that can be used to attract larger-scale investment.

18.4.1 Definition, Scope and Justification

The pilot-driven approaches often operate under grant conditions and rely on subsidized input, temporary labor and intensive support from project teams. Success is measured through outputs such as waste diverted, products manufactured, participants trained or positive evaluation reports. The central assumption is that successful pilots will naturally transition into sustainable enterprises, influence policy decisions or catalyze replication.

The pilot-driven approach derives from the linear innovation model prevalent in R & D policy, which conceptualizes change as a progression from prototype to pilot to scale. In the energy, water and waste sectors, this model occasionally produced successful transitions, reinforcing the belief that piloting is a reliable pathway for testing circular innovations (Andersen and Gulbrandsen 2020; Burnstine and Camargo 2025; Grönwall and Jonsson 2017). Philanthropic foundations and government innovation funds often favor pilot projects because they create visible evidence of activity, generate media attention and allow for experimentation without long-term financial commitments. The circular-economy discourse itself reinforces this model by celebrating "innovation pilots," "demonstration projects," and "testbeds," which naturally translate into projectized interventions. South Africa's development ecosystem, with its long history of donor-funded projects, further institutionalizes piloting as a conventional tool for delivering environmental and social interventions. Globally, circular-economy discourse also reinforces this project orientation, celebrating innovation labs, demonstration projects and testbeds as markers of progress.

18.4.2 Underlying Assumptions of the Pilot-Driven Approach

The pilot-driven model assumes that demonstrating technical feasibility under grant conditions is a reliable indicator of long-term commercial viability. It presumes that once a pilot proves successful, entities such as municipalities or private investors will step in to institutionalize or scale the model. This model assumes that the operational environment during a pilot, where inputs may be free, labor-subsidized and technical support abundant, closely resembles real-world conditions, allowing outcomes to be extrapolated. Furthermore, it assumes that organizations implementing pilots aim to transition into long-term operational roles. However, many organizations are structurally configured to optimize grant cycles rather than sustainable operations, creating institutional incentives to prioritize novelty over durability. Transition literature notes that niche experiments often function as incubators for new ideas, reinforcing the appeal of pilots as a mechanism for generating early-stage learning (Geels 2002). Transition scholars describe a "scaling fallacy" as the belief that niche performance under protected conditions can be extrapolated to regime-level viability (Turnheim and Sovacool 2020).

18.4.3 Challenges and Opportunities for Reframing the Pilot-Driven Approach

Pilot projects in South Africa often encounter a "post-pilot cliff," where operations collapse immediately after grant funding is withdrawn. Although pilots may succeed during the funded period, they often lack the necessary financial architecture, market commitments, governance structures or logistical systems required for long-term viability. The technical feasibility demonstrated in a pilot rarely accounts for the full cost structures, especially when inputs are subsidized or processes are simplified. Moreover, pilot conditions are unrepresentative of the challenges faced in real-world settings, including fluctuating feedstock availability, inconsistent quality, market volatility and resource constraints. The proliferation of isolated pilots also fragments learning and prevents cumulative system development, as each project operates in a silo with its own branding, metrics and stakeholders. Many organizations become locked in cycles of grant-seeking, repeatedly piloting without ever achieving scale.

The transition failure literature identifies similar patterns internationally, including overreliance on technological fixes, fragmented coordination and failure to build institutional capacity around niche innovations (Turnheim and Sovacool 2020). Textile recycling research highlights that technical feasibility often collapses when confronted with full cost structures, unstable feedstock, contamination challenges, market fragmentation and the high production costs of recycled fibers (Abrishami et al. 2024; Juanga-Labayen et al. 2022). Pilots that are not embedded within coordinated feedstock systems, procurement mechanisms, governance arrangements and business models tend to remain isolated successes rather than systemic precursors.

In addressing these challenges, the pilot-driven model may need to shift from using pilots as end points to using them as transitional phases within broader system-building pathways. Pilots should only be funded when they are explicitly linked to long-term revenue mechanisms such as EPR contracts, municipal service agreements or procurement commitments from industrial buyers. Stage-gated funding structures can support progression through pilot, early commercialization and scale phases, with clear criteria guiding continuation or termination. Pilots should focus on system integration rather than solely on technical outputs, testing the interactions between feedstock logistics, pre-processing, production, market engagement and governance. Evaluation processes must distinguish between outcomes that depend on temporary subsidies and those that reflect genuine viability. Learning from pilots must be consolidated and institutionalized rather than remaining confined to project reports. In this reframed model, pilots become catalysts for systemic change rather than isolated experiments.

18.5 THE INFORMAL-SECTOR-LED APPROACH

The informal-sector-led approach assumes that South Africa's extensive informal recycling economy can readily absorb new responsibilities related to textile collection, sorting, pre-processing and production. Waste pickers, reclaimers, small crafters and micro-entrepreneurs are envisioned as flexible, self-organizing actors who can quickly respond to emerging market signals with minimal support.

18.5.1 Definition, Scope and Justification

The informal-sector-led approach frames the informal sector as a low-cost labor reservoir that can deliver livelihood opportunities and circularity outcomes simultaneously (CSIR 2021; Biyada and Urbonavičius 2025; Schröder 2020). The underlying aim is to leverage the entrepreneurial energy and embedded networks of the informal sector while reducing public investment in infrastructure and support services.

The informal-sector-led model emerges from the empirical success of informal recyclers in managing other waste streams, particularly polyester (PET), metals and cardboard. Waste pickers have achieved high recovery rates despite operating under precarious conditions with limited institutional support. These successes have influenced national waste policy and international circular-economy discourse, which increasingly celebrates informal recyclers as "circularity champions" whose contributions should be integrated into formal systems. In the textile domain, small-scale examples of craft-based reuse, where artisans convert fabric offcuts into bags or decorative items, are often mistakenly interpreted as evidence that the informal sector can scale up its participation to match industrial recycling requirements. The approach, therefore, blends recognition and romanticization, projecting the proven competence of informal workers in one context onto a much more complex and low-value stream.

18.5.2 Underlying Assumptions of the Informal-Sector-Led Approach

The approach assumes that informal workers have the spare capacity to experiment with new streams despite operating under survivalist pressures and thin economic margins. It presumes that integration can largely be achieved through symbolic gestures, such as recognition, training or the provision of basic equipment on a business grant support basis, without addressing structural issues, including pricing, infrastructure, working conditions or governance. It further assumes that informal collectors can meet the technical requirements of textile-to-textile recycling, including fiber sorting, contamination management and traceability, without access to depots, sorting facilities or quality-control systems. Finally, it assumes the existence of downstream buyers willing to purchase the collected materials at fair prices, overlooking the fact that markets for recycled textiles are currently limited and underdeveloped. These assumptions diverge significantly from empirical evidence, which shows that textile recycling is far more complex than recycling PET or cardboard, due to the need for fiber-specific sorting, decontamination, disassembly and specialized machinery (Abrishami et al. 2024).

18.5.3 Challenges and Opportunities for Reframing the Informal-Sector-Led Approach

This approach shifts substantial risks onto the poorest actors in the value chain. Textile waste is bulky, lightweight, contaminated and of low value compared to metals or PET. Without funded storage, densification or transport, it is economically irrational for informal collectors to prioritize textiles. The sorting and quality requirements are far more complex than those of other recyclables, demanding specialized knowledge and infrastructure that are unavailable to informal workers. Expecting the informal sector to pioneer textile feedstock chains without guaranteed buyers or risk-sharing mechanisms effectively uses poverty as venture capital. Moreover, while policy documents increasingly reference "integration," formal mechanisms to ensure fair payment, co-governance or social protections remain limited or unenforced. As a result, informal participation becomes symbolic rather than substantive, reinforcing existing inequalities rather than addressing them.

Textile recycling globally faces significant technical and economic barriers that informal systems cannot overcome without substantial support. Large-scale recyclers report sorting hundreds of textile categories, which require labor-intensive and capital-intensive systems, specialized knowledge and consistent feedstock quality (Abrishami et al. 2024). Studies further highlight the absence of dedicated textile collection systems as a major barrier, making informal aggregation costly and inefficient (Juanga-Labayen et al. 2022). Without financed logistics, storage, equipment, safety standards or guaranteed markets, informal actors face disproportionate risks and limited value capture. Critical circular-economy scholarship also warns that informal labor is often used to externalize costs while formal actors capture benefits, reproducing unjust and extractive value chains (Korhonen et al. 2018). This dynamic contradicts just-transition principles centered on fairness, redistribution and structural change (Swilling and Annecke 2012; Grin et al. 2010).

A more just and realistic model treats informal recyclers as contractual partners with defined roles, rights and responsibilities within circular textile systems. This requires formalizing their participation through service-level agreements embedded in municipal tenders, EPR frameworks and PROs. Such agreements must specify floor prices, payment terms, quality requirements and social protections. Infrastructure investments are essential: community-based depots, sorting hubs and pre-processing centers equipped with balers, cutters, testing tools and data systems enable informal workers to meet quality standards and aggregate meaningful volumes. Capacity-building initiatives should expand beyond basic skills to include quality management, digital documentation and negotiation. In this reframed approach, informal actors participate within structured, supported and rights-based systems rather than through precarious and unsupported labor.

18.6 THE PRODUCTION-LED LIVELIHOOD APPROACH

The production-led livelihood model is grounded in the assumption that environmental value will translate directly into market value. It assumes that once recycled textile products are made, consumer or industrial demand will naturally materialize. This approach is rooted in the idea that recycled products possess inherent desirability due to their environmental attributes or alignment with sustainability trends, where recycling is framed as a product innovation strategy rather than a systemic transition (Eppinger 2022).

18.6.1 Definition, Scope and Justification

The production-led livelihood model encourages micro-entrepreneurs, artisans and community enterprises to produce as much as possible, expecting that markets will reward these efforts and generate livelihoods. The model draws on a broader socio-economic assumption prevalent in contemporary capitalism that individuals must secure their survival by creating and selling goods in competitive markets.

The production-led livelihood approach reflects a global context characterized by overproduction, consumer saturation and pressures on individuals to monetize creativity. In South Africa, where formal employment opportunities are limited and entrepreneurial activity is widely promoted as a solution to unemployment (Sonnenberg et al. 2024), the imperative to "make and sell" becomes particularly pronounced. Microenterprise development programs, social entrepreneurship initiatives and craft incubators often reinforce this belief by highlighting individual success stories of makers or artisans who have transformed production into livelihood opportunities. At the same time, circular-economy discourse can subtly reinforce the idea that recycling and reuse allow for continued production without environmental harm, thereby aligning with growth-oriented economic thinking. International examples of successful eco-brands and recycled-content products, often developed in high-income contexts with extensive consumer bases, are frequently misinterpreted as universally replicable models (Swilling and Annecke 2012; Reike et al. 2022).

18.6.2 Underlying Assumptions of the Production-Led Livelihood Approach

The production-led livelihood model assumes a direct line between production and income generation. The approach assumes elasticity in demand in that markets can absorb increased volumes of recycled products without becoming saturated. It also assumes that producers will be able to access the necessary inputs for production, including workspace, equipment, feedstock and capital. In addition, it assumes that artisans and microenterprises can compete meaningfully with low-cost imports and mass-produced goods, despite lacking comparable economies of scale. In addition, it is believed that environmental awareness will translate into consumer purchasing behavior and that consumers and retailers will prioritize recycled goods over cheaper virgin alternatives. It further assumes that recycled textile production is cost-competitive, despite global evidence indicating that recycled materials are often more expensive than virgin fibers due to contamination, sorting and processing costs (Abrishami et al. 2024). Finally, the model assumes that continuous production is both socially desirable and economically viable.

18.6.3 Challenges and Opportunities for Reframing the Production-Led Livelihood Approach

In practice, the production-led model falters on both economic and structural grounds. It fails to recognize that producing recycled goods is often far easier than selling them, particularly in an economy where consumer purchasing power is limited and price sensitivity is high. Despite increasing awareness of sustainability issues, consumers often prioritize cost, convenience and aesthetics over environmental attributes (Sonnenberg et al. 2014). Retailers and industrial buyers, faced with competitive pressures and stringent quality requirements, rarely adjust procurement practices without clear incentives or risk-sharing arrangements. The result is that micro-entrepreneurs and artisans produce goods that struggle to find markets, leading to unsold inventory, reduced margins and precarious incomes. Encouraging producers to generate more goods in an already saturated market intensifies competition, depresses prices and increases financial vulnerability. This misalignment between production and actual demand highlights the model's limitations.

Global textile circularity research indicates that less than 1% of textile waste is recycled back into textiles, reflecting limited demand for recycled fibers and insufficient price incentives (Juanga-Labayen et al. 2022). Recycled inputs are typically more expensive, markets are fragmented and procurement incentives are weak, making craft-scale recycled products uncompetitive against cheap imported goods.

Local markets for handmade goods are oversaturated, characterized by weak discretionary spending and homogenous product offerings. Without validated demand, micro-producers accumulate unsold inventory, experience low margins and face precarious incomes. Critical circularity scholarship highlights that relying on increasing production rather than sufficiency or service-based models risk reinforcing unsustainable material flows (Korhonen et al. 2018). Just transitions literature similarly caution against livelihood models that rely on expanded material throughput rather than system-level restructuring (Swilling and Annecke 2012).

Stronger and more sustainable framing requires shifting from production-first logic to demand-led and sufficiency-oriented models. Circular textile systems must be built around validated market demand rather than assumptions about consumer behavior. This may involve establishing procurement agreements with industrial buyers, retail partners or public entities before production begins. It also requires diversifying value pathways beyond product sales, including repair services, co-design workshops, collaborative making, remanufacturing and product-as-a-service models. Instead of encouraging thousands of micro-producers to compete in oversaturated markets, ecosystem-building strategies emphasize fewer but higher-quality products tailored to specific market niches. Shared branding, cooperative production and coordinated market development can reduce competition among vulnerable producers. The reframed model aligns circularity with principles of sufficiency, meaningful work and reduced overproduction, thereby supporting more resilient and just livelihood strategies.

18.7 THE ECOSYSTEM-BUILDING MODEL AS A VIABLE ALTERNATIVE APPROACH

The ecosystem-building systems model reframes textile circularity in South Africa as a complex socio-technical transformation that cannot be driven by isolated interventions in technology, training, entrepreneurship or informal participation. Instead of assuming that any single lever can catalyze the sector, this approach conceptualizes circularity as a multilayered system that requires the deliberate construction of interdependent infrastructures, institutions, relationships and financial flows.

18.7.1 Definition and Scope

Within this framework, technology, human capital, enterprise development, pilot projects, collection systems and market creation are all understood as mutually reinforcing components of a configuration that works (Papachristos 2014; Fischer and Herrmann 2011). The principal aim of this approach is therefore not the optimization of one element but the establishment of a minimum viable ecosystem that can sustain feedstock generation, pre-processing, production, demand stimulation and governance over time. Success within this model is measured not by isolated indicators such as the number of machines installed or the quantity of trainees certified, but by the overall stability, inclusiveness, coherence and adaptive capacity of the system (Geels and Schot 2007). The central question becomes whether the ecosystem has developed institutional arrangements, logistical pipelines, market linkages and governance mechanisms needed to ensure that circular textile practices can endure and evolve (Papachristos 2014).

18.7.2 Theoretical Foundations of the Ecosystem-Building Approach

The origins of the ecosystem-building approach reflect several strands of contemporary transitions theory that critique intervention logics focused on single solutions. Multilevel Perspective research views transitions as co-evolutionary processes across niche innovations, incumbent regimes and broader landscape pressures, where technologies alone cannot trigger systemic change without coordinated shifts in markets, infrastructures, regulations and cultural practices (Geels 2002; Papachristos 2014). Empirical analyses of transformation patterns similarly show that some socio-technical systems, especially highly globalized and institutionally entrenched sectors such as textiles, possess limited transformative potential unless multiple interdependent functions develop simultaneously (Miörner et al. 2021).

The approach also aligns with Deep Transitions theory, which argues that multiple industrial sectors share underlying meta-rules shaped by long historical waves of fossil-fueled, resource-intensive production (Kanger and Schot 2019; Fischer and Herrmann 2011). Within this framing, circular textile development in South Africa cannot be pursued as an isolated industrial modernization project; rather, it represents part of a broader shift away from the growth-through-volume logic that has governed industrial production for more than a century. An ecosystem-building strategy thus becomes a mechanism for institutional experimentation, foreground principles such as sufficiency, shared infrastructure, distributed capabilities and coordinated governance.

Failure-oriented transition studies address this problem by examining failure in greater detail, rather than focusing solely on isolated successful projects. Turnheim and Sovacool (2020) demonstrate that socio-technical transitions frequently fail when actors overestimate the catalytic power of technology, underestimate institutional dependencies or rely on isolated pilots that do not alter the underlying structural conditions. These failure modes mirror the structural weaknesses observed in the five earlier approaches analyzed in this chapter. The ecosystem-building model directly responds by embedding coordination, risk-sharing and long-term governance into its design logic, rather than treating them as downstream outcomes of technical success.

Finally, spatial transitions research stresses that transitions unfold unevenly across places, shaped by local industrial legacies, actor networks and governance capacities (Coenen et al. 2012). Evidence from Finland illustrates how even high-capacity contexts can fail to advance textile circularity without strong governance coherence, buyer alignment and standardized interventions (Salmenperä 2023). In South Africa, where institutional fragmentation, infrastructural gaps and market incoherence are more pronounced, the need for regionally grounded ecosystems, shared pre-processing hubs and deliberate coordination is even stronger (Ramirez 2021). The author's framework of transition contexts also supports the need for differentiated ecosystem strategies that respond to specific regional conditions (Berkhout et al. 2004).

Concepts such as "minimum viable ecosystems" and "system innovation" emphasize mission-oriented innovation systems interventions that cultivate demand, standards, capabilities and legitimacy simultaneously (Reike et al. 2022). In the South African context, the approach resonates with lessons from other sectors, such as renewable energy, smallholder agriculture and municipal waste integration, where ecosystem-building has proven more effective than isolated innovation, as it often incorporates a just-transition lens. This lens enables inclusive governance, redistributive mechanisms and decent work within emerging circular economies (Schröder 2020). The convergence of these empirical frustrations and theoretical insights has made the ecosystem-building model an increasingly compelling alternative.

18.7.3 Key Assumptions of an Ecosystem-Building Approach

Unlike the preceding five approaches, the ecosystem-building model is premised on the assumption that circularity is inherently complex, path-dependent and multi-actor. It rejects the idea that any individual intervention, whether technological, educational, entrepreneurial or informal, can independently drive sectoral transformation. Instead, it assumes that change will emerge through iterative experimentation, continuous learning and coordinated decision-making among actors with diverse interests, resource constraints and time horizons. This approach further assumes that early-stage ecosystem-building will be costly and difficult, requiring patient capital and a tolerance for slow progress. However, these investments lay the foundation for long-term stability and inclusiveness.

Another key assumption is that circular value chains must be designed to ensure the structural inclusion of marginalized actors (Schröder 2020; Swilling and Annecke 2012).

Informal collectors, community-based enterprises and small manufacturers are viewed not as peripheral beneficiaries but as indispensable partners who perform critical system functions (Calvo et al. 2020). Their participation must therefore be safeguarded through formal mechanisms such as service-level agreements, remunerated roles and representation in governance rather than through rhetorical commitments alone. The approach also assumes the necessity of risk-sharing arrangements. Because textile recycling markets are volatile, the model presumes that governance structures should distribute risks and benefits across stakeholders in ways that prevent corporate concentration of power and avoid the exploitation of vulnerable actors. Finally, it assumes that no single funding source can support all ecosystem functions. Instead, finance must be blended to align different funder mandates with specific system needs.

The ecosystem-building model rests on a different set of assumptions than the other five approaches. Instead of presuming that scale, training or informal flexibility will spontaneously create circularity, it assumes that system-level conditions must be deliberately constructed. This aligns with circular-economy critiques that describe the concept as "essentially contested" and warn against technocratic, firm-centered formulations that downplay questions of governance, justice and institutional design (Korhonen et al. 2018). In South Africa, where inequality and historical exclusion shape industrial participation, a viable circular textile system must therefore foreground sufficiency, redistribution, institutional coherence and meaningful participation, consistent with just transitions scholarship (Swilling and Annecke 2012; Swilling et al. 2016).

The approach further assumes that durable circular economies require stable feedstock, financed logistics, shared pre-processing capacity and coordinated market development before advanced technologies or small-scale entrepreneurship can succeed. Global textile recycling research strongly supports this assumption. Studies consistently identify fiber heterogeneity, contamination, lack of dedicated collection infrastructure and weak market demand as key barriers to viability (Juanga-Labayen et al. 2022; Abrishami et al. 2024; Biyada and Urbonavičius 2025). Without systems that address these structural barriers, both technology-centric and livelihood-oriented models continue to underperform.

18.7.4 Benefits of Adopting an Ecosystem-Building Approach

The ecosystem-building approach provides a more plausible and sustainable pathway because it directly addresses the structural weaknesses revealed by the technology-centric, training-led, pilot-driven, informal-sector and production-led livelihood models. By prioritizing feedstock logistics, standards and pre-processing infrastructure, it eliminates the technology-centric problem of constructing industrial plants without reliable material flows. It embeds training within multiyear support structures that include access to machinery, working capital, workspace and procurement, preventing the chronic underemployment that follows standalone skills programs. By designing pilots as explicit stages within long-term institutionalization pathways, it prevents the abrupt collapse of operations at the end of grant cycles and ensures that successful experiments transition into scalable systems.

The approach also tackles the vulnerabilities of informal-sector-dependent models by establishing the necessary physical infrastructure, payment guarantees and governance rights that enable informal actors to contribute under fair and secure conditions (Schröder 2020; Calvo et al. 2020). It resolves the production-led model's tendency toward oversupply by demanding validated demand, coordinated procurement and sufficiency-based production strategies before manufacturing begins (Burnstine and Camargo 2025). In doing so, it reorients circularity away from assumptions of infinite market absorption and toward realistic, socially just and environmentally responsible modes of economic participation. The ecosystem-building approach thus offers a more honest account of what it takes to establish a circular textile industry capable of delivering environmental benefits and meaningful livelihoods under South Africa's current macroeconomic conditions.

The ecosystem-building model is strengthened by theoretical and empirical insights into why transitions fail under fragmented or single-lever approaches. Turnheim and Sovacool (2020) identify overreliance on technological substitution, weak coordination, unrealistic expectations and neglect of institutional and cultural dynamics as core reasons for transition failure. These dynamics are visible in South Africa's textile circularity landscape, where machinery remains underutilized, pilots collapse after grant withdrawal, skilled artisans lack markets and informal recyclers face untenable burdens without compensation or infrastructure. An ecosystem-building strategy directly counters these failure modes by designing interdependencies explicitly rather than assuming they will emerge organically.

Empirical evidence on textile recycling further underlines the need for ecosystem coordination. Large-scale global recyclers report that the sorting and categorization of textile waste into hundreds of fractions are labor-capital- and technology-intensive, requiring stable feedstock, specialized machinery, trained operators and predictable markets (Abrishami et al. 2024). Other studies highlight the economic impracticality of recycling textiles when virgin materials are cheaper, markets are fragmented and regulatory frameworks are weak or inconsistent (Juanga-Labayen et al. 2022). These structural barriers cannot be resolved by training, pilots or small-scale entrepreneurial production alone. They require systemic interventions such as funded collection systems, hub-based pre-processing infrastructure, shared machinery platforms, long-term procurement commitments and governance models capable of coordinating dispersed actors (GreenCape 2023).

Reframing textile circularity through an ecosystem-building lens means approaching transition as a sequence of coordinated system functions rather than as a set of isolated projects. This aligns with innovation systems frameworks that identify core system functions—knowledge development, market formation, resource mobilization, legitimation and alignment—as prerequisites for stabilizing new industrial trajectories (Hekkert et al. 2007). It also echoes the principles of just transitions, which argue that sustainability cannot be achieved through efficiency alone but must involve institutional arrangements that redistribute risks and benefits while supporting meaningful, low-carbon livelihoods (Swilling and Annecke 2012).

The ecosystem-building model therefore prioritizes the construction of minimum viable ecosystems: funded and governed feedstock systems; shared pre-processing and makerspace facilities; contractual or policy-based market commitments;

blended financing structures; and cooperative, franchise or multi-actor governance mechanisms. Only once these foundational elements stabilize can more advanced processing technologies be integrated, as premature technological installation risks reproducing the failures documented in both South Africa and comparable international contexts (Salmenperä 2023). Through this reframing, the ecosystem-building model offers not merely a sectoral strategy, but a pathway aligned with deep global transition trends, critical circular-economy scholarship and the imperatives of justice-oriented socio-technical transformation. It provides a more realistic basis for building a circular textile economy capable of generating environmental benefits, institutional resilience and equitable livelihoods in South Africa.

18.7.5 Pathways to Practical Implementation

Translating the ecosystem-building model into practice requires a sequence of deliberate, context-sensitive actions. Implementation typically begins with a system-wide mapping exercise that identifies all existing and potential actors across the textile value chain from informal collectors and community organizations to designers, recyclers, manufacturers, brands, retailers, municipalities and funders (Hitchen et al. 2022). This mapping must clarify not only who is present but also what capacities, resources and constraints shape their contributions. Such analysis reveals the "missing middle" infrastructures, institutional voids and fragmented relationships that prevent actors from collaborating effectively (Swilling and Annecke 2012).

Following this, multi-stakeholder platforms can be convened to facilitate dialogue, align interests and co-develop a shared vision for the circular textile ecosystem. These platforms enable stakeholders to articulate roles, responsibilities, governance principles and priority investments. At the operational level, implementation often involves establishing community-based collection and pre-processing hubs equipped with sorting tables, balers, cutting equipment, testing tools and data systems. These hubs serve as foundational infrastructure, ensuring a reliable supply of feedstock for both industrial recyclers and artisanal producers. They can be embedded within cooperatives, social enterprises, municipal facilities or hybrid structures that combine community ownership with formal oversight.

Linked to these hubs, regional makerspaces or micro-factories can provide access to shared machinery, technical expertise, product development support and incubation services for enterprises seeking to create value from recycled textiles. In parallel, the ecosystem must invest in market development and demand creation, particularly through establishing forward orders, long-term procurement commitments or preferential procurement arrangements with institutional buyers such as retailers, hotels, interior designers and public entities. Governance structures, ranging from cooperative federations and social franchise networks to multi-actor oversight bodies, must be developed to manage shared assets, define standards, distribute revenue, resolve disputes and monitor impact.

Data and monitoring systems are crucial for tracking material flows, employment levels, income generation, environmental impacts and compliance with standards (Nahman et al. 2024). Over time, as the ecosystem stabilizes, it can integrate more advanced technologies, larger plants or specialized processes. However, these upgrades should only occur once feedstock supply, market demand and governance arrangements have reached adequate levels of maturity. In this way, the ecosystem-building model provides a stepwise, adaptive and inclusive pathway for developing a circular textile sector that generates both environmental benefits and economic advantages.

Last but not least, a central pillar of the ecosystem-building model is the strategic use of blended finance. The costs associated with developing collection systems, pre-processing infrastructure, shared makerspaces, enterprise incubation, data platforms and governance arrangements are substantial and cannot be carried out by any single institution. Philanthropic capital is well suited to early-stage experimentation, community mobilization and the establishment of shared physical spaces such as makerspaces, hubs or pre-processing centers. Skills-development funding, derived from B-BBEE obligations or Sector Education and Training Authority (SETA) budgets, can support training, stipends and long-term mentorship needed to integrate artisans, informal workers and small enterprises. CSI funding can be allocated to shared machinery, product development, branding and community enterprise support, particularly where the beneficiaries reside in historically disadvantaged areas.

Government grants and industrial policy instruments are essential for financing infrastructure upgrades, standards development, market development programs and job creation. EPR funds provide critical resources for feedstock logistics, quality-control, traceability systems and data platforms that track material flows and environmental outcomes. Concessional finance, such as low-interest loans, guarantees and first-loss capital, can support the acquisition of capital-intensive technologies once other parts of the ecosystem are functioning reliably. Blending these diverse funding streams creates a financial architecture in which each funder contributes according to its mandate, risk appetite and time horizon. The pooling of resources prevents overburdening any single mechanism, such as EPR levies, and ensures that the full range of ecosystem functions receives adequate support.

18.8 CONCLUSION

This chapter argued that South Africa's pursuit of circularity in the textile sector is unlikely to succeed through isolated, single-lever interventions. The persistent shortcomings of technology-led, human-capital-led, pilot-driven, informal-sector and production-oriented approaches underscore the challenges faced by fragmented efforts within the local context, which is characterized by weak infrastructure, incoherent markets and institutional gaps. In response, this chapter proposes an ecosystem-building model that treats circularity as a socio-technical transformation requiring coordinated development of feedstock systems, shared pre-processing capacity, inclusive governance and blended finance. This system-oriented reframing echoes the imperatives outlined in the UNEP (2023) Global Roadmap, which emphasizes that global textile circularity requires integrated policy environments, investment in shared infrastructure, enhanced traceability, inclusive governance and cross-value chain collaboration. Perhaps, we need to acknowledge that textile circularity is not so much a matter of scaling isolated innovations, but rather of creating enabling environments, including funded collection systems, harmonized standards, market development mechanisms and institutional coordination across public and private actors.

However, several limitations must be acknowledged. Foremost, this chapter provides only a conceptual view, which may underplay the context-specific variability and localized constraints across South Africa's diverse textile landscape. The critique directed toward conventional mainstream approaches, therefore, mostly relies on secondary literature. Hence, more empirical evidence from local pilot sites, recyclers and informal value chain participants is needed to strengthen conclusions. In this regard, the ecosystem-building model, while promising, remains a theoretical and strategic framework for now, requiring further practical implementation and testing, along with detailed costing and governance experimentation. Future research should therefore deepen empirical understanding of how circular textile waste management and recycling initiatives perform in practice. Such research may also want to explore regionally specific ecosystem-building pathways, assess the institutional and financial requirements of minimum viable ecosystems and investigate governance models capable of enabling inclusive participation. Greater attention would have to be devoted to market behavior, demand formation and the lived experiences of informal actors, as this will be essential in achieving the desired outcomes. Advancing these research directions will help refine the ecosystem-building approach, supporting a more just, resilient and context-responsive circular textile transition in South Africa.

REFERENCES

Abrishami, S., Baig, G. A., Moztarzadeh, F., Li, S. and Shahidi, S. 2024. Textile recycling and recovery: An eco-friend perspective on textile and garment industries challenges. *Textile Research Journal* 94(23–24): 2825–2844.

Andersen, A. D. and Gulbrandsen, M. 2020. The innovation and industry dynamics of technology phase-out in sustainability transitions: Insights from diversifying petroleum technology suppliers in Norway. *Energy Research and Social Science* 64(2020): 101447.

Berkhout, F., Smith, A. and Stirling, A. 2004. Socio-technological regimes and transition contexts. In *System Innovation and the Transition to Sustainability: Theory, Evidence and Policy* (Eds. Elzen, B., Geels, F. W., & Green, K.), 44(106), Cheltenham: Edward Elgar, 48–75.

Biyada, S. and Urbonavičius, J. 2025. Circularity in textile waste: Challenges and pathways to sustainability. *Cleaner Engineering and Technology* 24: 100905.

Burnstine, A. and Camargo, A. 2025. Fashioning the future: Bio-based textiles, circular innovation, and sustainability in emerging markets. *Cleaner and Circular Bioeconomy* 12: 100194.

Calvo, S., Morales, A., Núñez-Cacho Utrilla, P. and Guaita Martínez, J. M. 2020. Addressing sustainable social change for all: Upcycled-based social creative businesses for the transformation of socio-technical regimes. *International Journal of Environmental Research and Public Health* 17, 2527.

Cavusgil, S. T., Deligonul, S., Kardes, I. and Cavusgil, E. 2018. Middle-class consumers in emerging markets: 509 Conceptualization, propositions, and implications for international marketers. *Journal of International Marketing* 26(3): 94–108.

Cernansky, R. 2024. What Renewcell's bankruptcy says about next-gen fashion: Once a sustainable fashion darling, Renewcell's failure proves systemic change can't happen without significant industry overhaul. Vogue. https://www.vogue.com/article/what-renewcells-bankruptcy-says-about-next-gen-fashion [Accessed 10 January 2026].

Coenen, L., Benneworth, P. and Truffer, B. 2012. Toward a spatial perspective on sustainability transitions. *Research Policy* 41(6): 968–979.

Council for Scientific and Industrial Research (CSIR). 2021. The circular economy as a development opportunity. Pretoria, South Africa. https://www.circulareconomy.co.za/wp-content/uploads/2021/12/CSIR-2021-Circular-Economy-As-Development-Opportunity.pdf [Accessed 10 January 2026].

Department of Environmental Affairs. 2020. National waste management strategy. https://www.dffe.gov.za/sites/default/files/docs/nationalwaste_management_strategy.pdf [Accessed 15 January 2026].

Department of Trade, Industry and Competition. 2019. Retail–Clothing, Textiles, Footwear and Leather (R-CTFL) Masterplan 2030. Government of South Africa. https://www.thedtic.gov.za [Accessed 7 January 2026].

Department of Trade, Industry and Competition. 2021. Clothing, textiles, footwear and leather growth program (CTFLGP): Program guidelines. Government of South Africa. https://www.thedtic.gov.za [Accessed 5 January 2026].

Eitan, A. and Hekkert, M. P. 2023. Locked-in transitions? Towards a conceptualization of path-dependence lock-ins in the renewable energy landscape. *Energy Research and Social Science* 106: 103316.

Eppinger, E. 2022. Recycling technologies for enabling sustainability transitions of the fashion industry: Status quo and avenues for increasing post-consumer waste recycling. *Sustainability: Science, Practice and Policy* 18(1): 114–128.

European Commission. 2022. EU strategy for sustainable and circular textiles (COM (2022) 141 final). https://environment.ec.europa.eu/document/download/74126c90-5cbf-46d0-ab6b-60878644b395_en?filename=COM_2022_141_1_EN_ACT_part1_v8.pdf [Accessed 6 January 2026].

Fischer, G. and Herrmann, T. 2011. Socio-technical systems: A meta-design perspective. *International Journal of Sociotechnology and Knowledge Development* 3(1): 1–33.

Geels, F. W. 2002. Technological transitions as evolutionary reconfiguration processes: A multi-level perspective and a case-study. *Research Policy* 31(8–9): 1257–1274.

Geels, F. W. and Schot, J. 2007. Typology of sociotechnical transition pathways. *Research Policy* 36(3): 399–417.

Godfrey, L., Scott, D. and Trois, C. 2013. Caught between the global economy and local bureaucracy: The barriers to good waste management practice in South Africa. *Waste Management and Research* 31(3): 295–305.

GreenCape. 2019. Building a resilient clothing and textile industry in Cape Town: Resilience case study. GreenCape. https://greencape.co.za/library/resilience-case-study-building-a-resilient-clothing-and-textile-industry-in-cape-town/ [Accessed 8 December 2025].

GreenCape. 2023. Textile recycling and fiber recovery: Industry brief. GreenCape. https://greencape.co.za/wp-content/uploads/2023/06/Textile-recycling-and-fibre-recovery_industry_brief. [Accessed 8 December 2025].

Grin, J., Schot, J., Rotmans, J., Geels, F. W. and Loorbach, D. 2010. *Transitions to Sustainable Development: New Directions in the Study of Long-Term Transformative Change.* Routledge, New York.

Grönwall, J. and Jonsson, A. C. 2017. The impact of 'zero' coming into fashion: Zero liquid discharge uptake and socio-technical transitions in Tirupur. *Water Alternatives* 10(2): 602–624.

Hitchen, G., Ekmekcioglu, E. and Ozcinar, C. 2022. Mapping the virtual production ecosystem. Report of a research project undertaken by the Creative Research and Innovation Centre at Loughborough University, London.

Jayarathna, R. V., Saddler, J. B. and Clarke, M. 2025. Skills for a circular economy transformation: A regional manufacturing-level analysis. *International Journal of Training and Development* 29(1): 45–66.

Juanga-Labayen, J. P., Song, X. and Li, X. 2022. A review on textile recycling practices and challenges. *Textiles* 2: 174–188.

Kanger, L. and Schot, J. 2019. Deep transitions: Theorizing the long-term patterns of socio-technical change. *Environmental Innovation and Societal Transitions* 32: 7–21.

Klitkou, A., Bolwig, S., Hansen, T. and Wessberg, N. 2015. The role of lock-in mechanisms in transition processes: The case of energy for road transport. *Environmental Innovation and Societal Transitions* 16: 22–37.

Korhonen, J., Honkasalo, A. and Seppälä, J. 2018. Circular economy: The concept and its limitations. *Ecological Economics* 143: 37–46.

McKinsey and Company. 2020. The state of fashion 2020: Navigating uncertainty. https://www.mckinsey.com/industries/retail/our-insights/the-state-of-fashion-2020-navigating-uncertainty [Accessed 8 December 2025].

Miörner, J., Binz, C. and Fuenfschilling, L. 2021. Understanding transformation patters in different socio-technical systems: A scheme of analysis. GEIST: Geography of Innovation and Sustainability Transitions. *GEIST Working Series* 2021(11): 1–29.

Nahman, A., Goga, T., van Deventer, H., Oelofse, S., Haywood, L., Davies, S., Musvoto, C. and Stafford, W. 2024. Towards a South African circular economy monitoring system (SACEMS). https://www.circulareconomy.co.za/wp-content/uploads/2025/04/60-CSIR_Final_Technical-report_SACEMS_draft-indicators.pdf [Accessed 10 December 2025].

Nkatha, K. 2023. How fast fashion is fuelling the fashion waste crisis in Africa. https://www.greenpeace.org/africa/en/blog/54589/how-fast-fashion-is-fuelling-the-fashion-waste-crisis-in-africa/ [Accessed 10 December 2025].

Papachristos, G. 2014. Towards multi-system sociotechnical transitions: Why simulate. *Technology Analysis and Strategic Management* 26(9): 1037–1055.

Parliamentary Research Unit. 2025. Industrialization and employment: Supporting industrial development and expanding employment opportunities to benefit the youth. https://www.parliament.gov.za [Accessed 10 December 2025].

Prevent Waste Alliance. 2022. Development from a voluntary to a mandatory extended producer responsibility scheme in South Africa: Country report. Bonn, Germany.

Ramirez, P. 2021. Technological revolutions, socio-technical transitions and the role of agency: Värmland's transition to a regional bioeconomy. *Regional Studies* 55(10–11): 1642–1651.

Reike, D., Hekkert, M. P. and Negro, S. O. 2022. Understanding circular economy transitions: The case of circular textiles. *Business Strategy and the Environment* 32(3): 1032–1058.

Republic of South Africa. 2003. Broad-Based Black Economic Empowerment Act, No. 53 of 2003 (as amended by Act 46 of 2013). Government Gazette, 9 January 2004.

Salmenperä, H. 2023. Governing the transition to circularity of textiles: Finnish companies' expectations of interventions for change. *Circular Economy and Sustainability* 3: 1747–1767.

Schröder, P. 2020. *Promoting a Just Transition to an Inclusive Circular Economy*. London: Chatham House.

Snoek, S. 2017. Circular economy in the textile industry: Transition theory in Dutch start-ups towards a circular economy. MSc thesis, Wageningen University.

Sonnenberg, N., Jacobs, B. and Momberg, D. 2014. The role of information exposure in female university students' evaluation and selection of eco-friendly apparel in the South African emerging economy. *Clothing and Textiles Research Journal* 32(4): 266–281.

Sonnenberg, N. C., Patnaik, S. and Romo, G. M. 2024. Pursuing zero waste in the South African textile industry: A case study of SMME-scaled needle felting technology. In *Zero Waste Sustainable Apparel Industry* (Ed. Muthu, S. S.), Cham: Springer Nature Switzerland, 117–141.

South Africa. 2000. Preferential Procurement Policy Framework Act (Act No. 5 of 2000). https://www.gov.za/documents/acts/preferential-procurement-policy-framework-act-5-2000-03-feb-2000 [Accessed 10 December 2025].

Swilling, M. and Annecke, E. 2012. *Just Transitions in a Complex World.* UCT Press, Cape Town.

Swilling, M., Musango, J. and Wakeford, J. 2016. Developmental states and sustainability transitions: Prospects of a just transition in South Africa. *Journal of Environmental Policy & Planning* 18(5): 650–672.

Turnheim, B. and Sovacool, B. K. 2020. Exploring the role of failure in socio-technical transitions research. *Environmental Innovation and Societal Transitions* 37: 267–289.

United Nations Environment Program (UNEP). 2023. *Sustainability and Circularity in the Textile Value Chain: A Global Roadmap*. Paris: UNEP.

Whitfield, L. and Mkhabela, V. 2022. The business strategies of South African textile firms and global trends in 4IR and sustainability technologies. SARChI Industrial Development Working Paper Series WP 2023-01. University of Johannesburg.

19 Exploring Green Innovation Strategies and ESG Levels in the Green Supply Chain under Cost-Sharing

Sarat Kumar Jena

19.1 INTRODUCTION

Over the decades, garment firms consume a huge amount of packaging materials such as plastic, cartons, bags, tape, foam fillers, and cardboard boxes, because of the rapid development of e-commerce and online shopping, which generates a huge amount of waste and greenhouse gases that pollute the environment. The garment industry contributes significantly to India's 9.46 million tons of plastic waste each year, of which 40% remains uncollected and 43% is used for packaging, most of which is single use (PTI 2019). This excessive waste of resources and the resulting environmental pollution pose a significant social burden. To support low-carbon and sustainable development, reducing packaging waste is crucial (Li et al. 2021; Zhang et al. 2024).

To reduce garment packaging waste and promote sustainability, many garment firms have adopted initiatives within the green supply chain (GSC). Additionally, green innovation improves express packaging recycling value. To improve express packaging recycling value, some firms have adopted green innovations. For instance, H&M provides a garment collection program through which customers can return used clothing and packaging materials for recycling or reuse (Kavilanz 2020). Adidas has taken a similar approach, using recyclable packaging alternatives and reusable bags to reduce its environmental impact.

Sustainability practices are becoming more important, but implementing green innovations can be difficult. Despite their long-term benefits and enhancement of corporate reputation, these investments have high short-term costs, especially in industries with significant environmental and social impacts. Therefore, some downstream and upstream firms share the financial burden. Adidas, for instance, is collaborating with suppliers to develop recyclable packaging, such as reusable and collapsible bags. Garment firms and retailers can collaborate on green innovation. Despite green innovation's potential for reducing logistics or purchase volumes, increased

DOI: 10.1201/9781003689225-19

product traceability may result in sales losses. Firms must carefully balance green innovation's benefits and drawbacks.

Firms now face social and governance pressures beyond environmental concerns and profit maximization (Dmytriyev et al. 2021; Cerciello et al. 2023) that drive them to adopt environmental, social, and governance (ESG) principles. Integrating ESG into corporate decision-making means integrating sustainability, social justice, and governance. Due to issues such as poor labor conditions, unethical supply chains, and corporate scandals, firms must implement ethical labor policies and ensure transparent governance.

For instance, H&M ensures fair wages and good working conditions for its suppliers through fair trade and ethical sourcing policies (H&M Group 2025).

Studies that examine green innovations and social welfare practices within garment package delivery services are in their infancy, considering their objectives within garment firms' decision-making strategies. Currently, no studies have been carried out on how garment firms and suppliers can maximize profits while achieving ESG goals. Due to the aforementioned background, this study explores the following research questions: (1) How will retailers and garment firms' profit be affected under non- and no green innovation cost-sharing? (2) What is the impact of green innovation investment on supply chain profits?

For the purpose of answering the above questions, we devised a two-stage Stackelberg game model of GSC consisting of a garment manufacturer and a retailer, where the manufacturer produces sustainable garments at an ESG level and recycles the used products, whereas the retailer sells the garments (e.g., clothing). In this model, we examine two scenarios: (1) nongreen innovation cost-sharing and (2) green innovation cost-sharing.

19.2 LITERATURE REVIEW

This section describes how green innovation and ESG level impact the closed-loop supply chain in a packaging manufacturing firm. This study connects two main streams: GSC and ESG in supply chain.

19.2.1 Green Supply Chain

Many scholars have studied GSC by considering the impact of product greenness levels or customers' green preference behavior. As part of their study, Heydari et al. (2019) constructed three types of dual-channel GSCs: open, closed, and transitional, considering the manufacturer's green investment. Recently, researchers have examined how eco-labels influence customers' purchasing decisions. In their study, Wang et al. (2025) contend that eco-labels incorporating descriptive textual information enhance customers' cognitive fluency regarding label information and purchase intent more effectively than single-icon labels. According to Seman et al. (2019), GSCs, green innovation, and environmental performance are positively correlated. Furthermore, their study found that green innovation enhances environmental performance. Additionally, Le et al. (2022) conclude that firms' sustainable performance is influenced by their adoption of green innovation when looking at corporate social responsibility (CSR), sustainable corporate performance, and green innovation. Yang and Zhang (2023) examine the impact of emissions reduction efforts by manufacturers by considering a differential game model. For reducing emissions, collaboration strategies result in higher financial returns. A study by Pan et al. (2024) examines different policies for reducing carbon emissions using Stackelberg's game theory. The fluctuation in carbon prices affects manufacturers' and retailers' strategic decisions, which ultimately affect profitability. In addition, Li et al. (2024) found that digital finance can enhance green innovation in corporate supply chains, thereby improving supply chain efficiency. Zhao et al. (2025) report an information asymmetry in dual-channel GSCs when it comes to profits. While their study considered how consumer preferences influence supply chain decisions, it did not fully consider how channel selection affects supply chain decisions as well. According to Taleizadeh et al. (2025), dual-channel sales agreements and cost-sharing agreements can be used to balance economic performance and environmental sustainability.

There is a good deal of literature on the relationship between green innovation and environmental performance. However, to our knowledge, no study has assessed green innovation's ESG impact on supply chain profit. Therefore, we investigate green innovation's effects on supply chain profitability to decrease these gaps.

19.2.2 ESG in Supply Chain

The United Nations Global Compact introduced ESG in 2004 with "Who Cares Wins." It has since attracted government and regulatory attention (Eccles et al. 2020). Folqué et al. (2021) emphasize that ESG is increasingly concerned with environmental challenges, such as carbon emissions and biodiversity loss. Tan and Zhu (2022) argue that ESG principles align corporate goals with society's values and achieve sustainable economic growth. The "social" aspect of ESG is particularly important when it comes to supply chain financing, according to Baid and Jayaraman (2022). CSR is increasingly incorporated into supply chain strategies to improve sustainability (Kuo et al. 2023). Bade et al. (2024) examine the impact of German pharmaceutical companies' ESG maturity on supply chain security.

Further, all of these studies emphasize the importance of ESG factors in optimizing the sustainability and performance of supply chains. Despite this, none of the studies have examined how to achieve ESG levels while sustaining profits through green innovation. To maximize ESG performance for green companies in terms of environmental impact, consumer surplus, and profit, this study has been carried out.

19.3 MODEL

In this section, we discuss how ESG levels and green innovation can affect a garment manufacturer's GSC. According to this model, the optimal selling price, wholesale price, and green innovation effort are determined under both cost-sharing and non-cost-sharing scenarios. Model notations are provided in Table 19.1.

The model in this study involves a GSC system consisting of a garment manufacturer (he) and a retailer (she) who sells garments to customers. The manufacturer produces new packages at a unit cost of c and supplies them to the retailer at a unit

TABLE 19.1
Notations Used in This Study Model

Symbol	Description
M and R	Represent garment manufacturer and retailer, respectively
Model C	Green innovation cost-sharing between retailer and garment manufacturer
NC	Non-cost-sharing
i	Model C and Model NC
D	Market demand of the product
t	Garment manufacturer's degree of concern toward ESG $(0 \le t \le 1)$
α	The market size of the garment manufacturer
β	Price elasticity of the product
λ	Green innovation effort cost coefficient
η	Green innovation elasticity of package demand
θ	Environmental performance coefficient
π_M	Profit of the garment manufacturer
π_R	Profit of the retailer
π_T	Total profit of the GSC
EPI	Environmental performance index
CS	Consumer surplus
Decision Variables	
p	Selling price of the new green innovation package
S	Green innovation effort by the package supplier
w	Wholesale price of the new green innovation package

wholesale price of w. For simplicity, the manufacturing cost c is normalized to 0. Here, garment manufacturers invest in green packaging with the green effort formulated as a quadratic function $\frac{\lambda S^2}{2}$ which shows the increasing recycling times, where $\lambda(\lambda > 0)$ denotes cost coefficient (Ghosh and Shah 2015). For example, SF Express's "π-box" use polypropylene (PP) environmentally friendly honeycomb materials, and each box can be recycled for more than 70 times; this dramatically increases the recyclability of express packages.

The packaging manufacturing process must be optimized to maximize profit by determining the optimal selling price (p) and green innovation efforts (G). This optimization supports garment manufacturers' investment in green innovation to create environmentally friendly products and enhance their ESG performance. Following Kurata et al. (2007), Huang and Swaminathan (2009), and Jena and Singhal (2023), the demand function for the new green express package product is linear in the self-price and green innovation efforts, expressed as $D = (\alpha - \beta p + \eta S)$, where β is the sensitivity coefficient of package price and η is the sensitivity coefficient of green innovation level.

Furthermore, manufacturers of garments are encouraged to manage their supply chains sustainably. Green innovation requires integrating environmental and social concerns into governance and decision-making frameworks. Therefore, the study addresses ESG considerations while maximizing profit for garment manufacturers. Generally, firms' social responsibility is evaluated based on the surplus generated by their stakeholders (Panda 2014). "Consumer surplus (CS) is defined as the difference between the market price a customer pays for a product and the maximum price the customer is willing to pay" (Panda 2014). In this study, consumer surplus is modeled based on Jena and Singhal (2023), as shown in Equation (19.1):

$$\mathrm{CS} = \int_{p-\min}^{p-\max} D\,dp = \frac{D^2}{2\beta} \tag{19.1}$$

When a packaging supplier invests in green innovation, their environmental impact per unit decreases by θS, where $\theta > 0$ indicates the effectiveness of green innovation in improving environmental outcomes. Moreover, the GSC's environmental performance, measured by the environmental performance index (EPI), improves with the introduction of the new green express package. In this study, EPI is calculated using the indices proposed by Raz and Ovchinnikov (2015) and Yenipazarli (2017) as $\mathrm{EPI} = \theta \mathrm{SD}$. By integrating both EPI and consumer surplus (CS), ESG can be expressed as shown in Equation (19.2):

$$\mathrm{ESG} = t(\mathrm{EPI} + \mathrm{CS}) = t\left(\frac{D^2}{2\beta} + \theta SD\right), \tag{19.2}$$

where t represents the degree of ESG concern from the garment manufacturer.

19.3.1 Model No-Cost-Sharing (NC)

Generally, garment manufacturers and retailers work independently to maximize their profits. Therefore, the retailer places the order with the garment manufacturer, and then the garment manufacturer fulfills the order. In this scenario, however, the garment manufacturer invests in green innovation in order to minimize the amount of packaging waste. Additionally, garment manufacturers maximize their profit while maximizing the ESG level. In this scenario, it is assumed that the garment manufacturer acts as a leader and the retailer acts as a follower. The profit of the retailer and garment manufacturer can be calculated using Equations (19.3) and (19.4):

$$\max_{p} \pi_R^{NC} = D(p - w) \tag{19.3}$$

$$\max_{w,\,S} \pi_M^{NC} = Dw - \frac{\lambda S^2}{2} + t(\mathrm{CS} + \mathrm{EPI}). \tag{19.4}$$

π_R^{NC} is concave in p as $\dfrac{\partial\, \pi_R^{NC2}}{\partial^2 p} = -2\beta < 0$. By using backward induction method, we can obtain the equilibrium value of p.

$$p = \frac{\alpha + w\beta + S\eta}{2\beta} \tag{19.5}$$

Here, the equilibrium of retail price is put into Equation (19.3). Based on the first-order condition of (19.4), garment manufacturer's reaction function can be calculated:

$$w = \frac{(-2+t)(\alpha + S\eta) + 2St\beta\theta}{(-4+t)\beta} \tag{19.6}$$

$$S = -\frac{2w\beta\eta + t(\alpha - w\beta)(\eta + 2\beta\theta)}{t\eta(\eta + 4\beta\theta) - 4\beta\lambda} \tag{19.7}$$

Solving Equations (19.5), (19.6), and (19.7) simultaneously, the optimal equilibrium solution $\left(p^*, w^*, S^*\right)$ can be derived.

$$p^{NC*} = \frac{t\alpha\theta(\eta + t\beta\theta) + (-3+t)\alpha\lambda}{(\eta + t\beta\theta)^2 + (-4+t)\beta\lambda} \tag{19.8}$$

$$w^{NC*} = \frac{t\alpha\theta(\eta + t\beta\theta) + (-2+t)\alpha\lambda}{(\eta + t\beta\theta)^2 + (-4+t)\beta\lambda} \tag{19.9}$$

$$S^{NC*} = -\frac{\alpha(\eta + t\beta\theta)}{(\eta + t\beta\theta)^2 + (-4+t)\beta\lambda} \tag{19.10}$$

19.3.2 Model Cost-Sharing (C)

Generally, garment manufacturers and retailers work to maximize their profits. In Model C, we consider that green innovation for package is conducted by the garment manufacturer, and retailer shares the proportion ρ of the investment cost. The optimization problem of the retailer is

$$\max_{p} \pi_R^C = D(p - w) - \frac{\rho\lambda S^2}{2}. \tag{19.11}$$

The optimization problem of the manufacturer is as follows:

$$\max_{w,\,S} \pi_M^C = Dw - \frac{(1-\rho)\,\lambda S^2}{2} + t(\text{CS} + \text{EPI}). \tag{19.12}$$

π_R^C is concave in p as $\frac{\partial \pi_R^{C2}}{\partial^2 p} = -2\beta < 0$. By using backward induction method, we can obtain the equilibrium value of p.

$$p = \frac{\alpha + w\beta + S\eta}{2\beta} \tag{19.13}$$

Here, the equilibrium of retail price is put into Equation (19.12). We can calculate the garment manufacturer's reaction function based on the first-order condition (19.4):

$$w = \frac{-2\alpha + t\alpha - 2S\eta + St\eta + 2St\beta\theta}{(-4+t)\beta} \tag{19.14}$$

$$S = \frac{-t\alpha\eta - 2w\beta\eta + tw\beta\eta - 2t\alpha\beta\theta + 2tw\beta^2\theta}{t\eta^2 + 4t\beta\eta\theta - 4\beta\lambda + 4\beta\lambda\rho} \tag{19.15}$$

Solving Equations (19.13), (19.14), and (19.15) simultaneously, the optimal equilibrium solution (p^*, w^*, S^*) can be derived:

$$p^{C*} = \frac{\alpha\left(-3\lambda + t\left(\theta(\eta + t\beta\theta) + \lambda\right) - (-3+t)\lambda\rho\right)}{\eta^2 + 2t\beta\eta\theta + \beta\left(-4\lambda + t\left(t\beta\theta^2 + \lambda\right) - (-4+t)\lambda\rho\right)} \tag{19.16}$$

$$w^{NC*} = \frac{\alpha\left(-2\lambda + t\left(\theta(\eta + t\beta\theta) + \lambda\right) - (-2+t)\lambda\rho\right)}{\eta^2 + 2t\beta\eta\theta + \beta\left(-4\lambda + t\left(t\beta\theta^2 + \lambda\right) - (-4+t)\lambda\rho\right)} \tag{19.17}$$

$$S^{NC*} = -\frac{\alpha(\eta + t\beta\theta)}{\eta^2 + 2t\beta\eta\theta + \beta\left(-4\lambda + t\left(t\beta\theta^2 + \lambda\right) - (-4+t)\lambda\rho\right)} \tag{19.18}$$

19.4 NUMERICAL ANALYSIS

The numerical study of the model is presented in this section. Due to challenges in collecting real-time data for variables with multiple attributes, fictitious yet valid data were used. The models were validated by initializing most parameters based on the existing literature. Hypothetical data for parameters were sourced from previous studies (Chen et al. 2017; Jena and Singhal 2023). Parameters were set as follows: $\eta = 2, \beta = 5, \theta = 20, \lambda = 100, t = 0.1, \alpha = 400, \rho = 0.5$. Table 19.2 shows that wholesale prices are higher in the non-cost-sharing than in the cost-sharing strategy.

In the green innovation cost-sharing strategy, the retailer compensates for the additional costs of green innovation by raising the selling price. The retailer and manufacturer share some of the green innovation costs, which increases profits for

TABLE 19.2
Numerical Results of the Models

Model	W	P	S	D	ESG	EPI	CS	π_R	π_M	π_T
Model NC	38.91	59.64	0.932	10.36	15.57	48.30	107.46	214.89	414.49	629.38
Model C	38.85	59.80	1.89	10.47	20.85	98.74	109.76	210.61	419.10	629.71

the manufacturer and total supply chain. By investing more in green innovation, customers become more environmentally conscious and place more orders. Due to a higher EPI in Model C (cost-sharing) than in Model NC (no-cost-sharing), the Model C's ESG level is higher. Additionally, consumer surplus is higher in Model C. Therefore, it is crucial for garment manufacturers to work collaboratively with retailers to increase profits and overall supply chain profits.

19.4.1 Sensitivity Analysis

This section conducts a sensitivity analysis to evaluate the impact of varying key parameters—green innovation effort cost coefficient (λ), the degree of ESG concern (t), and cost-sharing (ρ)—on total GSC profit under Model NC and Model C.

19.4.2 Green Innovation Effort Cost Coefficient (λ)

The effect of the green innovation cost coefficient (λ) on the total profit of the GSC under cost-sharing and non-cost-sharing scenarios is shown in Figure 19.1. The figure illustrates that as the cost of green innovation increases, the overall profit of the GSC decreases. This occurs because higher investment in green innovation leads the garment

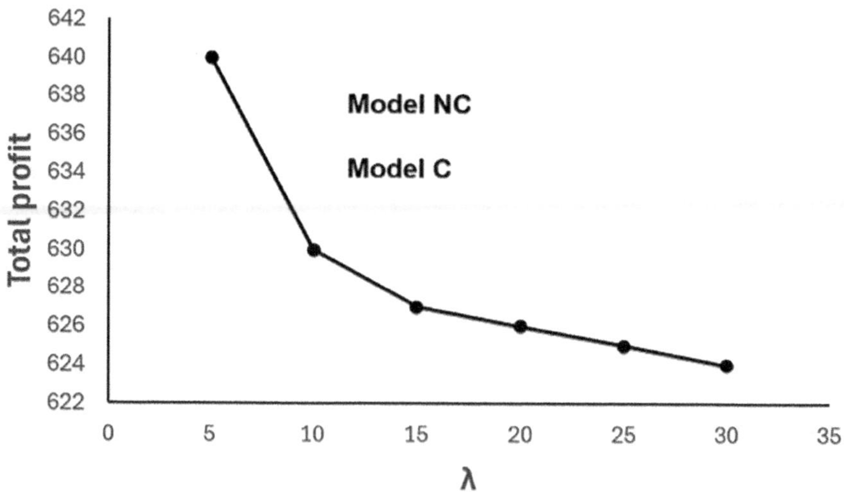

FIGURE 19.1 Impact of λ on green supply chain profit.

manufacturer to raise the wholesale price to offset the cost, which in turn increases the selling price and reduces product demand. Consequently, garment manufacturers should carefully consider what level of investment they should make in green packaging.

19.4.3 The Degree of ESG Concern (t)

Figure 19.2 shows the impact of ESG concern (t) on the total profit of GSC under two scenarios: cost-sharing and non-cost-sharing. In both models, when ESG concern (t) is increased up to a specific threshold value of 0.7, the GSC's total profit increases. As the t value rises above the threshold of 0.7, it decreases under Model C and increases under Model NC. The reason is lower ESG concern among consumers, retailers, and manufacturers coupled with higher green innovation costs. Consumers are becoming more concerned with ESG, thus requiring green packaging, causing manufacturers to invest in green innovation, which drives up costs. Moreover, both the retailer and the manufacturer share a portion of the costs. Therefore, both the retailer's and manufacturer's net profits decrease, resulting in an overall decrease in profits.

19.4.4 Cost-Sharing (ρ)

The effect of cost-sharing (ρ) on the manufacturer's profit, retailer's profit, and total profit of the GSC under cost-sharing is shown in Figure 19.3. It is observed that as cost-sharing increases, the manufacturer's profit increases, while the retailer's profit decreases. Consequently, supply chain profit decreases. When the retailer shares a proportion of the green innovation cost above a certain threshold value of 0.7, they compensate for this extra cost by increasing the selling price. This, in turn, negatively impacts product demand. Therefore, retailer experts must determine when and how much of the green and innovation costs should be shared.

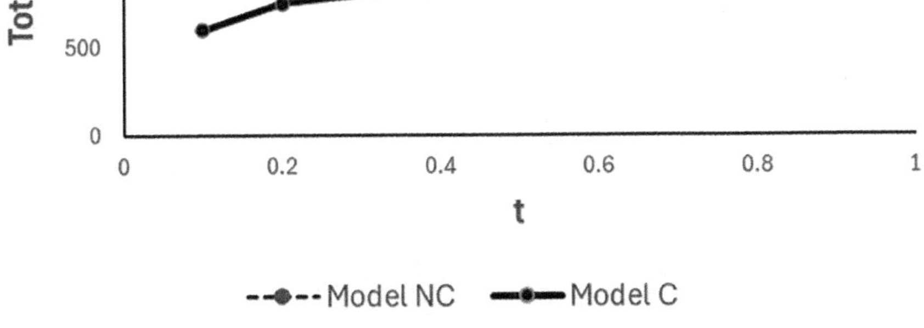

FIGURE 19.2 Impact of degree of ESG concern (t) on supply chain profit.

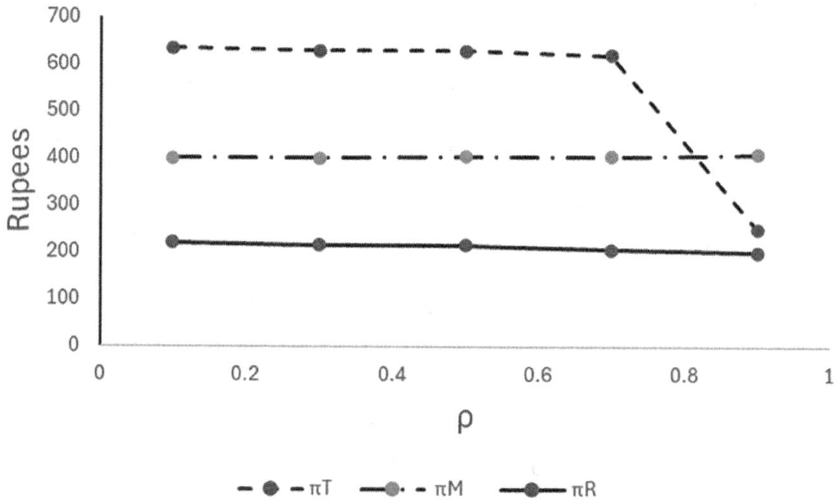

FIGURE 19.3 Impact of cost-sharing on different profits.

19.5 CONCLUSION, MANAGERIAL INSIGHTS, AND LIMITATION

Recently, a garment manufacturer invested in new, eco-friendly packaging and maximized profits by selling the green innovation package to a retailer. The retailer uses this green innovation package to sell garments. Due to ESG policies, the retailer also takes social and environmental concerns into account by sharing green innovation costs with the manufacturer. This study examines how green innovation and ESG scores impact GSC profit under cost-sharing and non-cost-sharing scenarios. The study found that total supply chain profit is higher in the cost-sharing scenario than in the non-cost-sharing scenario. However, it was also found that while the manufacturer's profit increases in cost-sharing scenario, the retailer's profit decreases compared to the non-cost-sharing scenario. The EPI and CS values are also higher in Model C than in Model NC, leading to a higher ESG score in Model C. In addition, the manufacturer invests more in green innovation packages in cost sharing than in non-cost sharing.

As mentioned in the previous section, sensitivity analyses can provide managerial insights into reducing costs and emissions. The first sensitivity analysis examines the impact of the allotted green innovation cost coefficient (λ), the degree of ESG concern (t), and the cost-sharing proportion (ρ) on the total GSC profit. It is found that as λ and t increase, the total cost rises, but the total GSC profit decreases under cost sharing. Thus, if a firm wishes to increase its total profit through green innovation, cost sharing should be adopted. Moreover, GSC profits also rise when green innovation, ESG scores, and CSR activities all increase at the same time. Accordingly, firms should implement green innovation packages while enhancing CSR and ESG programs.

Future research directions are highlighted in this study. Beyond green investments, advanced digital technology investments allow garment manufacturers, retailers, and logistic companies to track and use packages more efficiently. While this study concentrates on the green innovation investments of garment manufacturers, future research could also examine retailer investments in green innovation to better understand ESG impact. Furthermore, nonlinear and uncertain demand require further research to better reflect customer behavior. Lastly, future studies could gain valuable insights by including different supply chain members with dual ESG purposes, incorporating dual-channel modes, and incorporating exogenous commission rates.

REFERENCES

Bade, C., Olsacher, A., Boehme, P., Truebel, H., Bürger, L. and Fehring, L. 2024. Sustainability in the pharmaceutical industry: An assessment of sustainability maturity and effects of sustainability measure implementation on supply chain security. *Corporate Social Responsibility and Environmental Management* 31(1): 224–242.

Baid, V. and Jayaraman, V. 2022. Amplifying and promoting the "S" in ESG investing: The case for social responsibility in supply chain financing. *Managerial Finance* 48(8): 1279–1297.

Cerciello, M., Busato, F. and Taddeo, S. 2023. The effect of sustainable business practices on profitability. Accounting for strategic disclosure. *Corporate Social Responsibility and Environmental Management* 30(2): 802–819.

Chen, J., Zhou, Y.-W. and Zhong, Y. 2017. A pricing/ordering model for a dyadic supply chain with buyback guarantee financing and fairness concerns. *International Journal of Production Research* 55(18): 5287–5304.

Dmytriyev, S. D., Freeman, R. E. and Hörisch, J. 2021. The relationship between stakeholder theory and corporate social responsibility: Differences, similarities, and implications for social issues in management. *Journal of Management Studies* 58(6): 1441–1470.

Eccles, R. G., Lee, L.-E. and Stroehle, J. C. 2020. The social origins of ESG: An analysis of Innovest and KLD. *Organization and Environment* 33(4): 575–596.

Folqué, M., Escrig-Olmedo, E. and Santamaría, T. C. 2021. Sustainable development and financial system: Integrating ESG risks through sustainable investment strategies in a climate change context. *Sustainable Development* 29(5): 876–890.

Ghosh, D. and Shah, J. 2015. Supply chain analysis under green sensitive consumer demand and cost sharing contract. *International Journal of Production Economics* 164: 319–329.

Heydari, J., Govindan, K. and Aslani, A. 2019. Pricing and greening decisions in a three-tier dual channel supply chain. *International Journal of Production Economics* 217: 185–196.

Huang, W. and Swaminathan, J. M. 2009. Introduction of a second channel: Implications for pricing and profits. *European Journal of Operational Research* 194(1): 258–279.

H&M Group. 2025. https://www.hm.com/za/customer-service/sustainability/sustainability/ [Accessed 23 October 2025].

Jena, S. K. and Singhal, D. 2023. Optimizing the competitive sustainable process and pricing decision of digital supply chain: A power-balance perspective. *Computers and Industrial Engineering* 177: 109054.

Kavilanz, P. 2020. H&M will let people convert old clothing items into new ones at Stockholm store. CNN Business. [Accessed 25 October 2025].

Kuo, K.-C., Lu, W.-M. and Ganbaatar, O. 2023. Sustainability and profitability efficiencies: The moderating role of corporate social responsibility. *International Transactions in Operational Research* 30(5): 2506–2527.

Kurata, H. Yao, D.-Q. and Liu, J. J. 2007. Pricing policies under direct vs. indirect channel competition and national vs. store brand competition. *European Journal of Operational Research* 180(1): 262–281.

Le, T. T., Vo, X. V. and Venkatesh, V. G. 2022. Role of green innovation and supply chain management in driving sustainable corporate performance. *Journal of Cleaner Production* 374: 133875.

Li, G., Wu, H., Sethi, S. P. and Zhang, X. 2021. Contracting green product supply chains considering marketing efforts in the circular economy era. *International Journal of Production Economics* 234: 108041.

Li, G., Xue, J., Li, N. and Qi, Q. 2024. Does digital finance favor firms in supply chains? Roles of green innovation and bargaining power. *Transportation Research Part E: Logistics and Transportation Review* 183: 103431.

Pan, J., Wu, K.-S., Yang, C.-T., Lu, C.-J., and Lu, S. 2024. Using game theory to explore the multinational supply chain production inventory models of various carbon emission policy combinations. *Mathematics* 12(10): 1564.

Panda, S. 2014. Coordination of a socially responsible supply chain using revenue sharing contract. *Transportation Research Part E: Logistics and Transportation Review* 67: 92–104.

Press Turst of India (PTI). 2019. India generates 9.46 mn tonnes of plastic waste annually 40 percent of it remains uncollected. https://www.thehindubusinessline.com/news/science/india-generates-946-mn-tonnes-of-plastic-waste-annually/article29299108.ece [Accessed 30 September 2025].

Raz, G. and Ovchinnikov, A. 2015. Coordinating pricing and supply of public interest goods using government rebates and subsidies. *IEEE Transactions on Engineering Management* 62(1): 65–79.

Seman, N. A. A., Govindan, K., Mardani, A., Zakuan, N., Saman, M. Z. M., Hooker, R. E. and Ozkul, S. 2019. The mediating effect of green innovation on the relationship between green supply chain management and environmental performance. *Journal of Cleaner Production* 229: 115–127.

Taleizadeh, A. A., Adeli, M. and Abedsolta, H. 2025. Sustainability, pricing and marketing in a dual channel supply chain with different retailing scenarios and carbon emission reduction. *Journal of Cleaner Production* 496: 145149.

Tan, Y. and Zhu, Z. 2022. The effect of ESG rating events on corporate green innovation in China: The mediating role of financial constraints and managers' environmental awareness. *Technology in Society* 68: 101906.

Wang, J., Duan, K. and Zheng, Y. 2025. Green supply chain management, green technology innovation and firms' energy consumption intensity. *Energy Economics* 141: 108133.

Yang, X. and Zhang, P. 2023. Emission reduction of low-carbon supply chain based on uncertain differential game. *Journal of Optimization Theory and Applications* 199(2): 732–765.

Yenipazarli, A. 2017. To collaborate or not to collaborate: Prompting upstream eco-efficient innovation in a supply chain. *European Journal of Operational Research* 260(2): 571–587.

Zhang, C., Xiao, G. and Xu, L. 2024. Manufacturers' emission-reduction investments in competing supply chains with prisoner's dilemma: The economic and environmental impacts of retailer (s) capital constraint (s). *Transportation Research Part E: Logistics and Transportation Review* 187: 103602.

Zhao, S., Wang, M., Zhou, Q. and Xia, X. 2025. Managing manufacturer encroachment and product conflicts in a closed-loop supply chain: The case of information asymmetry. *Omega* 132: 103236.

20 Closed-Loop Supply Chain in the Clothing and Textile Industry

Kanti Prakash Brahma, Sarat Kumar Jena, and Bhavin J. Shah

20.1 WHAT IS A CLOSED-LOOP SUPPLY CHAIN (CLSC)

The concept of a circular economy (CE) focuses on moving away from the standard linear, unidirectional flow of materials, products, and waste into a system of circulation that decouples economic growth from resource use (Ellen MacArthur Foundation 2013; 2017). In the linear economy model based on the "take-make-dispose" pattern, products are manufactured using raw materials, then used, and ultimately disposed of as waste. Such a model does not account for the negative environmental impact of high resource use or the limitations imposed by the limited availability of finite resources. In contrast to the linear economy, the CE model advocates for a resilient system that considers the needs of business, people, and the planet. The core principles of CE include the elimination of waste and pollution, the efficient circulation of materials and products, and the sustainable regeneration of natural resources (Fashion for Good 2022). It reduces resource use and environmental impact by keeping materials and products in the economy for multiple life cycles through reuse, repair, remanufacturing or refurbishing, and recycling and has applications in a wide range of domains, including manufacturing, services, construction, resource sharing, and supply chain management (UNEP 2022).

For supply chain organizations, a CLSC system acts as a practical element in the transition to CE. CLSC "incorporates design, control, and operation of a system to maximize value creation over the entire life cycle of a product with dynamic recovery of value from different types and volumes of returns over time" (Abbey et al. 2024). It integrates the forward supply chain with the reverse supply chain, which involves waste collection and processing to circulate products and materials at their highest value. At a broad level, CLSC can be envisioned as a system involving three primary activities: Product/Material Acquisition Management, Reprocessing, and the Reprocessed Products Market (Figure 20.1).

The product/material acquisition activities include management of reverse supply chain, including collection, sorting, and grading of three different types of products/materials: (1) disposed used products, (2) unsold end-products or intermediate material, and (3) material waste during the production process. Collection, sorting, and grading play a pivotal role in enabling CLSC, as the quality and quantity of

DOI: 10.1201/9781003689225-20

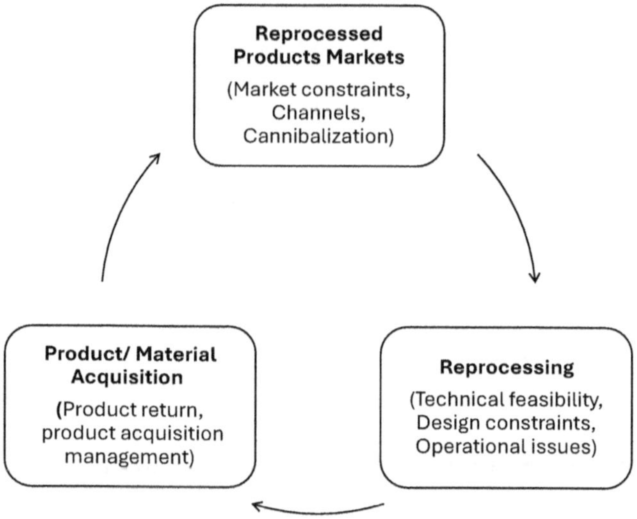

FIGURE 20.1 CLSC activities. (Adapted from Abbey et al. 2024.)

returned products/materials impact on the other two activities—reprocessing and the market for reprocessed products. The reprocessing activities entail processing collected materials and products so that they can be reintroduced into the closed loop. Reprocessing may involve assessing the technical feasibility of reprocessing, considering design constraints, making necessary modifications to processes, and addressing operational issues such as maintaining throughput and adjusting process parameters to ensure optimal performance. Reprocessed products market activities include addressing challenges such as pricing and promotion, deciding on sales channels, and ensuring that reprocessed products do not cannibalize sales of new products.

Although the CLSC activities mentioned in Figure 20.1 apply to supply chains across various industries, the details of application and operationalization may vary depending on the specific industry context. For example, the Clothing and Textile (C and T) industry faces unique challenges for waste collection, reprocessing, and disposal. These challenges originate from industry-specific characteristics like higher product and material variation, lack of high degree of standardization of the materials and processing parameters, degradation of strength of products during use. We have discussed C and T industry-specific CLSC activities in more detail later in this chapter.

20.2 NEED FOR CLSC IN C AND T

The C and T industry is notorious for its negative environmental impact caused by resource-intensive raw materials, polluting manufacturing processes, and the disposal of a high quantity of waste. The industry uses 93 billion m^3 of water for production and 5 trillion L for dyeing process. It releases up to 72 toxic chemicals into the water and contributes 4%–10% of global carbon emissions (Balchin 2023). Such a negative impact calls for action to promote sustainable production and consumption of C and T. CLSC is considered one of the avenues to advance sustainable production and consumption. Increasing production and consumption, coupled with the limitations of deriving growth from standard forward supply chain models, underscores the need for CLSC in C and T. The transition to CLSC is driven by limited resource availability, increasing raw materials volatility, and an evolving global regulatory landscape.

Production and consumption of C and T are increasing substantially due to demand and supply-side factors. On the demand side, an increasing population, disposable income, and fashion awareness contribute to C and T demand growth. At the same time, growing acceptance of the fast fashion business model, characterized by a higher number of short-lived products at lower prices, leads to increased consumption and production. Moreover, such short-lived products are often made with inferior materials, creating challenges for durability and reprocessing. On the supply side, the C and T supply chain is characterized by the presence of multiple stakeholders often spread across different countries. This leads to the risk of overproduction and higher inventory. Although it is challenging to predict fashion trends, organizations often predict trends and produce in advance in low-cost manufacturing countries, as they offer a substantial cost advantage. Advance production helps accommodate production and shipping lead times. The time lag between realizing a fashion trend and producing it often results in unsold inventory, as actual trends may differ from predicted ones. The increase in production and consumption demands higher resource use. Since resource availability is finite, there is a need to explore more sustainable models, such as CLSC.

The increasing cost and price volatility of virgin materials, as well as the limitations of raw material production in tandem with demand growth, call for the adoption of CLSC. For example, to cater for increasing consumption, it may not be feasible to increase the production of raw materials like cotton, as this would require more agricultural land. The cotton price is vulnerable to multiple factors, including weather conditions, changes in international trade policies, and demand for alternative crops. Similarly, synthetic fibers like polyester or polyamide prices are linked with petroleum-based raw materials that are vulnerable to global trade policies. Additionally, since these synthetics fibers shed microfibers during washing contributing to microplastic pollution, there is a need to restrict the use of these materials. Therefore, organizations are increasingly exploring alternative options, such as CLSC, where products and materials can be reprocessed, thereby disconnecting their business model from such volatility.

As regulations evolve, calling for more actions to address waste generation and reuse, organizations are expected to adopt CLSC with higher intensity. Changing regulatory landscape mandates C and T organizations to adopt CLSC. In Europe, regulations like the Extended Producer Responsibility (EPR) and Eco-design for Sustainable Products Regulation promote CLSC in C and T (Brown and Börkey 2024). Similarly, in the USA, several states have enacted regulations (e.g., Responsible Textile Recovery Act [RTRA] of 2024 in California) to promote apparel CLSC. These regulations force brands to take accountability for end-of-life products, and encourage investments in collection, sorting, and recycling systems. This, in turn, drives supply chain redesign by embedding circular material flows and enabling more traceable, resource-efficient operations. Moreover, adoption of CLSC in C and T industry mitigates price and procurement risks by reducing dependency on volatile raw material markets. It lowers reputation risk associated with waste and sustainability-related regulatory exposure. Less price volatility combined with stable and predictable resource availability increases supply chain resilience. Transformation of waste into useful resources, less supply chain risk, and better control over price have the potential to strengthen sustainable practices. The strengthening of sustainable practices accelerates the transition to CE and creates long-term value for organizations and their customers.

20.3 COMPARISON OF FORWARD SUPPLY CHAIN AND CLSC

This section explains the forward supply chain of C and T industry, discusses the types of waste generated in the supply chain, and highlights the activities of CLSC. It compares the forward supply chain and CLSC on basis of material flow, waste generation, value recovery, and market dynamics. The forward supply chain in C and T begins with virgin raw materials, which are converted into fiber form. In the case of synthetic fibers like polyester, raw materials such as purified terephthalic acid and mono-ethylene glycol, derived from petroleum-based sources, are polymerized to produce fibers; for natural fibers like cotton, seeds and impurities are removed to obtain pure cotton fiber. Fibers are converted to yarn through spinning, yarn to fabric through weaving or knitting, and fabric to apparel through garmenting. Apparels are packed and sent to distribution centers or warehouses, from which it reaches consumers through retail.

The forward supply chain generates waste at each stage of production and sales. Production waste includes spinning waste in the form of fiber waste, weaving or knitting waste in the form of yarn waste, and garmenting waste in the form of cut fabrics. Unsold fabric and apparel are also considered waste. The production waste and unsold fabric and apparel together constitute pre-consumer waste. The waste caused by the disposal of used products by consumers is referred to as post-consumer waste. The quality and characteristics of pre- and post-consumer waste vary substantially due to the difference in the stage at which waste is generated. Table 20.1 outlines the differences between the two types of waste.

As discussed earlier, CLSC combines reverse supply chain with forward supply chain and involves three broad-level activities: Product/Material Acquisition Management, Reprocessing, and the Market for Reprocessed Products.

In the C and T industry context, product and material acquisition involves collecting pre-consumer waste from production facilities and post-consumer waste from consumers. Since the origin and collection of pre-consumer waste occur in production facilities, manufacturing organizations provide waste for reprocessing, eliminating the need for a dedicated collection system. On the contrary, post-consumer waste is dispersed and spread across multiple locations, demanding a robust waste collection system. Different retailers have collection boxes in their stores where consumers can drop off their used garments for recycling. However, such collection mechanisms

TABLE 20.1
Differences between Pre- and Post-Consumer Waste

Waste Characteristics	Pre-Consumer Waste	Post-Consumer Waste
Source of waste	Production waste, unsold inventory.	Disposal of used C and T.
Homogeneity	Since the materials of different types, colors, and qualities are not mixed, homogeneity is high.	Collected waste contains a mix of different material types, colors, and quality.
Contamination with a non-textile component	Low for production waste, but unsold apparel inventory could still contain non-textile components, such as buttons and zippers.	High due to the presence of components like buttons, zippers, and the mixing of textile waste with other waste, like medical waste.
Ease of sorting	Sorting is easier owing to the homogeneity of the waste.	Difficult due to a lack of homogeneity.
Waste quality (fiber length and strength)	Fiber length and strength are higher.	Multiple uses and washing reduce fiber length and strength.
Material composition	The material composition (whether pure material or a blend of different materials) and exact material type (cotton, polyester) are known.	The material composition is unknown, as the collected waste is a mix of different materials.

are limited. It brings banks, municipal bodies, NGOs, charity donations, and individual waste collectors to contribute to a significant chunk of waste collection.

Reprocessing at C and T starts with sorting and grading the collected waste. Most of the collected waste, especially post-consumer waste, contains mixed or blended materials that require separation based on material type, color, and quality. The lack of uniformity in feedstock, due to variations in the color, composition, and quality of the collected waste, makes sorting time-consuming and inefficient. The collected garments usually do not carry tags or labels, making it difficult to determine their composition. There is a lack of technology to segregate mixed materials accurately or to separate high-quality recycled fibers from mono-composition materials. Contamination of feedstock because of medical waste, zippers, buttons, labels, sewing threads, dyes, and chemicals adds to sorting complications. Incorrect segregation of waste based on quality leads to challenges at the later stages. For example, material reprocessing technology varies depending on material composition; pure cotton can be mechanically recycled, while pure polyester cannot. Sorting and grading are predominantly manual processes carried out in fragmented and unorganized setups. Recently, infrared technologies were being deployed for sorting.

Sorting and grading help determine the level of reprocessing that waste should undergo at CLSC. Reprocessing of waste for CLSC in C and T is considered at three levels: product reprocessing, component reprocessing, and material reprocessing (Denizel and Schumm 2024). Apparel waste of relatively high quality is considered for product reprocessing. In product reprocessing, the apparel is mended or repaired and reintroduced to the market. Apparel waste is not suitable for product reprocessing, while fabric waste is considered for component reprocessing, which is similar to remanufacturing in the electronics or automobile industry. However, component reprocessing in C and T does not result in the replication of the original product using reprocessed components. It results in new products of the same or higher value through the upcycling of components. The waste that cannot be used in component reprocessing is subjected to material-level reprocessing, in which materials are disintegrated to the fiber level. Mechanical and chemical recycling are the two dominant material reprocessing processes. Mechanical recycling applies to natural fibers such as cotton and wool. This results in low-quality reprocessed fibers compared to virgin fibers (McKinsey 2022). Chemical recycling applies to synthetic fibers (such as polyester and polyamide) and man-made cellulosic fibers (like viscose), yielding a quality comparable to that of virgin fibers. Once disintegrated to the fiber level, the reprocessed material (recycled material) is reintroduced as a raw material for spinning. Table 20.2 illustrates the suitability of pre- and post-consumer waste for various levels of reprocessing at C and T CLSC.

Reprocessing at all levels will require consideration of technical feasibility, design constraints, and operational issues. Technical feasibility may include the feasibility of specific CLSC operations, such as product repair, disassembling components to reuse fabric, and material reprocessing through the selection of the right recycling technology. Since the reprocessed product and material in C and T have a lower strength compared to new ones, reprocessing requires design considerations for both materials and products. CLSC requires designers to consider product durability, recyclability, and the homogeneity of materials used during the design stage, before manufacturing. For example, plain or twill weaves are preferred over complex weaves for woven fabrics manufactured from recycled materials. Additionally, mono-materials (such as 100% cotton or 100% polyester) are preferred over blends of different fiber types. The operational issues of reprocessing could range from process issues like the need to lower production speed to accommodate relatively weaker materials, elimination of intermediate processes like combing in spinning, deciding the right mix of reprocessed and virgin material to increase the share of reprocessed material without compromising the performance (strength and appearance) of the reprocessed product.

The dynamics of the reprocessed product market are substantially different from those of a new product. In some industries, such as electronics and automobiles, reprocessed products are sold at a discount compared to new materials. However,

TABLE 20.2
Suitability of Pre- and Post-Consumer Waste for Reprocessing for CLSC

CLSC Activity	Pre-Consumer Waste	Post-Consumer Waste
Product reprocessing	Not applicable for production waste and unsold fabric, but high for unsold apparel	Depending on the quality of waste (high suitability for high-quality waste)
Component reprocessing	Not applicable for production waste, but high for unsold fabric and apparel	Depending on the quality of waste (high suitability for high-quality waste)
Material reprocessing	Relatively more due to higher material strength	Relatively less due to less strength

in C and T, the reprocessing cost could be higher than the retained value of the material or product, and the resultant selling price could be higher than that of the new product (Dissanayake and Sinha 2015), creating challenges for consumer acceptance. A reprocessed product channel can be identical to a new product or a distinct channel. The selection of the channel may vary depending on the CLSC activities in which an organization is involved. H&M Group has a separate brand named Sellpy, which is an online marketplace dedicated to buying and selling second-hand apparel (reprocessed products). The Group sells products made from recycled materials in H&M's stores and online platforms. The sales of reprocessed products may compete with those of new products from the same organization, leading to cannibalization. Not all CLSC activities have the potential to result in cannibalization. Reprocessing apparel with minor repairs may result in a product that is very similar to the new product, creating the risk of cannibalization. Since component reprocessing or material reprocessing yields different products, the risk of cannibalization is reduced in these activities. The risk of cannibalization could be mitigated by segregating sales channels and employing a distinct marketing mix for CLSC products. In a market characterized by a sustainability-conscious consumer segment, the risk of cannibalization negatively impacting profitability is reduced; instead, the organizations can leverage the consumer's willingness to pay for sustainability to price CLSC products higher than new products.

Supply chain traceability and transparency, product labeling, and sustainability certifications are critical for consumer acceptance of CLSC. These elements create consumer confidence and trust by providing verified information on material origins, recycling content, and end-of-life pathways. Certifications such as Global Recycled Standard (GRS) or Recycled Claimed Standards provide desired authenticity to recycled material-based products. Ultimately, these elements empower consumers to make informed choices and strengthen their involvement in C and T CLSC.

20.4 SELECT CASE STUDY/CASE STUDIES

This section presents three case studies detailing the operating models of three different organizations involved in four distinct CLSC activities. The case studies address three CLSC activities previously discussed: reselling (product reprocessing), component reprocessing, and recycling (material reprocessing).

20.4.1 THREDUP

Headquartered in California, USA, Thredup is one of the biggest online resale platforms for apparel, shoes, and accessories. It generated an annual revenue of US$ 260 million in 2024 and employed over 1,600 employees (Thredup 2025). The organization operates with a mission: "to inspire the world to think second-hand first." In 2024, the organization had 1.3 million active buyers. Active buyers are those who have created an account and purchased at least one product from the website within the last 12 months. Thredup's managed marketplace combines buyers and sellers of second-hand clothing. The key value proposition for the seller is the potential to unlock value for their used clothes. The buyers are attracted to the platform due to its ability to provide a wide variety of resale products from different price segments and brands at a price up to 90% lower than the retail price of the new product. The organization offers second-hand apparel from more than 55,000 brands, including J. Crew, Ann Taylor, Banana Republic, BCBG, and Lululemon. In 2024, the organization was featured in *Time* magazine's list of best inventions for its advanced AI search, which enhances the product search process by utilizing ultra-specific keywords.

The process of selling second-hand clothing includes the seller sending the used product to Thredup, inspection and listing by Thredup, and transfer of cash or credit to the seller upon sales realization. The organization accepts clean, undamaged, unaltered women's and kids' clothing for resale. The seller has the option to transfer ownership of the inventory to Thredup or to use Thredup as a marketplace without transferring the inventory. Thredup does not charge any upfront fees for sending products for resale. Once it receives products from sellers, it inspects the products for quality. If the products are deemed suitable, it takes photographs of them and lists them on its website. It features a proprietary algorithm to determinc the resale price. Upon realization of sales, the customer receives money after adjustments for costs related to shipping, inspection, product photography, and logistics.

20.4.2 RE/DONE

RE/DONE was founded in 2014 in Los Angeles, USA, with the mission to collect used Levi's denim jeans and reintroduce them with a new style and look through upcycling. In 2016, the organization started reprocessing sweatshirts. The business model of the organization involves five key steps: (1) sourcing of used products, (2) selection of used products for component reprocessing based on desired end-product, (3) dismantling of the components, (4) laying and cutting of the denim fabric, and (5) resewing of the components. Sourcing involves hand-picking-used products from rag houses around the USA based on pre-set criteria related to size and condition. The sourced products are sanitized. In the next step, the organization determines the size of the garment that needs to be reprocessed to achieve the desired size of the finished product. The size of the garment could be up to four sizes larger than the desired size of the reprocessed garment. The selected garment is then dismantled to segregate different components, including the waistband and pockets. The dismantled components are laid out according to the required pattern and sewn together to create a reprocessed product with a new style and fit. The dismantling and sewing activities take place in RE/DONE factories located in Mexico and Turkey. The factories are equipped with modern technologies to ensure reduced resource consumption and sustainable manufacturing practices. The reprocessed products are sold in over 350 retail stores across 30 countries, as well as on the online platform shopredone.com. Since the reprocessing process, other than initial disinfection, does not use water, each pair of upcycled jeans saves around 9 gallons of water (Redone 2025).

In 2016, in partnership with Hanes, RE/DONE introduced products made from reprocessed and low-impact materials, including recycled and organic cotton. Over time, the organization has expanded beyond jeans and diversified its operations to include the upcycling of leather and footwear.

20.4.3 Recover

Recover is among the biggest producers of mechanically recycled cotton globally. It contributes to CLSC in C and T through material reprocessing. Headquartered in Spain, the organization has manufacturing locations in Spain, El Salvador, Bangladesh, Pakistan, and Vietnam, with a combined capacity of 180,000 tons of recycled fibers per annum (Recover 2022). The organization aims "to close the loop in fashion by transforming textile waste into valuable raw materials for new garments." The organization collaborates with C and T recyclers and supply chain stakeholders to source cotton-based pre-and post-consumer waste. The waste used by Recover contains at least 95% cotton. Its recycling process converts textile waste into recycled fiber through a sequence of four steps: cutting, treatment, shredding, and packing. In the cutting process, cotton waste is cut into small pieces, and the non-textile components are removed from the waste. The organization has automated machines to remove buttons, trims, zippers, and other materials from the waste. During the treatment process, the smaller cut pieces are treated with antistatic agents to minimize the generation of static charge during subsequent processes. The shredding process separates fibers from the cut pieces, disentangling and aligning them so that they can be used as raw material for spinning. The aligned fibers are packed into bales, allowing for easy transportation to spinning units. Since mechanical recycling reduces fiber length owing to cutting and shredding processes, Recover recommends blending recycled cotton with virgin cotton to achieve the desired quality and performance.

Recover differentiated its recycled cotton products into two categories: RCotton, suitable for overdyeing, and RColorBlend, in which input waste of the same color is already blended to produce colored recycled fibers. Recover offers ten standard colors for RColorBlend. Since the organization maintains an inventory of RCotton, the lead time for it is not particularly important; however, for RColorBlend, the lead time could be 1–2 weeks. Recycled cotton from Recover is GRS-certified, meaning that independent third parties verify and validate the organization's claims regarding the recycled content and chain of custody. Recover adheres to global standards for sustainable practices, encompassing both social and environmental aspects.

Recover's contribution to CLSC and CE is reflected in the numerous recognitions and awards it has received. In 2021, it received the Ryan Young Climate+ Award in the Climate Leader category and the Drapers Award in the Disruptor category. It won the ITMF Sustainability and Innovation Award and the Green Product Award in 2022, as well as the Just Style Marketing Award in 2024.

20.5 CHALLENGES FOR CLSC IN C AND T

This section discusses the challenges to CLSC adoption in C and T, based on the three CLSC activities outlined in Figure 20.1. We have also discussed the challenges that organizations face during their CLSC transition journey, including tensions and paradoxes that arise.

20.5.1 Product/Material Acquisition Challenges

The challenges associated with product/material acquisition include a lack of consumer awareness about the importance of waste segregation, fragmentation of the reverse supply chain, complex sorting systems, and inadequate monitoring and tracking systems for waste.

Consumer awareness of the importance of waste segregation and proper waste disposal plays a crucial role in ensuring the availability of high-quality waste with minimal contamination. In the absence of consumer awareness, the municipality's waste collection system and take-back programs become ineffective, and the risk of contamination of textile waste with other waste increases, as consumers fail to segregate different kinds of waste and mix them. Regulations can play a pivotal role in creating awareness among consumers on the environmental impact of their waste disposal behaviors. Such awareness initiatives have the potential to yield favorable impacts on their approach to waste (Long and Nasiry 2022). Right media and communication are essential to enhance consumer awareness of waste disposal behaviors (Bianchi and Birtwistle 2012).

The reverse supply chain of the C and T industry is highly fragmented as consumers are dispersed geographically. No brand or retailer dominates the industry, as even the most prominent brands hold only a small market share. This high degree of fragmentation makes centralized waste collection challenging and leads to additional challenges related to scale and composition of waste. The collection, sorting, and recycling facilities are fragmented, and there is a lack of centralized control. Furthermore, there is a significant difference in waste collection system in developed economies compared to developing economies. The developed economies have high share of post-consumer waste owing to higher use of C and T compared to production. In post-consumer waste collection in developed economies, municipality bodies, NGOs, and charity donations play crucial roles. Additionally, brands and retailers encourage consumers to drop their used C and T in in-store collection boxes for reprocessing. In the developing countries, due to the presence of C and T manufacturing, both pre-consumer waste and post-consumer waste coexist. Collection of pre-consumer waste is easier, while post-consumer waste collection is dominated by individual waste collectors making the segment highly fragmented.

Limitations on the accuracy and scale of sorting act as barriers to the adoption of CLSC. Several emerging technologies for recycling demand cleaner feedstock. This requires accurate sorting, whereas the sorting process is predominantly manual, resulting in quality challenges in the sorted waste. C and T sorting requires attention to several criteria, including fiber content, composition, and quality differences that are not easily identifiable based on the number of uses. This requires training and knowledge sharing among workers in sorting facilities, which is often lacking due to the small scale of these facilities.

Since the reverse supply chain in C and T is relatively immature compared to the forward supply chain, there is a need for industry-wide waste tracking and monitoring systems. Such waste tracking and monitoring are crucial to creating visibility of the source, availability, and quantity of waste. It will help sorters and recyclers connect with their respective buyers, advancing business and trade. Although companies like Reverse Resources and Refashion Recycle have developed platforms to create visibility of waste, the initiatives are still in the early stages of implementation.

20.5.2 Reprocessing Challenges

The challenge of reprocessing activities lies in creating value from products and materials not designed for reprocessing, which involves a labor-intensive disassembling process, low production efficiency in downstream activities, and a lack of commonly accepted quality standards.

Product design and material selection are critical for enabling CLSC. Although reprocessing at the product level with minor repairs can be carried out for any waste products meeting the minimum quality requirements for reprocessing, the ease of component and material reprocessing depends on the design of the new product and material. The use of complex designs, textile components with multiple material types, and non-textile components with different fiber types, as well as unplanned selection of dyes and chemicals, hinders component and material reprocessing. For example, it is relatively more straightforward to produce recycled cotton from waste comprising 100% cotton fiber compared to a waste that is a blend of cotton and polyester fibers, as the fibers behave differently, and the segregation of the fibers from waste is expensive.

Dismantling for component reprocessing is labor-intensive and time-consuming, as it requires the removal of non-textile components, such as buttons, zippers, and embroidery, as well as the removal of sewing threads, so that the fabric components can be utilized. The lack of uniformity in product design, variation in style, the presence of multiple components in a single product, and the variation in component quality and type from one product to another make automation difficult for component reprocessing.

The forward supply chain is not designed to address the specific requirements of reprocessing. For example, garmenting in the case of component reprocessing requires a different setup compared to the regular garmenting designed for higher production quantities. The standard forward supply chain is primarily designed for mass production, where multiple garments can undergo the cut-and-sew process. However, in component reprocessing, the cut-and-sew process is less productive due to the irregular shape of the components. Similarly, in the case of material reprocessing, the recycled material is used as input in the standard spinning process, which is designed for virgin material, without considering the specific requirements of the recycled material, such as lower strength and shorter fiber length. As a consequence, the productivity of downstream manufacturing processes decreases for reprocessing.

The standard forward supply chain in the C and T clothing industry has established quality standards that enable the industry to communicate in a common language for procurement and sales. For example, a garment manufacturing organization prepares a fabric technical data sheet that contains specifications such as yarn count, yarn density, grams per square meter, and strength requirements (tensile, tear, and abrasion) as well as appearance requirements. Such quality standards are not designed for CLSC for C and T. The CLSC follows the quality standards designed for virgin material for reprocessed products and materials, which forces organizations to aim for quality requirements that are difficult to achieve. The lack of a customized quality standard for reprocessed products and materials acts as a barrier to advancing the transition to CLSC.

20.5.3 Reprocessed Product Market Challenges

The reprocessed product market is different from the more established standard forward supply chain market. The key challenges in the reprocessed product market include limited awareness of reprocessed materials and products, concerns about the cannibalization of new products, and both actual and perceived quality issues.

Consumer awareness of the environmental impact of reprocessed materials and products compared to new ones is relatively low. Even when consumers demonstrate awareness of sustainability and environmental impact, this awareness does not necessarily translate into purchasing behavior. Consumers often link reprocessed materials with inferior quality, lower value, and hygiene and safety issues. They expect reprocessed products to be priced lower than new products but match them in quality and appearance.

The sale of reprocessed C and T could impact the sales of new products from the same organization, leading to market cannibalization where existing consumers of new products shift from new products to reprocessed products. Cannibalization risk is particularly high for premium or luxury products, where the perceived price difference between new and reprocessed products is substantial. To avoid cannibalization, organizations should develop distinct marketing strategies for their reprocessed products, differentiating them from their new products. The differentiation could take the form of a different brand name, the use of a different channel, or the use of a third-party provider for CLSC products.

Quality is one of the most critical factors in consumers' buying decisions. C and T CLSC products face actual and perceived quality challenges. For example, mechanically recycled cotton faces a significant quality challenge, as the strength of mechanically recycled cotton fiber can be 30%–50% lower than that of virgin fiber due to physical wear and tear during the cutting and shredding processes in recycling, as well as exposure to chemicals during production and use. Lower strength and shorter fibers also lead to an inferior product appearance. On the contrary, chemically recycled cotton yields a fiber quality comparable to that of virgin cotton, but it still faces a perceived quality challenge, as consumers, unaware of the differences between various recycling technologies, tend to perceive all kinds of recycled materials similarly.

20.6 PARADOXICAL TENSIONS IN CLSC TRANSITION

The immature and evolving state of CLSC implementation in the C and T industry raises several contextual paradoxical tensions. Such tensions emerge as challenges for the conceptualization and implementation of CLSC in the C and T industry, requiring a deeper understanding and careful calibration to draft a balanced approach toward CLSC. Mutha et al. (2016) have discussed this paradoxical tension in the context of remanufacturing, and Dehghannejad et al. (2025) have explored it in the context of CLSC transition in the C and T industry.

The paradoxical tensions exist at various stages in the C and T supply chain, ranging from business model to product design to regulations. Examples of such tension include business model for flexibility vs. sustainability, designing for reprocessing vs. durability, and measuring financial performance vs. circularity. Business models like fast fashion are driven by consumer demand to change styles and appearance quickly, whereas CLSC requires more durable, high-quality products with extended product life. Product designing for remanufacturing and recycling in C and T conflicts with the requirements for designing for durability (H&M Group 2023). Designing for durability might require adding additional seams or blending different materials, which will create challenges for component and material reprocessing in subsequent product life cycles. Performance measurement parameters of organizations transitioning to CLSC can create paradoxical tensions due to incongruency in the measurement matrix used for different teams. While product designers might focus on using recycled materials, which could come at a price premium due to lower volume compared to virgin materials, the procurement team might view it as a negative impact on the cost of goods sold.

Similarly, governments face trade-offs and paradoxical tensions in formulating regulations for CLSC (Long and Nasiry 2022; Long and Gui 2024). Stringent regulation on the organization's waste disposal practices could lead to unintended consequences. If the cost of waste disposal or the penalty for leftover inventory is very high, C and T organizations could delay production decisions to produce closer to the actual trend. However, since production to actual trend comes at a higher cost, they might compromise on product quality to maintain profitability at the cost of the environment. Adaptive policy that can enable CLSC ecosystem to respond effectively to evolving conditions by incorporating continuous learning, feedback loops, and flexible mechanisms will be instrumental in handling such regulatory paradoxes. Since the CLSC in C and T is in a fluid state, and both the standard forward supply chain and the CLSC coexist, addressing the paradoxical challenges requires a focused approach and attention to detail.

20.7 OPPORTUNITIES FOR CLSC IN C AND T

Despite the challenges described in the previous section, several opportunities exist that serve as driving factors for the faster adoption of CLSC in C and T. The key opportunities discussed in this chapter include demand-linked opportunities, collection and reprocessing opportunities, data and technology, and the evolving regulatory landscape.

20.7.1 Demand-Linked Opportunities

Increasing consumer awareness has the potential to make CLSC sustainable in the long term. Since organizations in the price-sensitive C and T industry tend to satisfy consumer needs and retain consumers, consumer acceptance and willingness to purchase CLSC products will enable a sustainable transition to CLSC. Consumer education can be effective in encouraging organizations to adopt more sustainable CLSC models (Long and Nasiry 2022). There are analyses demonstrating the possibilities of profitable and self-sustainable CLSC in C and T once the practices achieve scale and maturity. However, since the initial transition to CLSC will require investment support for research, innovation, and learning, green funding shared by brands and consumers will be instrumental in this process. Several consumer surveys have revealed an increasing willingness among consumers to pay for CLSC (McKinsey 2022). Proper communication and branding of CLSC products have the potential to increase demand and acceptance.

20.7.2 Collection and Reprocessing Opportunities

Improvement in the collection and reprocessing of materials and products will drive the growth of CLSC in the industry. The European Union's Waste Framework Directive has brought the focus back to collection, sorting, and recycling (EU 1998). The framework mandates a separate collection mechanism for textile waste in all EU-27 countries and stresses the scaling up of sorting and recycling facilities to make them economically viable. The introduction of mandatory standardization for sorting and recycling, along with a uniform reporting structure, will support the creation of desired visibility and traceability of C and T waste. As the acceptance of CLSC increases, C and T machine manufacturing organizations have begun developing and introducing machines designed to address reprocessing. Switzerland-based spinning machine manufacturing organizations, Rieter (2023) and Saurer (2023), have introduced spinning machines for recycled fibers that could address the challenges associated with short fibers in mechanically recycled cotton. Designing for reprocessing is gaining momentum, as designers have begun to factor in the ease of remanufacturing and recycling, as well as the selection of dyeing and finishing methods, during the design stage. Italian fashion brand Napapijri designs products in which all components, such as fabric, lining, trimming, and buttons, are made of the same material to ensure ease of recyclability (Napapijri 2026). The brand encourages its consumers to return used clothes and incentivizes such returns through a discount on a new purchase.

20.8 DATA AND TECHNOLOGY

The proliferation of modern data-driven applications in the supply chain can help address some of the complex problems faced by CLSC. Visibility of waste quality and quantity in the reverse supply chain is a concern for sorting and recycling units. Technology can be leveraged to create simple, yet easily accessible databases linked to waste inventories, creating opportunities to address the issue. The industry has witnessed web-based platforms like Reverse Resources and Refashion Recycle for creating visibility into the quantity and source of pre-consumer waste. There are opportunities to create similar visibility for post-consumer waste. The source and quantity of reprocessed material have been a concern in C and T CLSC. The use of modern traceability solutions, such as blockchain technology, can ensure the proper chain of custody and quantity reconciliation of CLSC products and materials, thereby avoiding the scope for manipulation or counterfeiting.

Digital product passports (DPPs), as proposed by the EU, create opportunities to advance CLSC through higher visibility of waste composition, environmental impact, and source, among other things (Ospital et al. 2022). DPP has the potential to reduce complexity in sorting and recycling, as the absence of labels in disposed waste hinders the sorting process.

The rapidly growing innovation related to additive traceability is emerging as a promising option to ascertain the origin of reprocessed materials. Additive tracers, such as artificial DNA codes, rare inks, or other optical substances, can be added during reprocessing and detected in downstream processes to validate their presence.

20.9 EVOLVING REGULATORY LANDSCAPE

Considering the environmental consequences of C and T's forwarding supply chain, regulators worldwide are focusing on policies and regulations conducive to CLSC. In Europe, EPR holds the manufacturing organization responsible for the environmental impact of the entire product life cycle, forcing organizations to consider CLSC aspects at the design concept stage. In the USA, Massachusetts has banned the disposal of textiles and apparel that could be reused or recycled, forcing consumers and businesses to explore reuse and recycling options. One of the stated objectives of the ban is to encourage investment in the recycling ecosystem and promote infrastructure growth and employment generation through uninterrupted material supply to CLSC. The Textile Reuse and Recycling Act of New York requires businesses to separate and recycle textile waste. The Responsible Textile Recovery Act (RTRA) of 2024 in California holds manufacturing organizations accountable for the environmental impact of their products throughout the product life cycle.

The opportunities are linked with the challenges discussed in previous sections. These opportunities could be leveraged to address challenges for CLSC in C and T industry. For example, collection and reprocessing opportunities could address some elements of product and material acquisition challenges. Similarly, evolutionary regulatory landscape has the potential to influence all the three challenges discussed in Section 20.5, namely, product and material acquisition challenges, reprocessing challenges, and reprocessed market challenges.

20.10 CONCLUSION

CLSC in C and T offers unique challenges and opportunities specific to the industry. The transition CLSC in C and T industry requires a balanced approach between cost efficiency, operational feasibility, and sustainability ambitions. However, these challenges also act as catalysts for learning and innovation by encouraging various stakeholders to explore new technologies and collaborative models. As innovations in business models, processes, technology, and regulations gain momentum, adoption of CLSC in C and T industry will overcome many of the existing challenges. Ultimately, the multifaceted interplay between challenges, opportunities, and tensions will define the trajectory of CLSC development, influencing how quickly and effectively C and T industry can move toward circularity.

A successful transition to CLSC could create a strategic advantage for organizations by delinking their dependency on a volatile virgin raw material market and derisking them from reputational and regulatory compliance risks. However, success in this transition requires addressing challenges, harnessing opportunities, and balancing the contradictory tensions of the industry discussed in this chapter. Organizations that can blend their business strategies with CLSC strategies and execute them seamlessly are expected to lead the industry.

REFERENCES

Abbey, J. D., Guide Jr, V. D. R. and Sun, X. 2024. Closed-loop supply chains: A strategic overview. In *Sustainable Supply Chains: A Research-Based Textbook on Operations and Strategy* (Ed. Tang, C. S.), Springer Nature, Cham, Switzerland, 355–378.

Balchin, N. 2023. How to make textile and garment production more sustainable. World Economic Forum. https://www.weforum.org/stories/2023/11/textile-garment-industry-sustainability-policy-measures/ [Accessed 10 February 2026].

Bianchi, C. and Birtwistle, G. 2012. Consumer clothing disposal behavior: A comparative study. *International Journal of Consumer Studies* 36(3): 335–341.

Brown, A. and Börkey, P. 2024. Extended producer responsibility in the garments sector. OECD Environment Working Papers.

Dehghannejad, M., Pal, R., Dissanayake, K. and Sandberg, E. 2025. Exploring paradoxical tensions in scaling circular business models within the textile and clothing industry. *Sustainable Production and Consumption* 56: 579–592.

Denizel, M. and Schumm, C. Z. 2024. Closed loop supply chains in apparel: Current state and future directions. *Journal of Operations Management* 70(2): 190–223.

Dissanayake, G. and Sinha, P. 2015. An examination of the product development process for fashion remanufacturing. *Resources, Conservation and Recycling* 104: 94–102.

Ellen MacArthur Foundation. 2013. Towards the circular economy, vol. 2: Opportunities for the consumer goods sector. https://www.ellenmacarthurfoundation.org/towards-the-circular-economy-vol-2-opportunities-for-the-consumer-goods [Accessed 10 December 2025].

Ellen MacArthur Foundation. 2017. A new textiles economy: Redesigning fashion's future. https://www.ellenmacarthurfoundation.org/publications [Accessed 10 December 2025].

EU. 1998. Directive, E.U. 98/EC of the European Parliament (2026). https://eur-lex.europa.eu/legal-content/EN/TXT/PDF/?uri=CELEX:02008L0098-20180705 [Accessed 4 November 2025].

Fashion for Good. 2022. Wealth in waste. https://www.fashionforgood.com/report/wealth-in-waste/ [Accessed 10 December 2025].

H&M Group. 2023. Circulator guide. https://hmgroup.com/wp-content/uploads/2023/03/Circulator-Guide-2.pdf [Accessed 10 December 2025].

Long, X. and Gui, L. 2024. Waste not want not? The environmental implications of quick response and upcycling. *Manufacturing and Service Operations Management* 26(2): 612–631.

Long, X. and Nasiry, J. 2022. Sustainability in the fast fashion industry. *Manufacturing and Service Operations Management* 24(3): 1276–1293.

McKinsey. 2022. Scaling textile recycling in Europe: Turning waste into value. https://www.mckinsey.com/industries/retail/our-insights/scaling-textile-recycling-in-europe-turning-waste-into-value [Accessed 1 November 2025].

Mutha, A., Bansal, S. and Guide, V. D. R. 2016. Managing demand uncertainty through core acquisition in remanufacturing. *Production and Operations Management* 25(8): 1449–1464.

Napapijri. 2026. https://www.napapijri.com/en-gb [Accessed 2 November 2025].

Ospital, P., Masson, D. H., Legardeur, J. and Beler, C. 2022. Digital product passport to support product transparency and circularity. *Global Fashion Conference*, Biarritz, France.

Recover. 2022. Fashion revolution week: The fight for circular and equitable fashion for all. https://recoverfiber.com/newsroom/fashion-revolution-week-the-fight-for-circular-and-equitable-fashion-for-all [Accessed 1 November 2025].

Redone. 2025. Redone about us. https://shopredone.com/en-in/pages/about-us [Accessed 1 November 2025].

Rieter. 2023. Com4 Recycling spinning system: Closing the loop for good. https://www.rieter.com/products/system-applications/recycling-spinning-system [Accessed 2 November 2025].

Saurer. 2023. Autocoro: The recycling all-rounder for every purpose. *Saurer website*. https://saurer.com/en/news/autocoro-recycling-all-rounder [Accessed 2 November 2025].

Thredup Annual Report. 2025. https://ir.thredup.com/financials-filings/annual-reports [Accessed 2 November 2025].

UNEP. 2022. *Sustainability and Circularity in the Textile Value Chain: Global Stocktaking*. United Nations Environment Programme (UNEP), Nairobi, Kenya.

Index

C

D

E

S

T

U

V

W

Y

Z

For Product Safety Concerns and Information please contact our EU representative GPSR@taylorandfrancis.com Taylor & Francis Verlag GmbH, Kaufingerstraße 24, 80331 München, Germany

Batch number: 10761599

Printed by Printforce, the Netherlands